Electron Microscopy

Electron Microscopy

Uday Veer Singh

RANDOM

RANDOM PUBLICATIONS

NEW DELHI (INDIA)

Electron Microscopy

ISBN 978-93-5111-253-2

Published in 2014 in India by

RANDOM PUBLICATIONS

4376-A/4B, Gali Murari Lal, Ansari Road
New Delhi-110 002
Phone : +91-11-43580356, +91-11-23289044
e-mail: randomexports@gmail.com, sales@randompublications.com, info@randompublications.com

Reprinted 2026

Type Setting by : Keystoneprintads, Delhi-110051

Preface

Electron Microscopes are scientific instruments that use a beam of highly energetic electrons to examine objects on a very fine scale. The advent of electron microscopes has opened up new vistas in the field of science. An electron microscope (EM) is a type of microscope that uses an electron beam to illuminate a specimen and produce a magnified image.

The laws of electricity and magnetism were originally formulated in terms of the electric fluids rather than in terms of charged particles. All of the notions of electromagnetic phenomena, including Maxwell's equations, were conceived in terms of the density and the current of the positive and negative electric fluids. It was not until after the electromagnetic interaction was understood that it was realized that the density of the two electric fluids was the density of electrons and protons and that currents were generated by the movement of these particles.

Accelerated electrons in an SEM carry significant amounts of kinetic energy, and this energy is dissipated as a variety of signals produced by electron-sample interactions when the incident electrons are decelerated in the solid sample. These signals include secondary electrons (that produce SEM images), backscattered electrons (BSE), diffracted backscattered electrons (EBSD that are used to determine crystal structures and orientations of minerals), photons (characteristic X-rays that are used for elemental analysis and continuum X-rays), visible light (cathodoluminescence—CL), and heat. The transmission electron microscope (TEM) operates on the same basic principles as the light microscope but uses electrons instead of light. What you can see with a light microscope is limited by the wavelength of light. TEMs use electrons as "light source" and their much lower wavelength makes it possible to get a resolution a thousand times better than with a light microscope.

You can see objects to the order of a few angstrom (10^{-10} m). For example, you can study small details in the cell or different materials down to near atomic levels. The possibility for high magnifications has made the TEM a valuable tool in both medical, biological and materials research. The Transmission Electron Microscope (TEM) allows the user to determine the internal structure of materials, either of biological or non-biological origin. Aberration in optical systems, *i.e.* in lenses or mirrors or a series of them, may be defined as the non-concurrence of rays from the points of an object after transmission through the system; it happens generally that an *image* formed by such a system is irregular, and consequently the correction of optical systems for *aberration* is of fundamental importance to the instrument-maker. Reference should be made to the articles *Reflexion, Refraction,* and *Caustic* for the general characters of reflected and refracted rays.

The AFM analyses of irradiated films of poly(STY-*co*-BMA) in figure show the significant impact of the laser irradiation on the topography of the film, together with the influence of the infrared-dye concentrations.The lateral distribution of the thermally generated acid, detected by the acid sensitive fluorescent acridine probe, was analyzed with LSCM. The influence of the infrared dye on the fluorescence images of the irradiated film is presented in figure. This book would definitely create a feel for this subject particularly among those who want to use this technique in their research work.

I thank all members of my team who have helped in the preparation of the book. My special thanks go to "Random Publications" who have published the book.

– Uday Veer Singh

Contents

1

Methods of Electron Microscopy

INTRODUCTION

Electron Microscopes are scientific instruments that use a beam of highly energetic electrons to examine objects on a very fine scale. This examination can yield the following information:

Topography

The surface features of an object or "how it looks", its texture; direct relation between these features and materials properties (hardness, reflectivity...etc.)

Morphology

The shape and size of the particles making up the object; direct relation between these structures and materials properties (ductility, strength, reactivity...etc.)

Composition

The elements and compounds that the object is composed of and the relative amounts of them; direct relationship between composition and materials properties (melting point, reactivity, hardness...etc.)

Crystallographic Information

How the atoms are arranged in the object; direct relation between these arrangements and materials properties (conductivity, electrical properties, strength...etc.).

ELECTRONS: LAWS OF ELECTRICITY AND MAGNETISM

The laws of electricity and magnetism were originally formulated in terms of the electric fluids rather than in terms of charged particles. All of the notions of electromagnetic phenomena, including Maxwell's equations, were conceived in terms of the density and the current of the positive and negative electric fluids. It was not until after the electromagnetic interaction was understood that it was realized that the density of the two electric fluids was the density of electrons and protons and that currents were generated by the movement of these particles. Just as the notion of caloric fluid disappeared once it was realized that heat was due to the motion of the atom, so, too, the notion of electric fluids disappeared once it was realized that electric phenomena were due to the underlying electrical structure of atoms. We now turn to a description of the experiments which led to the discovery of the underlying electrical structure of the atom. Students are often baffled as to how the physicists know that an atom consists of a positively charged nucleus about which negatively charged electrons orbit. They become skeptical about the existence of the electron when they discover that the electron has not and can not be visually observed. In order to counteract this skepticism, we shall carefully describe a series of experiments stretching over a period of almost 100 years in which the existence, charge and mass of the electron were established and in which the underlying structure of the atom was discovered.

The first hint of any connection between the atom and electrical phenomena came from the observation of the process of electrolysis. We have illustrated a typical electrolysis experiment. The anode or positive electrode and the cathode or the negative electrode are two metal rods connected to the positive and negative terminals of a battery and are immersed in water. Note the presence of a meter in the circuit placed between the negative terminal of the battery and the cathode. This device permits one to measure the amount of electricity flowing in the circuit. As the current flows in this circuit, hydrogen and oxygen gas will collect at the cathode and anode respectively. The volume of hydrogen gas that collects is twice that of the oxygen gas and is proportional to the total amount of electricity that flows in the circuit.

The flow of electricity is possible because the water molecules, H_20, separate chemically into two positively charged hydrogen ions, H^+ and an oxygen ion, O^{--}. An ion is an atom which carries a charge. Because there is an excess of negative charge at the cathode, the H^+ ions collect there, become neutralized and form hydrogen gas. The oxygen ions drawn by the positive charge collect at the anode, are neutralized and form oxygen gas. The ratio of the volume of hydrogen gas to oxygen gas is two to one because there are two hydrogen atoms to one oxygen atom in the water molecule.

BEAM OF ELECTRONS – MMAGNETIC LENSES

Condenser Magnetic Lens

Electrons can be focused either by electrostatic forces or magnetic forces. Although electron lenses in principle behave the same as optical lenses, there are differences. Also, the quality of electron lenses is not nearly as good as optical lenses in terms of aberrations. The electron beam is divergent after passing through the anode plate and must be collimated by condenser lenses and apertures into a relatively parallel stream. A magnetic lens is formed from two circularly symmetric iron pole-pieces with a copper winding in-between with a hole in the centre through which the beam passes.

The pole pieces are separated by an "air-gap" in which the focusing actually takes place. The divergence of the magnetic flux along the optical axis imparts a force on electrons back towards the optical (Z) axis, resulting in focusing action. The magnetic field also causes a rotation of the electrons (and the image) about the Z axis in a corkscrew fashion. The condenser lens controls the amount of current that passes down the rest of the column by focusing the electron beam to variable degrees. The sharper the focus, the less of the beam intercepted by an aperture (small hole) located below the lenses and the higher the current. Some microprobes have a second condenser lens that is used to provide a better focus for SEM work. This lower condenser lens may be "decoupled" (not used) for microprobe work and "coupled" to the upper lens to make one long lens for SEM work.

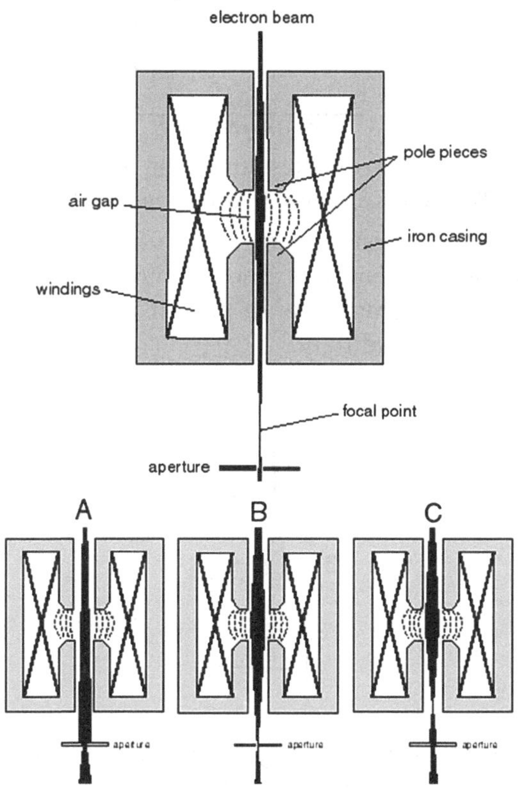

Fig. Controlling the sample beam by changing the current through the windings of a condenser lens: (a) no current, electron beam is unfocused; (b) small current, lens has long focal length, (c) larger current, lens has a shorter focal length and a smaller cone of electrons is transmitted through the aperture. Dimensions have been greatly exaggerated for clarity.

The beam characteristics are different for SEM and microprobe work. Probe beams are typically 1 to 5 μm in diameter with currents of 10^{-9} (1 nanoamp = 1 nA) to 10^{-7} amps (100 nA). There is no reason for the electron beam diameter to be smaller than the diameter of the excited region, which is about 2 μm. In contrast, because SE are produced only in the upper 10 A of the sample's surface, the beam size is not limited by the excitation area and much smaller beam diameters are possible. SEM uses much lower currents (about 10^{-12} amp = 1 picoamp = pA) and a smaller final aperture to limit the beam diameter. The beam regulator, which is located between the upper two condenser lenses, is used to maintain a stable beam current.

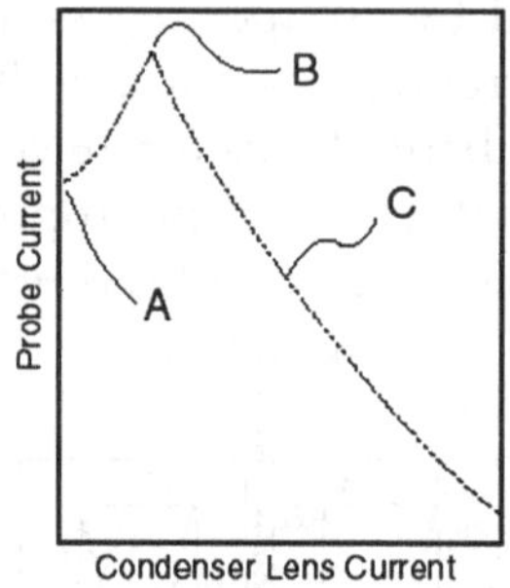

Fig. Change in Probe Current Resulting from Varying the Condenser Lens Current. The Points a, b and c Correspond to Diagrams in Figure.

The emission from a tungsten filament can vary by up to 1% in 10 minutes, an unacceptable amount for analysis. The beam regulator operates by sensing changes in the current impinging on a pair of apertures and adjusting the condenser lens current accordingly.

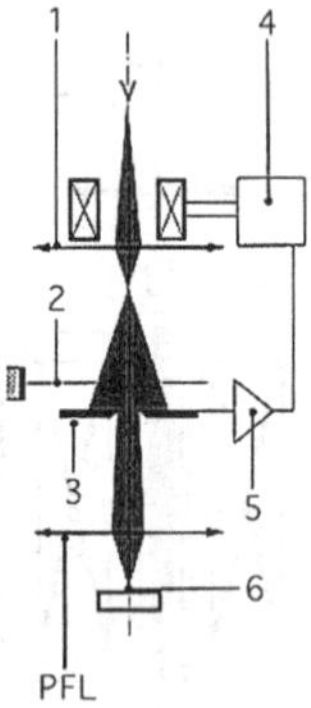

Fig. The electron beam is shown (greatly expanded) in red. The beam regulation system has two apertures: limiting (2) and probe-forming lens or PFL (3). The amount of current impinging on the PFL is amplified (5) and used to control the condenser lens power supply (4). The effects of the condenser lens (1) and PFL (objective lens) are shown schematically by divergent black arrows. The probe sample point is located at 6.

Electron microscopic tomography can provide a complete three-dimensional range of information about subcellular structures. Applied to mitochondria, it has shown that the conventional textbook model of this organelle is incorrect. The infoldings of the inner membrane (called cristae) are connected by narrow tubular regions to each other and to the outside, suggesting that internal diffusion of ions, metabolites and proteins may be restricted. Tomographic reconstruction of mitochondria in situ indicates these organelles may occur in clusters with stacks of endoplasmic reticulum, forming extended structures.

ELECTRON DIFFRACTION (ED)

Electron diffraction is a collective scattering phenomenon with electrons being (nearly elastically) scattered by atoms in a regular array (crystal). This can be understood in analogy to the Huygens principle for the diffraction of light. The incoming plane electron wave interacts with the atoms, and secondary waves are generated which interfere with each other. This occurs either constructively (reinforcement at certain scattering angles generating diffracted beams) or destructively (extinguishing of beams).

As in X-ray diffraction (XRD), the scattering event can be described as a reflection of the beams at planes of atoms (lattice planes). The *Bragg law* gives the relation between interplanar distance d and diffraction angle Θ:

$$n\lambda = 2d \sin \Theta$$

Bragg's Law of Diffraction:

The scattering of electrons, X-rays and neutrons by crystals can be described as a reflection of the beams at planes of atoms (lattice planes), independent of the actual physical reason causing the diffraction event. If the incident plane wave hits the crystal at an arbitrary angle, the interference of the reflected waves can be either destructive or constructive.

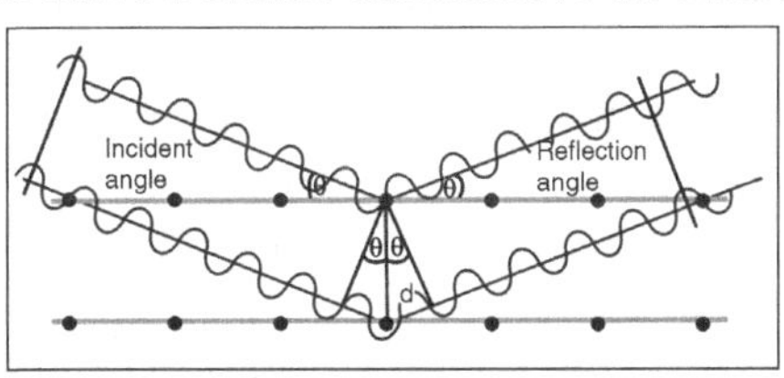

Destructive interference of reflected waves (in the reflected waves, maximum and minimum of the wave amplitude are superimposed).

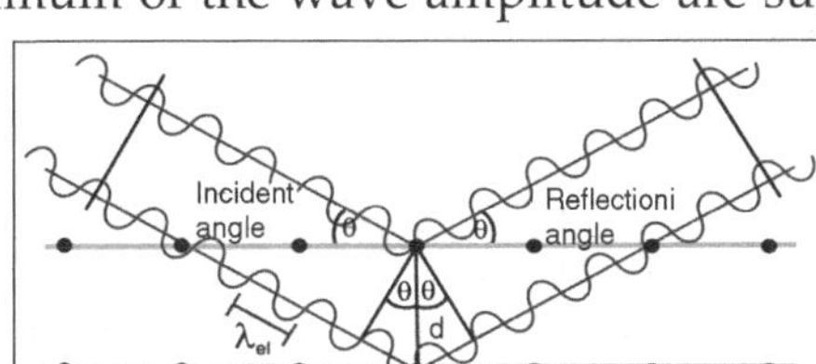

Constructive Interference of reflected waves (reflected waves in phase, i.e., maxima are superimposed). To obtain constructive interference, the path difference between the two incident and the scattered waves, which is $2d\sin\Theta$, has to be a multiple of the wavelength λ. For this case, the Bragg law then gives the relation between interplanar distance d and diffraction angle Θ:

$$n\lambda = 2d\sin\Theta$$

Since the wavelength λ of the electrons is known, interplanar distances can be calculated from ED patterns. Furthermore, information about crystal symmetry can be obtained. Consequently, electron diffraction represents a valuable tool in crystallography.

Estimate of scattering angles

λ_{el} = 0.00197 nm (1.97 pm) for 300 kV electrons. A typical value for the interplanar distance is d = 0.2 nm.

If these values are put in the Bragg law, then the scattering angle is: Θ = 0.28°. As a rule, the scattering angles in ED are very small: $0 < \Theta < 1$.

SCANNING TRANSMISSION ELECTRON MICROSCOPY (STEM)

Utilized signals: *elastically* scattered electrons - direct and diffracted beams (BF/DF) and incoherent scattering (Z contrast)

Methods	*Information*
BF/ADF-STEM	Detection of crystalline areas, defects and grain boundaries, phase analysis, particle size
HAADF-STEM	Detection of small clusters or even single atoms of a heavy metal in a matrix of light elements (Z-contrast), direct visualisation of structures and defects

In STEM, a tiny, convergent electron beam is scanned over a defined area of the sample.

At each spot, the generated signal is simultaneously recorded by selected detectors, building up an image.Furthermore, such a convergent beam is used to gain a highly localized signal from the specimen in analytical TEM (e.g.,

EDXS, EELS), and thus the combination of *STEM with analytical methods* is a main application in practical work.

Point Analysis in STEM

In *STEM,* an electron beam is scanned over a defined area of the sample. The beam can be localized on a certain point in the image and used to measure an *EDX* and/or *EEL* spectrum at there.

Moreover, line scans and mappings can be obtained by this powerful technique.

Example: Molybdenum Tungsten Oxide

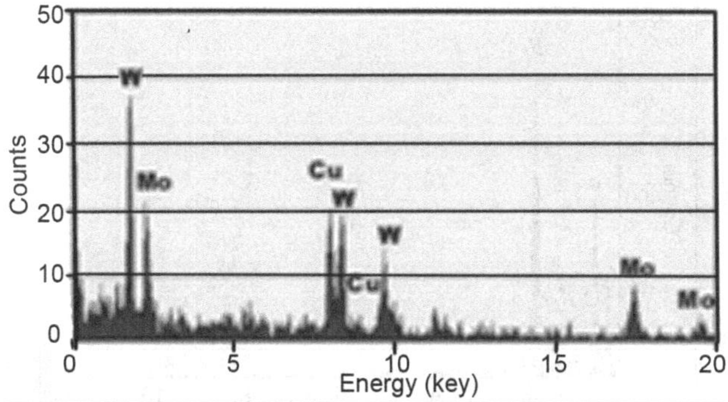

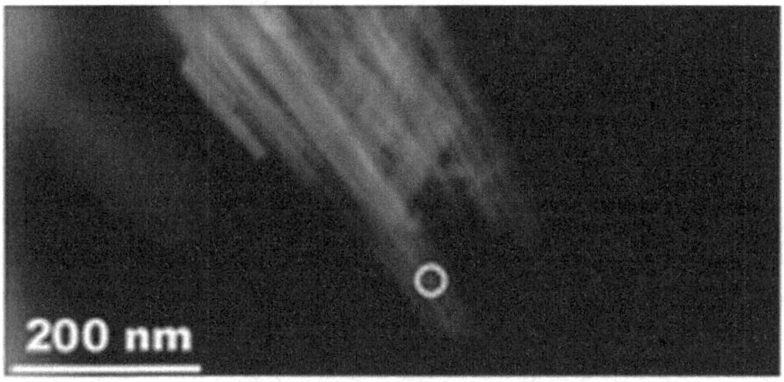

To solve the question whether the nanorods consist of a mixed Mo-W oxide or of the separated binary oxides MoO_3 and WO_3, spot analyses were performed. The EDX spectrum obtained at the position encircled in the HAADF-STEM image shows both metals and thus proves the presence of a mixed oxide.

STEM Detectors

Three types of detectors are used to obtain STEM images:

1. *BF detector:*It is placed at the same site as the aperture in *BF-TEM* and detects the intensity in the direct beam from a point on the specimen.
2. *ADF detector:* The annular dark field (ADF) detector is a disk with a hole in its centre where the BF detector is installed. The ADF detector uses scattered electrons for image formation, similar to the *DF* mode in TEM. The measured contrast mainly results from electrons diffracted in crystalline areas but is superimposed by incoherent Rutherford scattering.
3. *HAADF detector:* The high-angle annular dark field detector is also a disk with a hole, but the disk diameter and the hole are much larger than in the ADF detector. Thus, it detects electrons that are scattered to higher angles and almost only incoherent Rutherford scattering contributes to the image. Thereby, *Z contrast* is achieved.

X-RAY SPECTROSCOPY

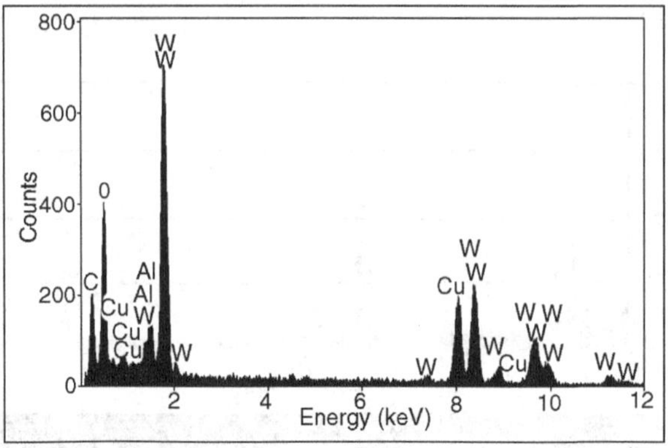

Fig. X-ray spectrum of a aluminum tungsten oxide on a carbon foil supported on a copper grid.

X-ray spectroscopy is a valuable tool for qualitative and quantitative element analysis. Each element has characteristic peak positions corresponding to the possible transitions in its electron shell.

The presence of copper, for example, is indicated by two K peaks at about 8.0 and 8.9 keV and a L peak at 0.85 eV. In heavy elements like tungsten, a lot of different transitions are possible and many peaks are therefore present.

TEMs are almost exclusively equipped with energy-dispersive spectrometers (energy-dispersive X-ray spectroscopy EDXS).

ELECTRON ENERGY LOSS SPECTROSCOPY (EELS)

A typical electron energy loss spectrum is shown below. It consists of three parts:

1. ***Zero-loss peak at 0 eV:*** It mainly contains electrons that still have the original beam energy E0, i.e., they have only interacted elastically or not at all with the specimen. In thin specimens, the intensity of the zero-loss beam is high, so that damage of the CCD chip can occur. Since there is no useful information in it, the zero-loss beam is often omitted during spectrum collection.
2. ***Low-loss region (< 100eV):*** Here, the electrons that have induced *plasmon* oscillations occur. Since the plasmon generation is the most frequent inelastic interaction of electron with the sample, the intensity in this region is relatively high. Intensity and number of plasmon peaks increases with specimen thickness.
3. ***High-Loss region (> 100eV):*** For the ionization of atoms, a specific minimum energy, the critical ionization energy EC or ionization threshold, must be transferred from the incident electron to the expelled inner-shell electron, which leads to ionization edges in the spectrum at energy losses that are characteristic for an element. Thus,

EELS is complementary to *X-ray spectroscopy*, and it can be utilized for qualitative and quantitative element analysis as well. In particular, the detection of light elements is a main task of EELS.

Compared to the plasmon generation, the inner-shell ionization is a much less probable process, leading to a low intensity of the peaks. In the high-loss region, the amount of inelastically scattered electrons drastically decreases with increasing energy loss, thus small peaks are superimposed on a strongly decreasing background (s. spectrum). Because of the low intensity, the representation of the high-loss region is often strongly enhanced (here: intensity gap at about 220 eV).

EELS of vanadium oxide nanotubes

The critical ionization energy EC is sensitive to the chemical situation of the element: e.g., the L edge of Cu metal and CuO are shifted in respect to each other (chemical shift). Moreover, the ionization process may take more energy than EC, and therefore there also is intensity located after the corresponding edge. Actually, this region, designated as ELNES (Energy-Loss Near-Edge Structure), mirrors the DOS and provides information about the bonding situation. Modulations further away from the ionization edge contain information about interatomic distances and coordination (EXELFS, Extended Energy-Loss Fine Structure).

A lot of information is present in a electron energy loss spectrum. The energy resolution (below 1 eV) is much better than in X-ray spectroscopy, and as a result more structural information can be obtained from the fine structure in EELS. For many questions, it is important to get this information with a high local resolution. Such mappings can be done with two different methods:

1. *STEM with EELS:* In STEM mode, an region of the sample is selected, and an EELS is measured on each spot of the defined grid (serial measurement).
2. *EFTEM using an energy filter:* An energy range is selected, and an energy-filtered image is recorded with electrons of this energy (parallel measurement).

ENERGY-FILTERED TRANSMISSION ELECTRON MICROSCOPY (EFTEM)

Energy filtering is a valuable tool for materials characterization. It can be applied for various problems:

1. *Zero-loss Filtering:* By using only the zero-loss beam, all inelastically scattered electron are omitted (reduction of noise). That leads to increased contrast in TEM images (important for specimens with low contrast, e.g. biological specimens) or in diffraction patterns.
2. *Electron Spectroscopic Imaging:* By using electrons with a well-defined energy loss (ionization edge), elemental distribution maps can be generated.

The energy filter (Gatan imaging filter, *GIF*), which is installed below a TEM, allows to record electron energy loss spectra (*EELS*) and element specific images (elemental maps) by means of the electron spectroscopic imaging (ESI) technique. Mostly, the so-called three-window method is applied for mappings (see figure): an image is taken after a suitable ionisation edge of the corresponding element (post-edge image, ΔE3), two additional images (pre-edge 1, ΔE1, and pre-edge 2, ΔE2) are recorded at energy losses smaller than the ionisation edge. Only the electrons passing through the selected energy slit contribute to these images. Specimen drift during recording of these images is compensated for by a cross-correlation algorithm implemented in the software of the GIF.

The pre-edge images are used for an approximate determination of the unspecific background which is then subtracted from the post-edge image leading to an elemental map with enhanced contrast. Alternatively, the post-edge image can be divided by a pre-edge image (ratio method).

Recording and calculating an elemental map by the three-window method is demonstrated by the corresponding images obtained at the K edge of carbon in a *vanadium oxide nanotube*. The TEM image shows the cross-section of a single VO_x-NT. In the post-edge image, the carbon distribution is clearly visvible (bright contrast at the C-containing sites): C is present around the tube due to the organic resins used during the *cross-sectional preparation*, in the centre of the tubes, and between the VO_x layers.

Due to the absence of C there, the VO_x layers appear with dark contrast in the ESI and post-edge image, whereas they are bright in the two pre-edge images since these in fact are post-edge images of the V_M edge.

TRANSMISSION ELECTRON MICROSCOPY (TEM)

Utilized Signals: elastically scattered electrons - direct and diffracted beams.

Methods	Information
BF/DF-TEM	Detection of crystalline areas, defects and grain boundaries, phase analysis, particle size
HRTEM	Direct visualisation of structures and defects
EFTEM	Zero-loss filtering
ED	Determination of lattice parameter and crystal symmetry, crystal orientation, phase analysis

A main advantageous characteristic of a *transmission electron microscope* is the possibility to obtain information in real space (imaging mode) and reciprocal space (diffraction mode) almost *simultaneously*.

An introduction into the function principles of a TEM with animations can be found on Goodhew's Matter *page* and into theory and practice of TEM on the web site of *Rodenburg*. The potential of HRTEM in combination with electron diffraction for structure determination is demonstrated on example of a new *Zr/Nb/W oxide.*

Bright Field (BF) Mode of the TEM

In the bright field (BF) mode of the TEM, an aperture is placed in the back focal plane of the objective lens which allows only the direct beam to pass. In this case, the image results from a weakening of the direct beam by its interaction with the sample. Therefore, mass-thickness and diffraction *contrast* contribute to image formation: thick areas, areas in which heavy atoms are enriched, and crystalline areas appear with dark contrast. It should be mentioned that the interpretation of images is often impeded by the simultaneous occurence of the contrast-forming phenomena.

Dark Field (DF) Images

In dark field (DF) images, the direct beam is blocked by the aperture while one or more diffracted beams are allowed to pass the objective aperture. Since diffracted beams have strongly interacted with the specimen, very useful information is present in DF images, e.g., about planar defects, stacking faults or particle size.

Lattice Images

To obtain lattice images, a larger objective aperture has to be selected that allows many beams including the direct beam to pass. The image is formed by the interference of the diffracted beams with the direct beam (phase contrast). If the point resolution of the microscope is sufficiently high and a suitable sample oriented along a zone axis, then high-resolution TEM (HRTEM) images are obtained. In many cases, the atomic structure of a specimen can directly be investigated by HRTEM.

The incident parallel electron beam, ideally a plane wave, interacts elastically while passing through the specimen, and the resulting modulations of its phase and amplitude are present in the electron wave leaving the specimen.

The wave here, the object exit wave $0(r)$, thus contains the information about the object structure.

Unfortunately, the objective lens is not an ideal but has aberrations (astigmatism, spherical C_s and chromatic C_c aberration) that reduce image quality. The intensity distribution of the exit wave function is decribed by the contrast transfer function (*CTF*).

Contrast Transfer Function

The contrast transfer function CTF (*detailed treatment*), sin(k), gives the phase changes of diffracted beams with respect to the direct beam.

$$X(k) = \pi\lambda \, \Delta f \, k^2 + 1/2\pi \, C_s \, \lambda^3 \, k^4$$

The complicated curve sin(*k*) strongly depends on C_s (spherical aberration coefficient, i.e. the quality of the TEM objective lens), on lambda (electron wavelength defined by *accelerating voltage*), on defocus and on the spatial frequency *k*. While this function is zero at the origin, it becomes positve for intermediate values of *k*. In this region of *k*, all information is transferred with positive phase contrast, i.e. the scattering centers (atom positions) appear with dark contrast Therefore, the information in HRTEM images is consequently directly interpretable till the point resolution. The point resolution of a TEM corresponds to the point where the CTF first crosses the *k* axis at Scherzer defocus. Both the Scherzer defocus Δf_s and the point resolution Δx are functions of C_s and λ:

$$D_S = \sqrt{\frac{4}{3} C_S \lambda}$$

$$D_x = 0.66 \sqrt[4]{C_S \lambda^3}$$

Depending on the defocus, the CTF may oscillate strongly. At larger *k*-vectors, it is strongly damped mainly due to the effect of chromatic aberration, focus spread and energy instabilities. Since the defocus is variable and can be adjusted at the microscope, an adequate value can be chosen to optimize the imaging conditions. CTF of a CM30 TEM (LaB6 cathode; ST lens with Cs = 1.15 mm, Cc = 1.5 mm) close to Scherzer defocus. The CTF crosses the k axis at about 5.1 nm^{-1} which corresponds to a resolution of about 1.9 E.

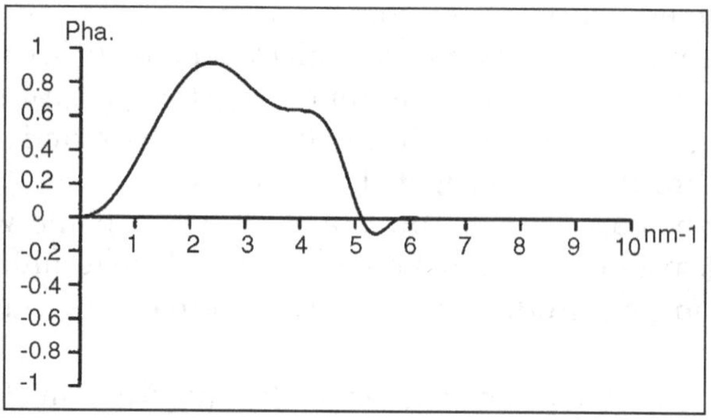

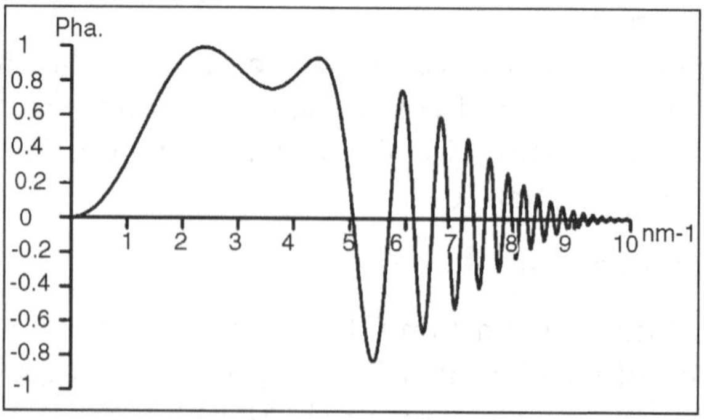

CTF of a 300 kV field-emission TEM (Cs = 1.15 mm, Cc = 1.5 mm) close to Scherzer defocus. In the coherent electron beam of the FEG, the effect of damping is reduced compared to the LaB_6 gun. As a result, there is additional information beyond the point resolution in the oscillating CTF till the information limit at about 1.0 E.

Modern HRTEM Methods

In TEMs with field emission guns, the damping of the CTF is less drastic than in TEMs with a LaB_6 cathode. Thus, there is information available in the oscillations beyond the point resolution till the information limit that can be used to achieve data of higher resolution. A up-to-date method for this is the exit wave function reconstruction from defocus series *(A. Thust, Jülich).*

According to Scherzer, a correction of spherical (C_s) and chromatic (C_c) aberrations cannot be achieved with rotationally symmetric electron lenses. Using asymmetric ray paths, it is possible to overcome this resolution limit and enter the sub-Engström range (*examples*). Such aberration correctors (*CEOS*), as first proposed by *H. Rose*, are now available in commercial TEMs (*FEI, JEOL, Zeiss*). Much HRTEM work has been done on complex oxides with a short and two long axes. If these are observed along the short axis, the a structural model is directly accessible by evaluation of the HRTEM image (examples: *block structures, Nb-W oxides*).

A structure determination of a new *bronze-type oxide* was performed by HRTEM and ED. Of course, the interpretation of HRTEM images has to be confirmed by *image simulation* (e.g., EMS programme of *P. Stadelmann* at the *CIME* (Centre Interdisciplinaire de Microscopie Electronique), EPFL Lausanne,).

DEVELOPMENT OF ELECTRON MICROSCOPES

Electron Microscopes were developed due to the limitations of Light Microscopes which are limited by the physics of light to 500x or 1000x magnification and a resolution of 0.2 micrometers. In the early 1930's this theoretical limit had been reached and there was a scientific desire to see the fine details of the interior structures of organic cells (nucleus, mitochondria...etc.). This required 10,000x plus magnification which was just not possible using Light Microscopes. The Transmission Electron Microscope (TEM) was the first type of Electron Microscope to be developed and is patterned exactly on the Light Transmission Microscope except that a focused beam of electrons is used instead of light to "see through" the specimen.

It was developed by Max Knoll and Ernst Ruska in Germany in 1931. The first Scanning Electron Microscope (SEM) debuted in 1942 with the first commercial instruments around 1965. Its late development was due to the electronics involved in "scanning" the beam of electrons across the sample. An excellent article was just published in *Scanning* detailing the history of SEMs and we would encourage those interested to read it.

HOW DO ELECTRON MICROSCOPES WORK

Electron Microscopes (EMs) function exactly as their optical counterparts except that they use a focused beam of electrons instead of light to "image" the specimen and gain information as to its structure and composition. The basic steps involved in all EMs:

1. A stream of electrons is formed (by the Electron Source) and accelerated toward the specimen using a positive electrical potential.
2. This stream is confined and focused using metal apertures and magnetic lenses into a thin, focused, monochromatic beam.
3. This beam is focused onto the sample using a magnetic lens.
4. Interactions occur inside the irradiated sample, affecting the electron beam.

These interactions and effects are detected and transformed into an image. The above steps are carried out in all EMs regardless of type. A more specific treatment of the workings of two different types of EMs are described in more detail:

- Transmission Electron Microscope.
- Scanning Electron Microscope.

FACILITY OF ELECTRON MICROSCOPE

In electron microscopy a beam of electrons are focused upon a sample. The interaction of the beam with the sample produces a number of measurable effects, most commonly:

Secondary Eectrons

These produce an image of the surface morphology of a sample. This is perhaps the most common form of electron microscope image used in the media (as well as scientific journals of course !). As well as producing superior magnifications than conventional light microscopy the technique also offers a far greater depth of field (focusing range).

Backscatter Electrons

These electron images identify materials of differing atomic density (atomic number contrast).

Cathodoluminescence

Some materials produce light in the visible spectrum when exited by an electron beam. Differences in the trace element concentrations in these materials may then be imaged.

X-rays

These are also produced with interaction of an electron beam and a sample. By measuring the wavelength of the x-rays it is possible to determine qualitatively which elements are present. If a flat, polished sample is used the x-rays measured may be compared to standards to produce quatitative analysis. Most conventional electron microscopes operate under vacuum and sample size is restricted to the size of the sample chamber and stage. Samples are normally coated (with a precious metal such as gold or with carbon) in order to produce an electrically conductive surface layer necessary for this work.

INSTRUMENTS

JEOL JSM-5410LV Scanning Electron Microscope

The *JEOL JSM-5410LV* scanning electron microscope (SEM) allows high resolution photography of samples. Magnification ranges from 35 to 200,000. In addition, this instrument allows samples to be analysed in low vacuum without coating. Thus archaeological artefacts, semi-viscous liquids (molasses), type samples, crystals, waxes and more can be examined.The different photographic modes are

Surface topology

Microstructural characteristics of solid samples - morphology of insects,plants,marine specimens and biological crystals

Atomic Number Contrast Mode

Differentiates between materials of different density - grains in sand - cartilage in animals - heavy metals in plants or tissue Cathodoluminescence - certain impurities in minerals cause visible light to be emitted.

JEOL JXA-8200 Superprobe

The JEOL JXA-8200 superprobe combines the features of high magnification, high resolution imaging with elemental analysis. It allows for the non-destructive analysis of very small amounts of solid materials. This is a state-of-the-art instrument.

JEOL 840A Electron Microprobe

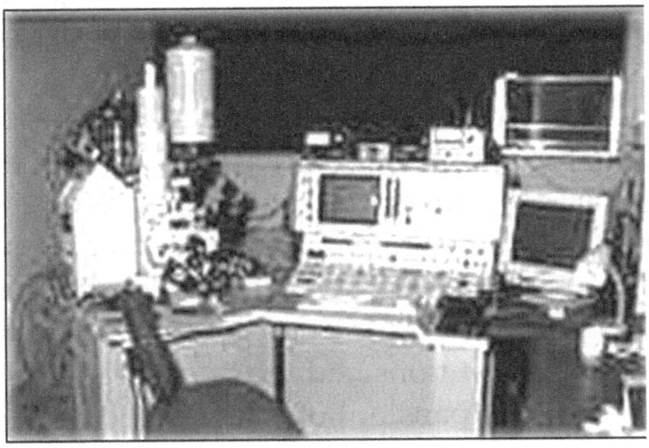

The JEOL 840A electron microprobe (EMPA) is a high resolution SEM and a WD/ED Combined Electron Probe Microanalyzer (EPMA). It has 3 wavelength dispersive X-ray spectrometers (WDS) and an energy dispersive X-ray spectrometer (EDS). The analytical components are controlled by Moran Scientific hardware and software.

The applications for the electron microprobe include analysis of a wide range of sample types inclluding geological, biological, archaeological and metalurgical materials. Chemical analyses may be obtained to provide

- Qualitative elemental analysis of individual features (particularly for identification of small, accessory phases).
- Quantitative element analysis of individual features (requires a flat, polished sample).
- X-ray line profiles and maps, (provides qualitative digital images showing the elemental distribution and quantitative data).

HIGH VOLTAGE ELECTRON MICROSCOPY STUDIES OF AXOPLASMIC TRANSPORT IN NEURONS: A POSSIBLE REGULATORY ROLE FOR DIVALENT CATIONS

ME Stearns

Light and high voltage electron microscopy (HVEM) procedures have been employed to examine the processes regulating saltatory motion in neurons. Light microscope studies demonstrate that organelle transport occurs by rapid bidirectional saltations along linear pathways in cultured neuroblastoma cells.

HVEM stereo images of axons reveal that microtubules (Mts) and organelles are suspended in a continuous latticework of fine microtrabecular filaments and that the Mts and lattice constitute a basic cytoskeletal structure mediating the motion of particles along axons. We propose that particle transport depends on dynamic properties of nonstatic microtrabecular lattice components. Experiments were initiated to determine the effects of changes in divalent cation concentrations (Ca^{2+} and Mg^{2+}) on:

(a) The continuation of transport.

(b) The corresponding structural properties of the microtrabecular lattice. We discovered that transport continues or is stimulated to a limited extent in cells exposed to small amounts of exogenously supplied Ca^{2+} and Mg^{2+} ions (less than 0.1 mM).

Exposure of neurons to increased dosages of Ca^{2+} and Mg^{2+} (0.2-1.0 mM) stimulates transport for 2-4 min at 37 degrees C, but after a 5- to 20-min exposure the saltatory movements of organelles are observed gradually to become shorter in duration and rate particle motion ceases to occur. HVEM observations demonstrated that Ca^{2+} - and with the cessation of motion. Ca^{2+}-containing solutions produced contractions of the microtrabecular filaments, whereas Mg^{2+}-containing solutions had the opposing effect of stimulating an elongation and assembly (expansion) of microtrabeculae.

On the basis of these observations we hypothesize that cycles of Ca^{2+}/ Mg^{2+}-coupled contractions and expansions of the microtrabecular lattice probably regulate organelle motion in nerve cells.

THE SCATTERING OF ELECTRONS BY ATOMS

Elastic scattering

We will study the *elastic scattering* of two isolated spinless particles. The only interaction we take into account is the interaction between the two particles. We ignore the internal structure of the particles. If they have internal structure we assume that it does not influence the elastic scattering process. If we ignore the centre of mass motion, we can study the scattering of particles m_1 and m_2 with interaction energy $V(r_1-r_2)$ by studying the scattering of a fictitious particle of reduced mass

$$m = \frac{m_1 m_2}{m_1 + m_2}$$

by a potential (energy) $V(r)$.

In a typical scattering experiment a target is struck by a beam of monoenergetic particles. Let F_i be the incident flux, i.e. the number of particles per unit area per unit time. $F_i = n_p v$, where n_p is the number of particles per unit volume. Typically n_p is very small and we can neglect any interaction between different incident particles.

We measure the number ΔN_p of particles scattered per unit time into a solid angle $\Delta\Omega$ about the direction defined by the spherical coordinates q and f.

We expect $\Delta N_p \mu F_i$, and $\Delta N_p \mu \Delta\Omega$. We define $\Delta N_p = \sigma_t(\theta, \phi) F_i \Delta\Omega$. Here σ_t (θ, ϕ) is the *differential scattering cross section of the target.* It has the units of an area. Commonly used units are cm^2 and barn=$10^{-24}cm^2$.

Often the target is made up of a large number N_t of scattering centers. If it is sufficiently thin, so that we can neglect multiple scattering, and if the distances between the scattering centers are large compared to the wavelength of the incident particles, so that we may neglect coherence between waves scattered from different centers, we have $\Delta N_p \mu N_t$.

Here we assume that we have a big beam and a small target, so that the beam covers the target completely. We then write $\Delta N_p = \sigma(\theta, \phi) N_t F_i \Delta\Omega$, where $\theta(\theta, \phi)$ is the *differential scattering cross section of a target particle.* The total scattering cross section is given by

$$\sigma = \int \sigma(\theta, \phi) d\Omega$$

In quantum mechanics we describe the scattering of a particle by a potential $V(r)$ by describing the time evolution of the wave packet $y(r, t)$ representing the state of the particle. Assume that the particle approaches the target from the left, along the z-axis. Assume that at $t \to -\infty$ the characteristics of the wave packet are known. The subsequent evolution can be calculated using the Schroedinger equation

$$-\frac{\hbar}{2m}\nabla^2\psi(r,t)+V(r)\psi(r,t), i\hbar\frac{\partial}{\partial t}\psi(r,t)$$

or

$$H\psi(r,t)=i\hbar\frac{\partial}{\partial t}\psi(r,t)$$

The eigenstates of the Hamiltonian form a basis for the state space E, $\{|E_k>\}$, the basis of stationary states. An arbitrary state vector may be expanded in terms of these basis vectors.

$$|\psi>=\sum_k a_k|E_k>$$

The time evolution of the eigenstates of H can be easily found.

$$H\psi k(r,t)=i\hbar\frac{\partial}{\partial t}\psi k\,(r,t),\ \psi k(r,t)=\phi_k(r)e^{\frac{E_k}{\hbar}}.$$

If we solve the eigenvalue equation $Hf_k(r)=E_kf_k(r)$, then we can obtain the evolution of any wave packet by expressing it as a superposition of stationary states. We will therefore first attempt to find the stationary state solutions of the scattering problem. We are trying to solve the equation

$$\left(-\frac{\hbar^2}{2m}\nabla^2+V(r)\right)\phi_k(r)=E_k\phi_k(r),$$

for $E_k>0$. Let

$$E_k\frac{\hbar^2k^2}{2m} \text{ and } V(r)=\frac{\hbar^2}{2m}U(r).$$

Then we may write

$$\left(\nabla^2+k^2-U(r)\right)\phi_k(r)=0.$$

For each value of k an infinite number of solutions exist. We must choose from them the one solution that corresponds to the physical problem to be studied.

To do this, we look at the *asymptotic behaviour* of the solutions. We assume that we have a finite range potential, *i.e.* that $U(r)\to 0$ as $r\to\infty$ faster than r^{-1}. Note that this excludes the Coulomb potential. Then

$$\left(\nabla^2+k^2\right)\phi_k(r)=0 \text{ as } r\to\infty$$

Plane waves of the form exp($ik.r$) moving into an arbitrary direction are solutions to this equation. In addition outgoing and incoming spherical waves of the form

$$\frac{e^{ur}}{r} \text{ and } \frac{e^{-ur}}{r}$$

are solutions as $r\to\infty$.

In our example, at large negative times $t \to -\infty$ the particle is a free particle described by a wave packet build from plane waves moving towards the origin along the z-axis from z=-∞. We therefore look for a plane wave term e^{ikz} in the asymptotic stationary state solution.

For large positive values of $t \to \infty$, the particle is described by transmitted wave packet build from plane waves and a scattered wave packet. We expect the scattered wave packet to be a superposition of outgoing stationary states. We therefore look for a term $\frac{e^{ur}}{r}$ in the asymptotic stationary state solution.

Since, in general, scattering is not isotropic, the amplitude of the scattered wave can be a function of q and f. The asymptotic form of the eigenfunctions of H is therefore given by

$$\phi_k(r, \theta, \phi) \to \mathrm{A}\left[e^{ikz} + f_k + (\theta, \phi)\frac{e^{ur}}{r} \right] \text{ as } r \to \infty$$

Here $f_k(q,f)$ is defined as the *scattering amplitude.*

We will now check that a wave packet with the proper asymptotic behaviour can indeed be build out of energy functions of this form. Let

$$\psi(r, t) = \int_0^\infty dkg(k)\phi_k(r)\, e^{\frac{E_t}{\hbar}}, \text{ with } E_k = \frac{\hbar^2 k^2}{2m}.$$

Let $g(k)=|g(k)|e^{ia(k)}$. This is the most general form. For simplicity let us assume that $a(k)$ is zero and $g(k)$ is real.

Note: By labeling our eigenfunctions with a single index k, we have neglected the degeneracy of the energy eigenvalues. Plane waves with the same energy, traveling in different directions should really be labeled with an index k. Limiting ourselves to an index k in building the wave packet means taking care of the energy dispersion, (the spread of the wave packet along the z-axis), but not of the angular dispersion (the spread of the wave packet perpendicular to the z-axis).

The *asymptotic* wave packet is

$$\psi(r, t) = \int_0^\infty dkg(k)e^{-i\frac{E_k}{\hbar}i} + \int_0^\infty dkg(k) f_k(\theta,\, \phi)\frac{e^{ikr}}{r}e^{-i\frac{E_k}{\hbar}t}.$$

Assume $g(k)$ has a pronounced peak at $k= k_0$ and is non zero only in a small interval Dk about k_0. We want to find the maximum of each of the two packets making up the asymptotic wave packet. Let

$$\beta(k) = -\frac{E_k t}{\hbar},\ \beta(\mathrm{k}) = \beta(\mathrm{k_0}) + (\mathrm{k} - \mathrm{k_0})\frac{\partial\beta}{\partial k}|k = k_0,$$

i.e. we assume β varies little in the small interval Δk. For the first wave packet we then have

$$\int_0^\infty dkg(k)\, e^{ikz}e^{-i\frac{E_k}{\hbar}} = e^{i(k_ez + \beta(k_0))}\int_0^\infty dkg(k)e^{i(k-k_0)(z-z_0)}.\ \text{where } z_0 = -\frac{\partial\beta}{\partial k}|k = k_0.$$

If $|z - z_0| >> \frac{1}{\Delta k}$, then the period of $e^{i(k-k_0)(z-z_0)} = e^{i\frac{2e}{\tau}(k-k_0)}$ is much smaller than Δk, and the function oscillates rapidly in the integration interval. The integral then yields approximately zero. If $|z-z_0| \gg 0$, then the period of $e^{i(k-k_0)(z-z_0)} = e^{i\frac{2e}{\tau}(k-k_0)}$ approaches infinity, and the function hardly oscillates at all in the integration interval. The integral then yields its maximum value. The first wave packet has its maximum value at $z_M = z_0 = \frac{\partial\beta}{\partial k} | k = k_0$.

$$-\frac{\partial\beta}{\partial k} | k = k_0 = \frac{t}{\hbar}\frac{\partial E_k}{\partial k} | k = k_0 = \frac{t}{\hbar}\frac{\hbar^2 2k_0}{2m} = vgt,\ v_g = \frac{\hbar k_0}{m}.$$

Consider the second wave packet.

$$\int_0^\infty dkg(k) f_k(\theta, \phi) - t\frac{e^{ikr}}{\hbar} e^{\frac{E_k}{\hbar}t} = \frac{| f_k(\theta, \phi)|}{r} e^{i(k_0r + \beta(k_0))} \int_0^\infty dkg(k)e^{i(k-k_0)(r-r_0)},$$

where $\beta'(k) = \sigma_\kappa(\theta, \phi) + \beta(k)$, $r_0 = -\frac{\partial\beta'}{\partial k} | k - k_0$, and $a_k\,(\theta, \phi)$ is the phase of the scattering amplitude. For a given q and f the second wave packet has its maximum value at

$$r_M\,(r, \theta, \phi) = -\frac{\partial\sigma_\kappa(\theta, \phi)}{\partial k}|_{k-k_0 + v_g t}.$$

For large negative values of t, r_M is negative, there is no scattered wave packet. All we find is the incident wave packet. For large positive values of t both packets are present. One packet moves along the z-axis and the other diverges in all directions. We can therefore build a proper asymptotic wave packet out of our asymptotic energy eigenfunctions and conclude that they are well suited for the description of the scattering process.

The energy eigenfunctions $\{\phi_k(r)\}$ form a basis for the space of square integrable functions L^2, but for $E_k > 0$ the functions $\phi_k(r)$ are not square integrable, i.e. $\phi_k(r) \notin L^2$. The basis can only be "orthonormalized in the Dirac sense". To find the scattering cross section we therefore do not look at probabilities, but probability currents. We compare the incident and the scattered probability current. The probability current density is given by

$$J = \frac{\hbar}{2mi}[\psi^{\#}\vec{\nabla}\psi - \psi\vec{\nabla}\psi^{\#}] = \frac{1}{m}\mathrm{Re}\left[\psi^{\#}\frac{\hbar}{i}\vec{\nabla}\psi\right].$$

In the asymptotic region the plane wave term represents a wave with incident current density

$$J_i = \frac{1}{m}\mathrm{Re}\left[e^{ikz}\frac{\hbar}{i}\frac{\partial}{\partial z}e^{ikz} \right] = \frac{\hbar k}{m}.$$

To find the current density represented by the term $f_k(\theta, \phi) = \frac{e^{ur}}{r}$ we evaluate

$$\mathrm{v}\left(f_k(\theta, \phi)\frac{e^{ikr}}{r} \right) = \left| \hat{r}\frac{\partial}{dr} + \hat{\theta}\frac{1}{r}\frac{\partial}{\partial\theta} + \hat{\phi}\frac{1}{r\sin\theta}\frac{\partial}{\partial\phi} \right| \left(f_k(\theta, \phi)\frac{q^{ikr}}{r} \right).$$

Retaining only the lowest order terms in r, we find that $f_k\ (\theta, \phi)\frac{e^{ur}}{r}$ represents an outgoing wave of current density

$$J_s = \frac{\hbar k}{m} \mid f_k(\theta, \phi) \mid^2 \frac{1}{r^2}.$$

In practice, interference terms are negligible except near the forward direction, since in an actual experiment the detector is generally placed in such a way that it can only be reached by scattered particles. We then have

$$dN_p = CJ_s.\ dS_s = CJ_s r^2 d\Omega = C\frac{\hbar k}{m} \mid f_k(\theta, \phi) \mid^2 d\Omega \ \ \mathrm{Fi} = C\frac{\hbar k}{m},\ dN_p$$

$$= F_i \mid f_k\ (\theta, \phi) \mid^2 d\Omega.$$

But $dN_p = s_k(q,f)F_i dW$. The scattering cross section is equal to the square of the scattering amplitude.

$$\sigma_k(\theta, \phi) = \mid f_k(\theta, \phi) \mid^2.$$

To calculate the cross section we need to know the asymptotic form of the stationary state wavefunction $\phi_k(\boldsymbol{r})$. To this effect we have to solve the eigenvalue equation $(\nabla^2 + k^2 - U(r))\,\phi_k(r) = 0$ given the potential $U(\boldsymbol{r})$.

Scattering by a central potential $V(r)$

A basis $\{\phi_{klm}(r, \theta, \phi)\}$ of eigenfunctions of H, L^2, and L_z exists. The eigenvalues are $\frac{\hbar^2 k^2}{2m}$, $l(l + 1)\hbar^2$, respectively. We call these wave functions *partial waves*.

For large r the partial waves will resemble the common eigenfunctions of H_0, L^2, and L_z, where H_0 is the Hamiltonian of a free particle. These eigenfunctions have a well defined energy and magnitude and z-component of angular momentum. We denote these eigenfunctions by $\phi^0_{klm}(r, \theta, \phi)$ and call them *free spherical waves*. We assume that the *asymptotic behaviors* of $f_{klm}(r, \theta, \phi)$ and $\phi^0_{klm}(r, \theta, \phi)$ are identical except for a difference in phase, since the solutions satisfy the same differential equation as r goes to infinity. We will show that for a fixed k the value of this phase shift for all l is all we need to calculate the scattering cross section.

2

Fundamental Principles of Scanning Electron Microscopy

Accelerated electrons in an SEM carry significant amounts of kinetic energy, and this energy is dissipated as a variety of signals produced by electron-sample interactions when the incident electrons are decelerated in the solid sample. These signals include secondary electrons (that produce SEM images), backscattered electrons (BSE), diffracted backscattered electrons (EBSD that are used to determine crystal structures and orientations of minerals), photons (characteristic X-rays that are used for elemental analysis and continuum X-rays), visible light (cathodoluminescence—CL), and heat.

Secondary electrons and backscattered electrons are commonly used for imaging samples: secondary electrons are most valuable for showing morphology and topography on samples and backscattered electrons are most valuable for illustrating contrasts in composition in multiphase samples (i.e. for rapid phase discrimination). X-ray generation is produced by inelastic collisions of the incident electrons with electrons in discrete ortitals (shells) of atoms in the sample.

As the excited electrons return to lower energy states, they yield X-rays that are of a fixed wavelength (that is related to the difference in energy levels of electrons in different shells for a given element). Thus, characteristic X-rays are produced for each element in a mineral that is "excited" by the electron beam. SEM analysis is considered to be "non-destructive"; that is, x-rays generated by electron interactions do not lead to volume loss of the sample, so it is possible to analyse the same materials repeatedly.

SCANNING ELECTRON MICROSCOPE

The path of the electron beam within the scanning electron microscope differs from that of the TEM. The technology used is based on television techniques. The method is suitable for the depiction of preparations with conductive surfaces. Biological objects have thus to be made conductive by coating with a thin layer of heavy metal (usually gold is taken).

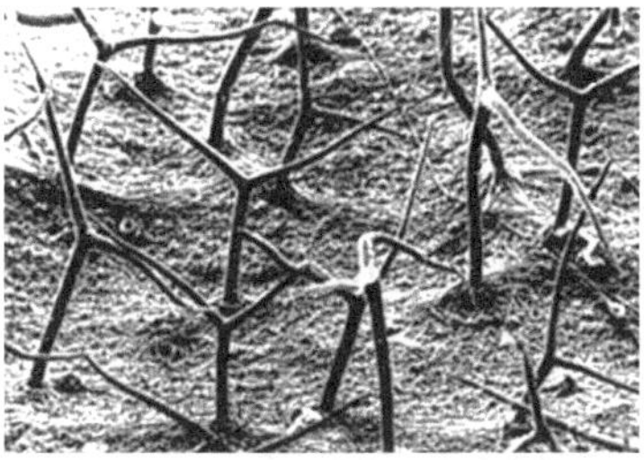

The power of resolution is normally smaller than in transmission electron microscopes, but the depth of focus is several orders of magnitude greater. Scanning electron microscopy is therefore also well-suited for very low magnifications. Numerous examples will be given in the following.

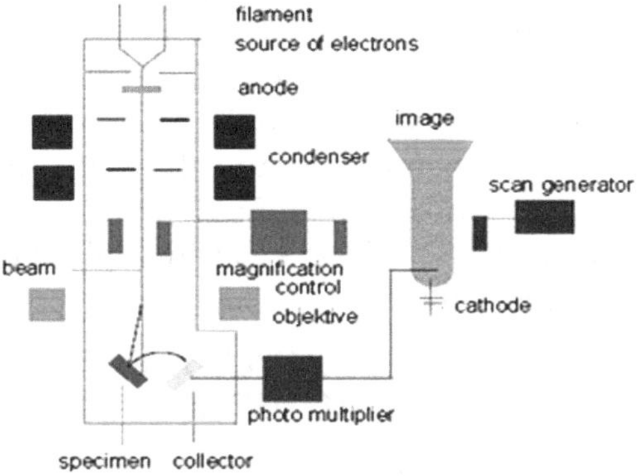

The surface of the object is scanned with the electron beam point by point whereby secondary electrons are set free. The intensity of this secondary radiation is dependent on the angle of inclination of the object's surface. The secondary electrons are collected by a detector that sits at an angle at the side above the object. The signal is then enhanced electronically. The magnification can be chosen smoothly (depending on the model) and the image appears a little later on a viewing screen.

WHAT IS SCANNING ELECTRON MICROSCOPY (SEM)

A typical SEM instrument, showing the electron column, sample chamber, EDS detector, electronics console, and visual display monitors.

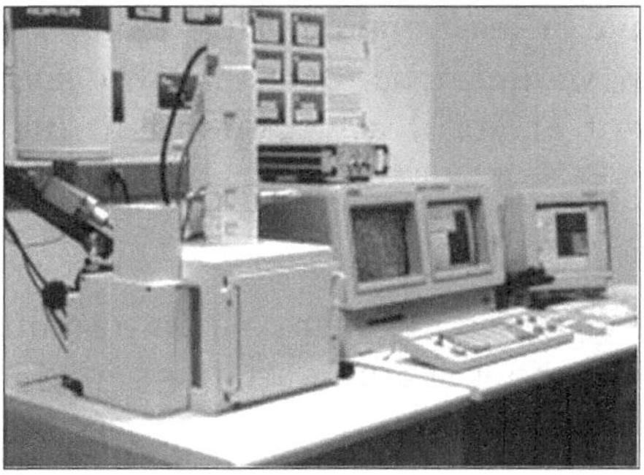

The scanning electron microscope (SEM) uses a focused beam of high-energy electrons to generate a variety of signals at the surface of solid specimens.

The signals that derive from electron-sample interactions reveal information about the sample including external morphology (texture), chemical composition, and crystalline structure and orientation of materials making up the sample.

In most applications, data are collected over a selected area of the surface of the sample, and a 2-dimensional image is generated that displays spatial variations in these properties. Areas ranging from approximately 1 cm to 5 microns in width can be imaged in a scanning mode using conventional SEM techniques (magnification ranging from 20X to approximately 30,000X, spatial resolution of 50 to 100 nm).

The SEM is also capable of performing analyses of selected point locations on the sample; this approach is especially useful in qualitatively or semi-quantitatively determining chemical compositions (using EDS), crystalline structure, and crystal orientations (using EBSD). The design and function of the SEM is very similar to the EPMA, and considerable overlap in capabilities exists between the two instruments.

SCANNING ELECTRON MICROSCOPY (SEM) INSTRUMENTATION

Essential components of all SEMs include the following:

- Electron Source ("Gun")
- Electron Lenses
- Sample Stage
- Detectors for all signals of interest
- Display/Data output devices
- Infrastructure Requirements:
  - Power Supply
  - Vacuum System
  - Cooling system
  - Vibration-free floor
  - Room free of ambient magnetic and electric fields

SEMs always have at least one detector (usually a secondary electron detector), and most have additional detectors. The specific capabilities of a particular instrument are critically dependent on which detectors it accommodates.

Applications

The SEM is routinely used to generate high-resolution images of shapes of objects (SEI) and to show spatial variations in chemical compositions (usually EDS, also BSE and CL).

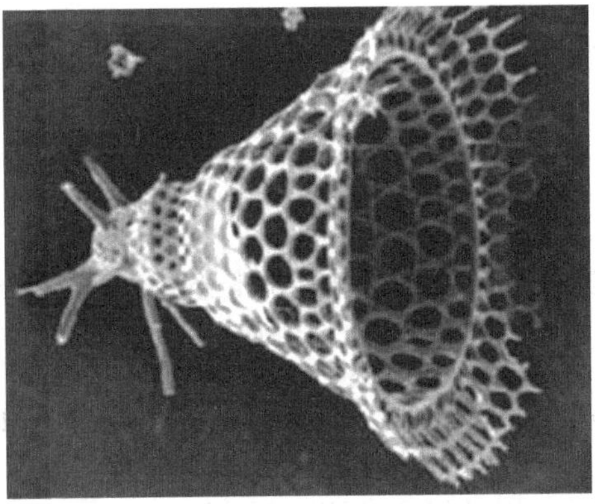

This instrument is also widely used to identify phases based on qualitative chemical analysis and/or crystalline structure. Precise measurement of measurement of very small features and objects down to 50 nm in size is also accomplished using the SEM.

Backescattered electron images can be used for rapid discrimination of phases in multiphase samples. SEMs equipped with diffracted backscattered electron detectors can be used to examine microfabric and crystallographic orientation in many materials.

STRENGTHS AND LIMITATIONS OF SCANNING ELECTRON MICROSCOPY (SEM)

There is arguably no other instrument with the breadth of applications in the study of solid materials that compares with the SEM. The SEM is critical in all fields that require characterization of solid materials. While this contribution is most concerned with geological applications, it is important to note that these applications are a very small subset of the scientific and industrial applications that exist for this instrumentation.

Most SEM's are comparatively easy to operate, with user-friendly "intuitive" interfaces. Many applications require minimal sample preparation. For many applications, data acquisition is rapid (less than 5 minutes/image for SEI, BSE, spot EDS analyses.) Modern SEMs generate data in digital formats, which are highly portable.

LIMITATIONS

Samples must be solid and they must fit into the microscope chamber. Maximum size in horizontal dimensions is usually on the order of 10 cm, vertical dimensions are generally much more limited and rarely exceed 40 mm. For most instruments samples must be stable in a vacuum on the order of 10^{-5} - 10^{-6} torr. Samples likely to outgas at low pressures (rocks saturated with hydrocarbons, "wet" samples such as coal, organic materials or swelling clays, and samples likely to decrepitate at low pressure) are unsuitable for examination in conventional SEM's.

However, "low vacuum" and "environmental" SEMs also exist, and many of these types of samples can be successfully examined in these specialized instruments. EDS detectors on SEM's cannot detect very light elements (H, He, and Li), and many instruments cannot detect elements with atomic numbers less than 11 (Na). Most SEMs use a solid state x-ray detector(*EDS*), and while these detectors are very fast and easy to utilize, they have relatively poor energy resolution and sensitivity to elements present in low abundances when compared to wavelength dispersive x-ray detectors (WDS) on most electron probe microanalyzers (EPMA).

An electrically conductive coating must be applied to electrically insulating samples for study in conventional SEM's, unless the instrument is capable of operation in a low vacuum mode.

SAMPLE COLLECTION AND PREPARATION

Sample preparation can be minimal or elaborate for SEM analysis, depending on the nature of the samples and the data required. Minimal preparation includes acquisition of a sample that will fit into the SEM chamber and some accommodation to prevent charge build-up on electrically insulating samples. Most electrically insulating samples are coated with a thin layer of conducting material, commonly carbon, gold, or some other metal or alloy.

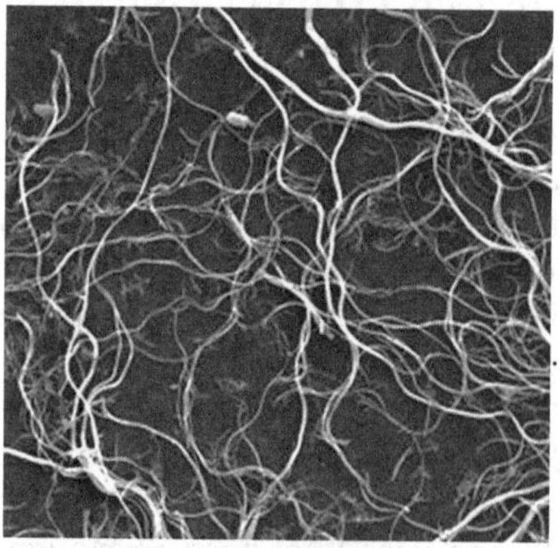

UICC Asbestos Chrysotile 'A' standard

The choice of material for conductive coatings depends on the data to be acquired: carbon is most desirable if elemental analysis is a priority, while metal coatings are most effective for high resolution electron imaging applications. Alternatively, an electrically insulating sample can be examined without a conductive coating in an instrument capable of "low vacuum" operation.

Data Collection, Results and Presentation

Representative SEM images of asbestiform minerals from the *USGS Denver Microbeam Laboratory*

Fig. Tremolite asbestos, Death Valley, California

Fig. Anthophyllite asbestos, Georgia

SCANNING ELECTRON MICROSCOPY (SEM)

Fig. Winchite-richterite asbestos, Libby, Montana

- Major components are the electron column consisting of an electron gun and the electron lenses, and the control console consisting of a cathode ray tube viewing screen and the scanning and control electronics for the electron beam.

- The three dimensional appearance of the image is due to the large depth of field.
- A scanning electron microscope (SEM) employs a probe lens to focus the electron beam into a fine probe and scanning coils are used to scan the probe over the sample.
- Resolution of the SEM is controlled by the probe lens. It is the inelastically scattered electrons that provide information. Usually < 30 keV electrons are used in SEM.
- In optical microscopy and TEM, information is collected continuously over the full field of view (from all image points simultaneously) and focused by suitable lenses to form a magnified image. In SEM, information is collected sequentially.

Electron Sources

The filament is resistively heated to 2000 – 2700K by applying a high voltage and a small amount of current to a point that valence electrons are released from its tip in what is called a space charge cloud. The amount of energy needed to cause electrons to leave the filament is called the work function. The electrons are released in all directions. The Wehnelt cap/grid has a slight negative potential (charge) - or excess of electrons - to create negative lines of force that focus the emitted electrons and control their emission.

The filament or cathode is supplied with a high negative voltage, e.g. -20,000 volts. The anode, a metal plate with a hole in it, is at ground potential (0 volts) and is greatly positive with respect to the cathode. This potential difference accelerates the electrons toward the anode. Along the route from the cathode to the anode, the paths of the individual electrons cross each other. This is called the point of crossover with diameter d_0. It is considered the real source in the electron gun, and its image is projected onto the specimen surface. After the crossover, the electron beam diverges with a divergence angle a_0. The condenser and objective lenses then produce a demagnified image of the crossover on the specimen with an image diameter d_p. The anode has a hole in it. This hole allows only a fraction of the electrons to continue down the column toward the lenses. The remaining electrons are collected on the anode and returned via the ground to the voltage supply.

Hot Cathode Guns: Tungsten filaments and Lab-six – LaB_6 crystals, producing thermionic emission of electrons.

Thermionic emission formula:

$$\text{Current density } I = AT^2 \exp(-E/kT)$$

A = Richardson's constant depending on the source material,

T = emission temperature K (C+273),

E = work function or the energy required to escape from the filament into the vacuum, k = Boltzman's constant.

Tungsten	LaB6
stable beam current	
stable beam current	
short life	10 times longer life
large tip	smaller tip
large emitting area (probe diameter)	smaller emitting area (probe diameter)
low brightness	higher brightness (10 times higher current)
high work function	lower work function
high evaporation rate	medium evaporation rate
lower vacuum (10^{-6} Torr)	higher vacuum (10^{-7} Torr)
low resolution	higher resolution (due to thinner beam)
2700 K	1700 K

Field Emission

This is based on the fact that electrons can be drawn off from a material by applying a high voltage in ultra high vacuum conditions:

- Stable beam current if heated, unstable if cold
- Very long life
- More monochromatic
- Small tip
- Very small emitting area (small probe diameter)
- Very high brightness (1000 times compared to hot guns)
- Concentrated electric field can tunnel through energy barrier
- Low evaporation rate
- Very high vacuum (10^{-10} Torr)
- Very high resolution (electrons are emitted from a very fine tip)
- Three types – cold, thermal, Schottky

Electron Beam

- The electrons that go through the hole in the anode and continue down the column to the specimen make up the electron beam.
- The flow of these electrons in called the beam current.
- The electron beam is manipulated by lenses and apertures.

Electromagnetic Lenses and Focusing

- Since electrons have a charge, their direction of travel can be altered by an electromagnetic field. An electron traveling in off-axis to a uniform magnetic field follows a helical path. Electrons can be brought to focus by engineering the electrostatic and/or magnetic fields. In the region of electron gun, the beam is influenced by the electrostatic field. All the subsequent focusing is achieved by electromagnetic lenses.

- The condenser lens concentrates (or demagnifies) the beam of electrons into a small area called a spot. It is like the condenser on an optical transmission microscope which concentrates light from the light source (e.g,. a light bulb) into a small area as it passes through the specimen slide. The size of the spot can be adjusted. It can be a large area or it can be adjusted to be a small area. Condenser or spot size control is one of the most important controls on an SEM.
- In the electromagnetic lens, the intensity of the field (the magnetic flux) causes a radial vector along the optical axis, so when an electron is accelerated through the pole pieces, it takes a helical path through the lens. The rotational force is the product of the electron velocity and the density of the magnetic flux. This vector interaction also results in focusing as the strength of the lens is changed.
- Focal length of the electromagnetic lens is controlled by varying the lens current. The focal length is approximately proportional to V/ $(NI)^2$ where V is the accelerating voltage, N is the number of turns in the magnet coil and we are the current.
- Electron lenses demagnify the image of the beam crossover (with d_0 ~ 50mm for a heated electron gun) in the electron gun to the final spot size on the specimen of ~10 nm. This represents a demagnification by 5000 times.

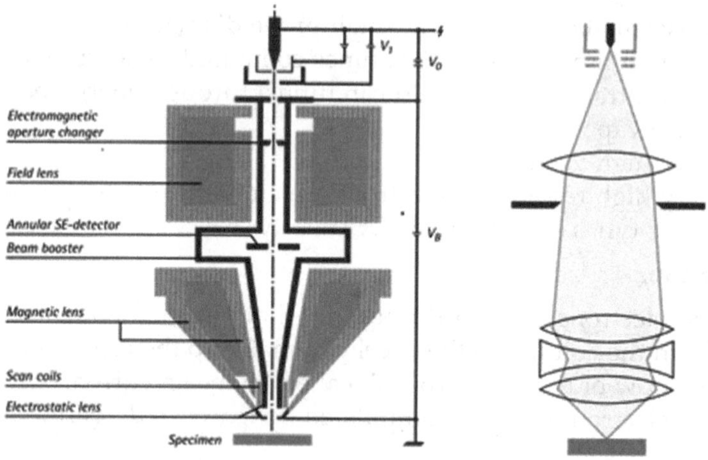

Operating priciple of the LEO 982 with GEMINI column.
V_1 - extractor voltage at first anode
V_0 - accelerator voltage at second anode
V_B - booster voltage.

- These lenses are created with high precision and even a hairline scratch can distort their magnetic field and will have to be replaced. Most electromagnetic lenses are cooled with water to prevent extra heating.

- Their functions are similar to optical lenses. A condenser lens can condense electrons; an objective lens can focus electrons on the specimen, and a projector lens can project an image onto a screen.

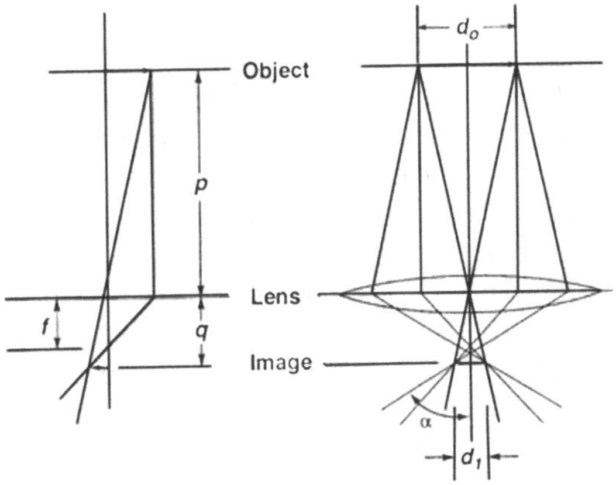

Geometric optics for a thin lens operated in demagnifying mode. (Left) Geometric ray diagram showing the graphical construction for determining the focal length *f*. (Right). The geometric demagnification of the gun crossover d_0 for forming the intermediate crossover d_1 after the first condenser lens. Note that the image is inverted compared to the object.

$$1/f = 1/p + 1/q$$

Demagnification m = $p/q = d_0/d_1$

- Like in optical microscopy, spherical aberration makes peripheral electrons to be deflected more than electrons that are closer to the centre. Electrons in the beam with slightly different velocities or wavelengths causes chromatic aberration.
- Both types of aberrations results in a disk rather than a point where all the rays converge. Both types of aberrations make the image blurred and reduce resolution.
- Diameter of the disk d_s in the case of spherical aberration

$$d_s = \tfrac{1}{2} C_s \alpha^3$$

Where C_s is the spherical aberration coefficient (usually few millimeters) directly related to the accelerating voltage and the focal length of the lens and a is the aperture angle.

- Diameter of the disk in the case of chromatic aberration

$$d_c = \Delta E/E\ C_c \alpha$$

Where ΔE/E is the fractional variation in electron beam voltage, C_c is the chromatic aberration coefficient related to the focal length of the lens and a is the aperture angle.

- Increasing the working distance produces a larger spot size at the specimen and a consequent degradation of the image resolution.
- Increasing the condenser lens strength increases the demagnification of each lens by reducing the probe size d_p.

Neglecting chromatic aberration, the dependence of probe size d_p on probe current i_p can be written as:

$$d_p = KC_s^{1/4}\lambda^{3/4}\left(\frac{i_p}{\beta\lambda^2}+1\right)^{3/8},$$

where constant K ~ 1. The value of d_p can be obtained in nm when C_s and l are in nm, i_p in Amperes and brightness in A/cm^2 sr.

- Depending on the type of electron source, and its inherent brightness, the focused beams can have sizes ranging from nanometers to micrometers carrying currents ranging from picoamperes to microamperes. The thermionic guns have a minimum probe size of ~ 10 nm. Field emission guns have the smallest probe diameters ~ 1.2 nm suitable for higher resolution SEM measurements.
- Resolution can be improved by increasing the accelerating voltage. But one has to worry about radiation damage at such high voltages.
- The lenses in a SEM are generally electromagnets, coils of copper wire around a hollow iron core through which the electrons pass as they are accelerated down the column. By applying a direct current through the copper coil, we can create a magnetic field in the hollow of the lens that will slightly change the path of the electron beam, and we can control that path by varying the current in the coil, which in turn changes the strength of the electromagnetic field.

The electromagnetic lens is a notoriously poor lens when compared to the glass lens that focuses light. Some aberrations cannot be corrected as they can in glass lenses.

ENVIRONMENTAL SCANNING ELECTRON MICROSCOPY

The ability of the ESEM to condense and evaporate liquids has enabled the in situ dynamic study of condensation, evaporation and transport of water inside carbon nanotubes. It has been possible to see liquid menisci inside straight, CVD-fabricated carbon nanotubes (CNTs) having disordered walls. From the measured contact angles, it is clear that these CNTs are hydrophilic. Complex meniscus shapes and slow liquid dynamics due to water confinement and strong interaction with tube walls have been observed.

ELECTRON MICROSCOPY STUDY OF WATER IN CARBON NANOPIPES

Since the discovery of carbon nanotubes (CNTs), the scientific community has been fascinated by their outstanding properties and potential applications, including use in nanofluidic devices and composites. The behaviour of water inside CNT channels is of interest for a number of reasons. First, a fundamental understanding of liquid flow in nanochannels is still lacking. Additionally, CNTs may be used as nanopipes in nanofluidic devices, and control of liquid transport through CNTs is needed to develop functional devices. Finally, there is an ongoing discussion in the literature regarding the wetting of CNT surfaces. Mostly hydrophobic behaviour is anticipated, but hydrophilic behaviour has also been shown for CNTs under special conditions. The *in situ* study of liquids of practical interest (aqueous solutions and biofluids) inside CNTs is therefore extremely important.

Transmission electron microscopy (TEM) studies have shown that closed CNTs can retain fluids trapped during synthesis. However, TEM is limited to imaging closed CNTs only, and the autoclave treatment used for filling the CNTs in these studies restricts the choice of liquids. Recent advances in environmental scanning electron microscopy (ESEM) have enabled scientists to image hydrated materials at low pressures. Unlike the conventional scanning electron microscope (SEM), the ESEM operates by circulating a gas, generally water vapour or nitrogen, through the chamber while imaging.

Furthermore, by means of several pressure-limiting apertures, the ESEM can be used at pressures of up to 20 Torr, which are much higher than the pressures allowed by the conventional SEM. With the ability to manipulate the pressure and temperature by means of a Peltier cooling stage within the ESEM chamber, it is possible to induce the condensation and evaporation of organic and aqueous fluids. In the case of distilled water, if the temperature is maintained constant at approximately 4 °C, water droplet formation will initiate around 5.8 Torr.

Interactions of CNT and carbon nanofibers with fluids have been observed with the ESEM. However, due to the low depth of penetration of the electron beam, these studies were limited to surface interactions only. The penetration depth of electrons depends on the atomic weight and density of the material, as well as the accelerating voltage. A higher voltage causes a deeper penetration of the electrons. Thus, thin walls and relatively high accelerating voltages are required to see through a CNT. The objective of this work was to conduct *in situ* fluidic experiments at the nanoscale in the ESEM, by using the ability to see through the CNT walls, and to study the liquid flow, condensation, and evaporation inside CNT.

Experimental Section.

The CNTs used in these experiments were obtained by chemical vapour deposition (CVD) of carbon in alumina membranes with pores ~250 nm in diameter from a 30% ethylene/helium gas mixture at 670 °C for 6 h in a process described in ref 17.

Sonication in 1 M NaOH for 90 min was used to dissolve the alumina and release the CNTs. Tubes of 200-300 nm in diameter have been fabricated. These carbon nanopipes have wide channels and very thin, smooth, and straight walls, similar to water pipes but orders of magnitude smaller, as shown in figure. A great majority had at least one end open and more than 50% of the tubes were open on both sides after release from the membrane.

High resolution TEM imaging shows that the CNTs have disordered walls. It is also likely that graphene sheets in the walls are hydrogen-terminated because of the synthesis from the C-H environment. Due to the CNTs' thin walls (12-15 nm), they are transparent under the ESEM electron beam at electron voltages as low as 10 kV, though optimal imaging was done at 25 kV or higher. To observe the CNTs under the ESEM in environmental mode, they were placed on a type-303 stainless steel sample holder surface, which was in turn positioned on a Peltier cooling stage. A gaseous secondary electron detector was employed for imaging.

There are two possible ways to induce condensation of water within the chamber: either a droplet of water can be placed on the sample holder surface and/or the cooling stage, or the moisture from the water vapour within the chamber can be used. However, it is easier to control the liquid when it is placed within the chamber initially than if the moisture from the chamber is condensed. The chamber was evacuated at an intermediate pressure (around 5 Torr) in order to evaporate the water droplet initially placed on the cooling stage. As a result, the CNT were initially dry. The temperature was set constant, generally to 4 or 5 °C. The pressure was then increased at a rate of about 0.1 Torr for approximately every 15 s. Since condensation of water within the ESEM chamber occurs suddenly, this slow pressure increase rate allowed for better control of liquid phase formation.

The contact angle for water on the sample holder was measured optically in air. The contact angle for water in nanotubes was measured from SEM images and the estimates were done with understanding that the tube surface was not flat and that the environment was a low-pressure water vapour rather than air.

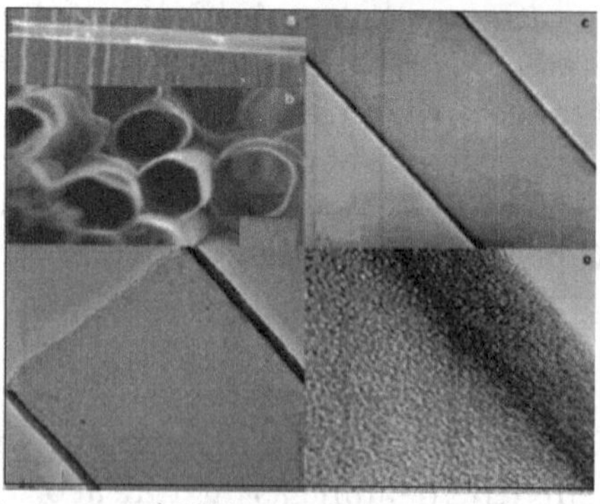

SEM and TEM images of CVD carbon nanopipes. Individual tubes. A bundle of tubes after dissolution of the membrane shows an asymmetric shape and polygonization due to thin and flexible walls. Open tip of a tube. The tapered morphology of the tip indicates it is the as-grown end of the nanotube, not the broken end. High-resolution TEM image showing that the tube wall has a disordered structure. The thickness of the tube wall is 12 nm.

Results and Discussion

Since the principles of image forming and contrasts in ESEM are different from the TEM or the conventional SEM, interpretation of images is not as straightforward. In TEM, the liquid is an obstacle to electron transmission and liquid plugs in nanotubes appear dark. In the field emission ESEM, the vapour phase may appear darker because there is less obstruction of the electron beam, the metal sample holder surface underneath is more visible.

Conversely, liquids look lighter because they impede electron penetration and the sample holder surface underneath is less discernible. Furthermore, analysis of liquid motion inside the tube, which was recorded on video, clearly indicates that the liquid is the lighter phase inside the CNT. A thin (tens of nanometers) layer of water is transparent to the electron beam at the maximum voltage used and some tube features can be observed through the water film.

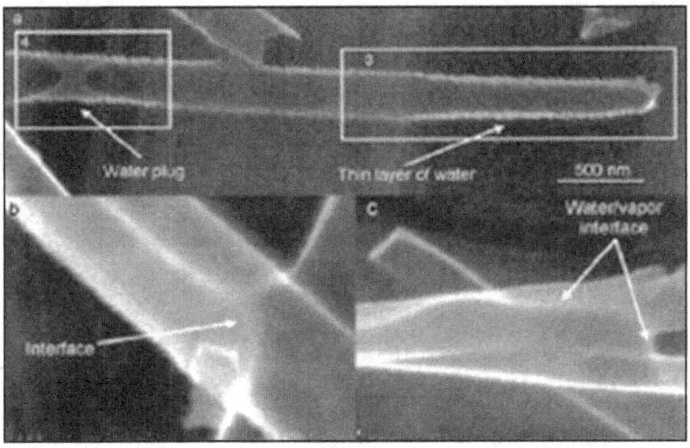

ESEM images showing the condensation of water inside a carbon tube

(a) and underneath or around the tubes. Water initially condenses inside open tubes when they are placed on a steel substrate. Condensation and formation of a water bridge may occur between a tube and the substrate when the tube is slightly elevated above the substrate

(b) or between two tubes which are close to each other

(c) Framed areas 3 and 4 show the sections of the tube presented in figures, respectively.

Before the first indication of water condensation on the sample holder surface (marked by the formation of small droplets), menisci began to become evident within the CNTs as pressure increased above 5 Torr at 4 °C, as seen in figure. This shows that the carbon tubes were more hydrophilic than steel (contact angle ~55°) and attracted water vapour from the environment. In addition to the hydrophilic walls, surface tension and thermocapillarity are the forces that may cause a liquid to become drawn into a CNT.

When water pressure increases further and the open tubes get fully or partially filled, condensation of water begins around the tubes, especially when several tubes are in close contact as shown in figure. Condensation experiments on branched, closed CNTs showed no meniscus formed because water did not penetrate inside the tube. A sequence of ESEM images taken under different pressures from the closed end of the tube shown in figure shows condensation of water inside the tube. Surface tension after the formation of a thin film of water inside the tube applies a force to the walls that is sufficient to change the tube shape slightly.

Small bubbles may eventually be trapped in the liquid. The process of condensation and evaporation of water inside the tubes is fully reversible and can be repeated multiple times. Figure shows the change in meniscus shape and size in response to pressure at a constant temperature of 4 °C. The dynamics of this process for the water plug with a volume of approximately 10^{-17} liters (10 attoliters) was recorded on video and can be quantitatively analyzed. Measured contact angles of the water with the CNT internal walls shown in Figure 4a are comparable to the ones predicted for H-terminated graphite.

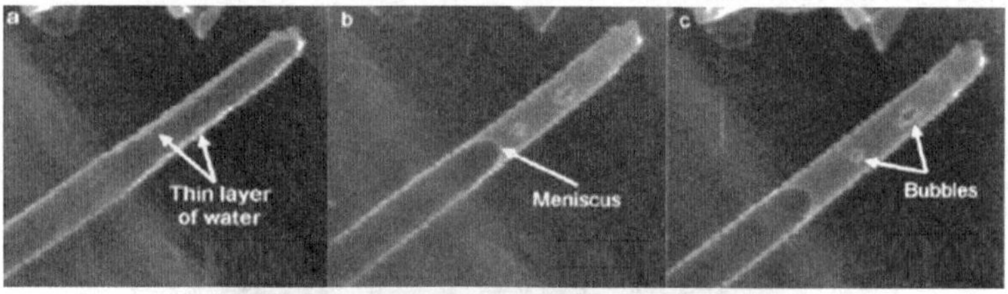

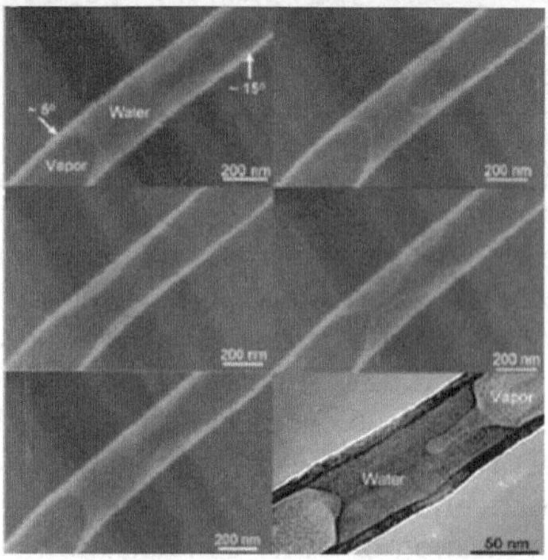

ESEM images showing the condensation of water at the closed end of the tube shown in figure. Condensation starts at the walls of the tube and the volume of liquid grows with increasing time or pressure. Two bubbles can be seen trapped inside the liquid.

ESEM images showing the dynamic behaviour of a water plug close to the open end of the tube shown in figure. The meniscus shape changes when, at a constant stage temperature, the vapour pressure of water in the chamber is changed

(a) 5.5 Torr,
(b) 5.8 Torr,

(c) 6.0 Torr,
(d) 5.8 Torr
(e) 5.7 Torr, where themeniscus returns to the shape seen in figure. The asymmetrical shape of the meniscus, especially the complex shape of the meniscus on the right side in figure, is a result of the difference in the vapour pressure caused by the open left end and closed right end of the tube.
(f) TEM image showing a similar plug shape in a closed CNT under pressure.

Figure was obtained at a scan rate of 33.2-66.4 ms for image clarity. As a result, there is a lag in the imaging of these dynamic processes, and the liquid phase seems to decrease as pressure increases. In addition, although the cooling stage is maintained at a constant temperature of 4-5 °C, CNTs have a high thermal conductivity and are quickly heated by the electron beam under the prolonged, continuous observation at high magnification and slow scan rates in the ESEM. Because of CNT heating under the electron beam, it is possible that water evaporation can occur even when the conditions dictate condensation. It is evident in figure that the menisci observed within the CNT are asymmetric.

This could be attributed to the fact that the pores in the A12O3 membrane are not perfectly circular, as shown in figure. This gives the CNT a nonuniform cross-sectional shape and could be the reason for the distorted shape of the meniscus. The slow scan rate at which the image was taken could also contribute to the distortion of the meniscus. However, the distinct difference in the meniscus shape on the left and right sides, which is reproducible from cycle to cycle, suggests that the closed end of the tube and presence of liquid at the tube tip led to different pressures on different sides of the water plug.

Pinning of the plug to the hydrophilic tube walls has been demonstrated for hydrothermal nanotubes and can explain why the plug deforms instead of just moving under the pressure gradient. In the case of hydrothermal nanotubes, the rapid heating of liquid inclusions by the electron beam caused the formation of a complex interface between the liquid and the gas inside the nanotube.

Movement of this interface occurred as a result of bubble formation in the bulk liquid. Basically, liquid inclusions were pinched-off between the gas bubbles, creating a wavelike pattern within the nanotubes. In addition, interface deformation by peeling of the graphite layers on the inner wall of the nanotubes was observed.

This behaviour was most likely caused by the strong attraction between the hydrophilic walls of the hydrothermal nanotubes and the liquid inclusion. A complex shape of the meniscus was also for a water plug in a hydrophilic nanochannel between two plates. According to this study, the meniscus shape is attributed to the negative pressure induced inside the nanochannel as a result of tensile capillary forces.

It has been argued that the results obtained on hydrothermal nanotubes were inherent to that specific system used, and pure water in a nanopipe may behave differently. However, the interfaces observed in ESEM are similar to the complex liquid-gas interfaces recorded in the experiments in hydrothermally produced tubes (50-100 nm inner diameter) with a $H_2O/CO/CH_4$ fluid at high pressures as shown in figure. The kinetics of membrane pinch-off and liquid propagation along the walls is also similar in order of magnitude.

The fact that liquid inclusions in two distinct types of CNT (highly ordered hydrothermally produced tubes having a conical scroll structure and disordered CVD tubes) under significantly different conditions (pressure above 760 Torr in TEM experiments and 4-6 Torr in ESEM experiments) behave in such a similar fashion strongly suggests that such a complex interface is intrinsic to the behaviour of water in nanochannels, at least those with hydrophilic walls.

It is believed that the forces responsible for liquid becoming drawn into a hydrophilic nanotube are surface tension and thermocapillarity. Studies on capillary condensation of a fluid in a cylindrical pore, which have been done by the scientific community throughout the past thirty years, could serve to explain some of the behaviour reported here. In 1972, Everett and Haynes first proposed their theory, which has been confirmed recently by Vishnyakov and Neimark. According to the Everett-Haynes scenario, a metastable, vapour-like fluid in a cylindrical pore with attractive walls will form a bump or undulation that will grow into a liquid-like bridge. As the liquid-like phase grows, the vapour-like bubble begins to contract until the entire cylinder is filled with condensed liquid.

This behaviour follows the Laplace equation

$$P_{cap} = \gamma(r_1^{-1} + r_2^{-1}) = \text{constant}$$

where ç is the surface tension of the liquid and r_1 and r_2 are the main radii of curvature. The observed behaviour indicates an attraction between the CNT inner walls and the water, which demonstrates that the CNTs are hydrophilic. An explanation for the behaviour seen in figures can be explored by using this model. In parallel, Cole and Saam studied the adsorption of liquid films in cylindrical pores. In this theory, the condensation of a liquid film in a cylindrical cavity does not form the liquid bump between the vapour phases.

In their argument, Cole and Saam state that a liquid bridge collapses so that only meniscus behaviour similar to that observed in figure occurs. In this case, the behaviour of the liquid/gas interface follows the equation where *pv* and *pl* are the equilibrium vapour and liquid pressures, *a* is the film radius, and ç is surface tension. The behaviour observed in figure has not been clearly hypothesized by one particular theory, since what has been observed varies from what is expected from different individual theories. However, it is also evident, by comparing the observations in figure, that this behaviour is innate of hydrophilic cylindrical pores irrespective of parameters such as pore diameter or external pressure.

The next step in studies such as those presented here, therefore, is to carefully analyse theories of capillary condensation presented thus far to explain the observed phenomena.

Conclusions.

Dynamic fluidic experiments with pure water inside the CNTs show that interface dynamics and complex shapes are controlled by the small tube diameter rather than wall structure, fluid composition, or pressure. Complex interfaces previously reported from the TEM study on hydrothermal nanotubes containing pressurized fluid have been observed for pure water in CVD tubes at low pressure.

This work shows that CVD CNTs with disordered walls are hydrophilic and form contact angles with water from 5 to 20 degrees. It is also shown that water condenses preferentially inside the CNT, which demonstrates that such tubes will easily imbibe aqueous fluids. This could indicate that CNTs can be used as probes to guide fluids to and from specific locations, such as cells, and that they could provide an efficient way to collect attoliter to picoliter amounts of liquids.

CONCEPT OF SCANNING TUNNELING MICROSCOPY (STM)

The Scanning Tunneling Microscope (STM) was introduced by G. Binnig and W. Rohrer at the IBM Research Laboratory in 1982 which was honoured by the Noble Prize in 1986. It has become widely used as an important instrument for real space analysis in surface science. The basic idea is to bring a fine metalic tip in close proximity (a few E) to a conductive sample. By applying a voltage (U4V) between the tip and the sample a small electric current (0.01nA-50nA) can flow from the sample to the tip or reverse, although the tip is not in physical contact with the sample.

This phenomenon is called electron tunneling. The exponential dependence of the tunneling current on the tip to sample distance results in a high vertical resolution. By scanning the tip across the surface and detecting the current (one can also use the current as a vertical positioning signal for the tip) a map of the surface can be generated with a resolution in the order of atomic distances.

It has to be mentioned that the image cannot just be interpreted as a topographic map as the tunneling current is influenced by the lateral and vertical variation of the electronic state density at the surface. The lateral resolution is about 1E whereas a vertical resolution up to 0.01E can be achieved. The STM can be used in ultra high vacuum, air or other environments.

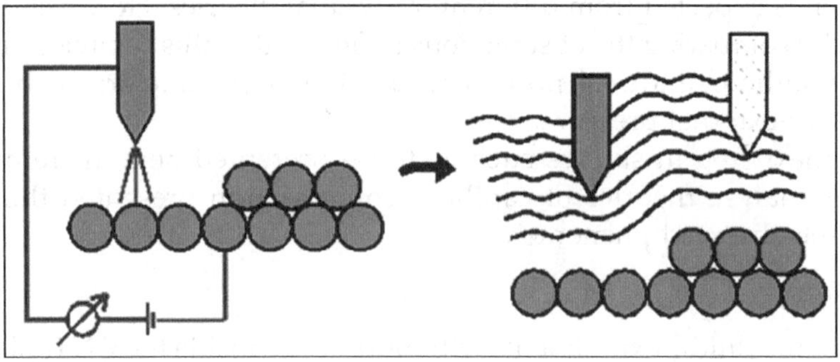

Theory

Electron Tunneling

In classical physics an electron cannot penetrate into or across a potential barrier if its energy E is smaller than the potential Φ within the barrier. A quantum mechanic treatment predicts an exponential decaying solution for the electron wave function in the barrier. For a rectangular barrier we get.

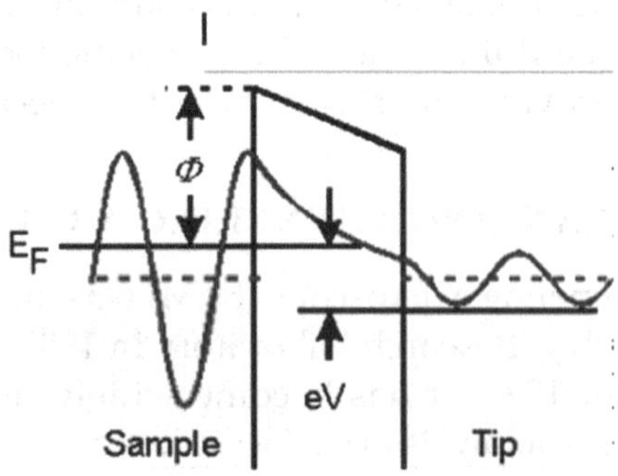

$$\Psi(d) = \Psi(0)e^{-kd} \text{ where } k = \frac{\sqrt{2m(\Phi - E)}}{\hbar}$$

The probability of finding an electron behind the barrier of the width d is

$$W(d) = |\Psi(d)|^2 = |\Psi(0)|^2 e^{-2kd}$$

In scanning tunnneling microscopy a small bias voltage V is applied so that due to the electric field the tunneling of electrons results in a tunneling current I. The height of the barrier can roughly be approximated by the average workfunction of sample and tip.

$$\Phi = 1/2\ (\Phi_{sample} + \Phi_{tip})$$

If the voltage is much smaller than the workfunction eV <<, the inverse decay length for all tunneling electrons can be simplified to

$$k \approx \frac{\sqrt{2m\Phi}}{\hbar}$$

The current is proportional to the probability of electrons to tunnel through the barrier:

$$I \propto \sum_{E_n=E_F-eV}^{E_F} |\Psi_n(0)|^2 e^{-2kd}$$

By using the definition of the local density of states for $\varepsilon \to 0$

$$\rho(z,E) \equiv \frac{1}{\varepsilon} \sum_{E_n=E-\varepsilon}^{E} |\psi_n(z)|^2$$

the current can be expressed by

$$I \propto Vr_{sa}(0, E_F)e^{-2kd} \; (*)$$

$$» \; Vrsa\,(0, E_F)e\,-1.025\sqrt{\Phi d} \text{ where } [d] = \text{Å};\ [\Phi] = \text{eV};$$

With 5eV as typical example for a workfunction value a change of 1E in distance causes a change of nearly one order of magnitude in current. This facilitates the high vertical resolution.

Also

$$I \propto V\rho_{sa}(d, E_F).$$

which means that the current is proportional to the local density of states of the sample at the Fermi energy at a distance d, i.e. the position of the tip.

A more exact calculation of the current density of the square barrier problem requires the Schrödinger's equation to be solved in the three regions: before, in and behind the barrier. The coefficients have to be adapted so that the overall solution is continually differentiable. Defining the transition probability as

$$T = \frac{j_T}{j_\iota}$$

yields

$$\frac{16E(V-E)}{V^2} e^{-2ed}; \quad k = \sqrt{\frac{2m(V-E)}{\hbar^2}}$$

with the approximation kd >> 1.

The current density itself is defined as

$$j_z = -\frac{i\hbar}{2m}\left(\psi^*(z)\frac{d\psi}{dz} - \psi(z)\frac{d\Psi^*}{dz}\right)$$

For a nonsquare potential the WKB method must be used. This is more adequate as the potential is changed by the applied voltage and influenced by the image force on the electron. The WKB method yields a transition probability of

$$T(E) \propto \exp\left(-\frac{2}{\hbar}\int_0^1 \sqrt{2mV(z) - E}\,dz\right)$$

Bardeen Approach

Another way of describing electron tunneling comes from Bardeen's approach which makes use of the time dependent perturbation theory. The probability of an electron in the state Ψ at E_Ψ to E_Ψ tunnel into a state χ at E_x is given by Fermi's Golden Rule

$$w = \frac{2\pi}{\hbar} |M|^2 \delta(E_\Psi - E_\chi)$$

The tunneling matrix element is given by an integral over a surface in the barrier region lying between the tip and the sample:

$$M = \frac{\hbar}{2m} \int_{z=z_0} \left(\chi^* \frac{\partial \Psi}{\partial z} - \Psi \frac{\partial \chi^*}{\partial z} \right) dS$$

Applying a bias voltage V and approximating the Fermi distribution as a step function ($kT << \Delta E_{resolution}$), the current is

$$I = \frac{4\pi e}{\hbar} \int_0^{eV} \rho_{sa}(E_F - eV + \varepsilon)\rho_{tip}(E_F + \varepsilon)|M|^2 d\varepsilon \ (**)$$

Hence the current is given by a combination of the local densities of states of the sample and the tip, weighted by the tunneling matrix element M.

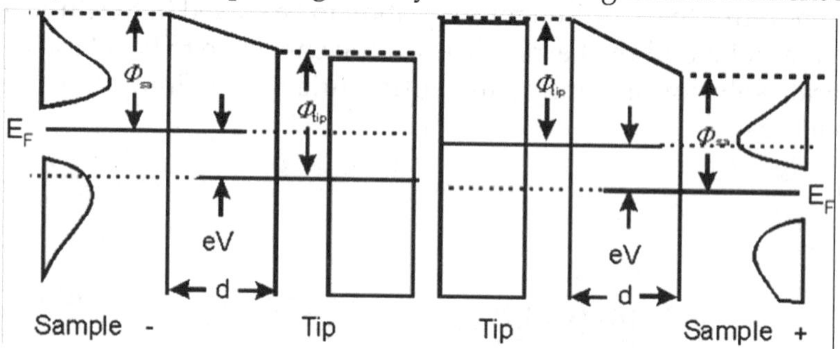

Schematic of electron tunneling with respect to the density of states of the sample.

d ($E_Y - E_c$) means that an electron can only tunnel if there is an unoccupied state with the same energy in the other electrode (thus inelastic tunneling is not treated). In case of a negative potential on the sample the occupied states generate the current, whereas in case of a positive bias the unoccupied states of the sample are of importance.

Therefore, as shown below, by altering the voltage, a complete different image can be detected as other states contribute to the tunneling current. This is used in tunneling spectroscopy. It should finally be mentioned that the probability of tunneling (expressed by M^2) is larger for electrons which are close to the fermi edge due to the lower barrier.

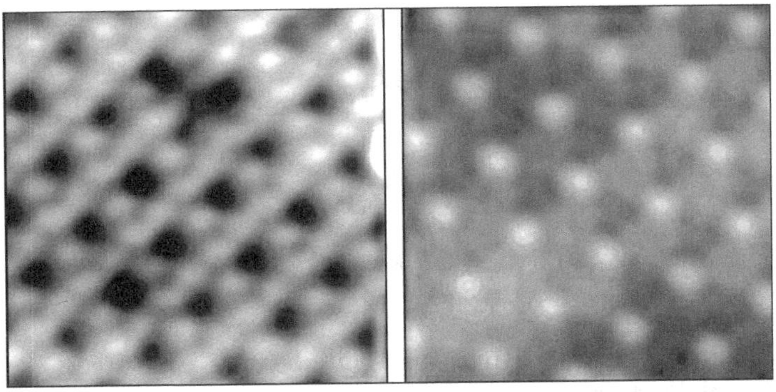

Imaging the occupied states of SiC(000)3 × 3 | Imaging the unoccupied states of SiC(000)3 × 3

Lateral Resolution

The lateral resolution of STM can not be understood in terms of a Fraunhofer diffraction resolution. The corresponding wave length of the tunneling electron would be > λ > Å.

Therefore the STM works in the near-field regime. The overall geometric curvature of the tip with a radius of curvature of e.g. 1000Å and x = 1 $Å^{-1}$ would give rise to a resolution of about 50Å. The actual atomic resolution can only be understood in a quantum mechanical view: The most prominent model in this respect is the s-wave-tip model. The tip is regarded as a protuding piece of Sommerfeld metal with a Radius of curvature R.

It is assumed that only the s-wave solutions of this quantum mechanical problem (spherical potential well) are important. Thus, at low bias the tunneling current is proportional to the local density of states at the centre of cuvature of the tip r_0:

$$I \propto \sum_{E_k=E_r-eV}^{E_r} |\Psi_\mu(r_0)|^2 = eV\rho_{sa}(r_0, E_F)$$

In this model only the properties of the sample contribute to the STM image which is quite easy to handle. But it cannot explain the atomic resolution.

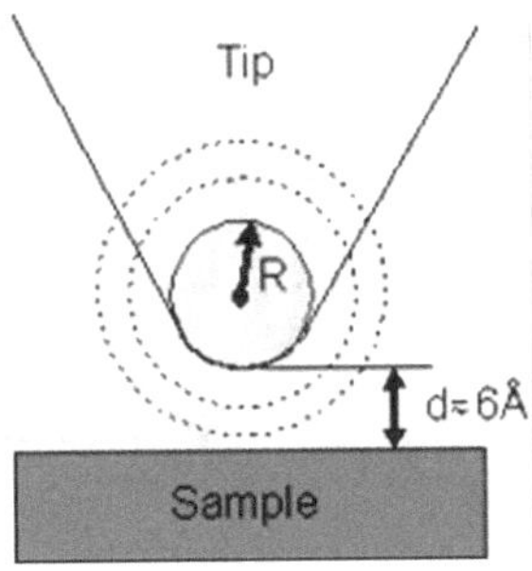

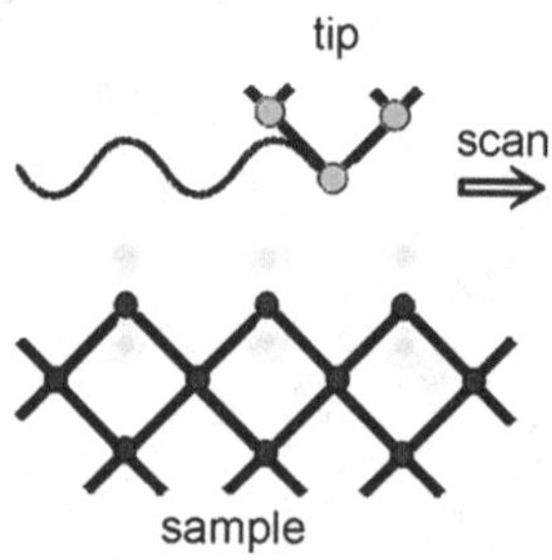

Interaction which Causes a High Corrugated Tunneling Distribution

Calculations and experiments showed that there is often a d_z^2 like state near the fermi edge present at the apex atom which also predominantely contributes to the tunneling current. It is understood that this state (and also the p_z like state) is advantageous for a „sharp„ tip.

Since the tunneling current is a convolution of the tip state and the sample state, there is a symmetry between both: By interchanging the electronic state of the tip and the sample state, the image should be the same (reciprocity principle). This can also explain the fact that the corrugation amplitude of an STM image is often larger than that of the LDOS of the sample (measured by helium scattering). In this case the tip traces a fictitious surface with a d_z^2 like state.

The state of the tip atom is dependent on the material and the orientation. As the tip is quite difficult to handle, it is one of the most difficult problems in a STM experiment.

Modes of Operation

Constant Current Mode

By using a feedback loop the tip is vertically adjusted in such a way that the current always stays constant. As the current is proportional to the local density of states, the tip follows a contour of a constant density of states during scanning. A kind of a topographic image of the surface is generated by recording the vertical position of the tip.

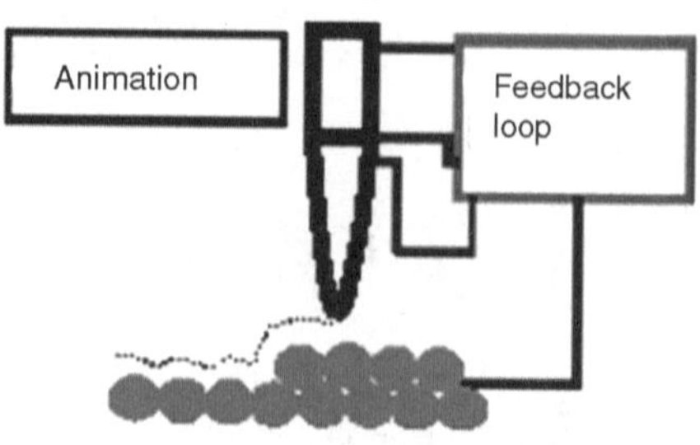

Constant Height Mode

In this mode the vertical position of the tip is not changed, equivalent to a slow or disabled feedback. The current as a function of lateral position represents the surface image. This mode is only appropriate for atomically flat surfaces as otherwise a tip crash would be inevitable. One of its advantages is that it can be used at high scanning frequencies (up to 10 kHz). In comparison, the scanning frequency in the constant current mode is about 1 image per second or even per several minutes.

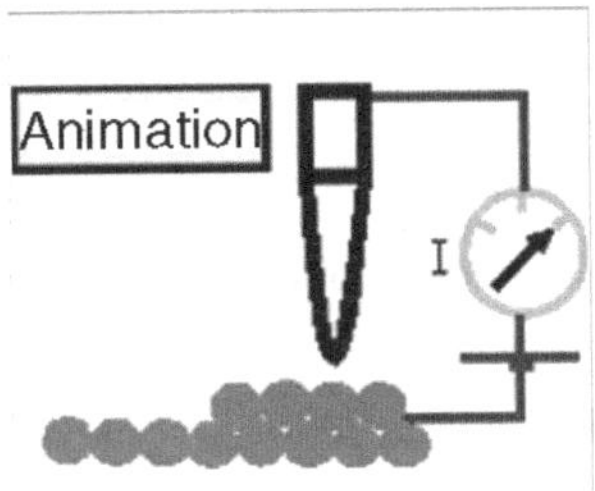

Barrier Height Imaging

Up to now homogoneous surfaces were considered. If there is an inhomogeneous compound in the surface the work function will be inhomogeneous as well. This alters the local barrier height. By using the two modes described above we would get a virtual hole or adatom. But this can further be explored:

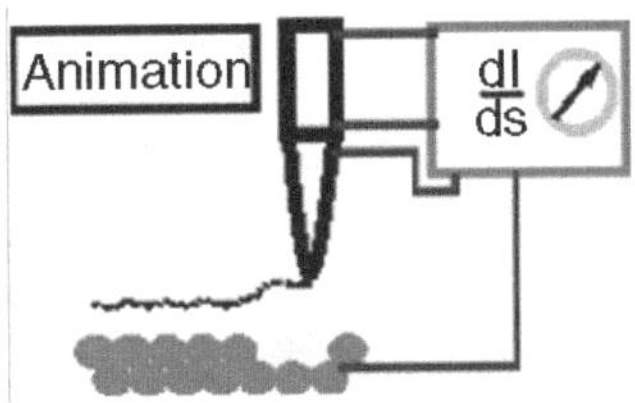

Differentiation of (*) yields

$$\frac{d(\ln I)}{ds} \propto \sqrt{\Phi}$$

Thus the work function can directly be measured by variing the tip-sample distance, which can be done by modulating the current with the feedback turned on.

Scanning Tunneling Spectroscopy

If the matrix element and the density of states of the tip is nearly constant, the tunneling current (**) can be estimated to

$$I \propto \int_0^{eV} \rho_{sa}(E_F - eV + \varepsilon)d\varepsilon$$

Differentiation yields the density of state,

$$\frac{dI}{dV} \propto \rho_{sa}(E_F - eV)$$

The density of states can be deduced by:

- Modulation of the bias voltage.
- Current-Imaging Tunneling Spectroscopy (CITS): The tip is scanned in the constant current mode to give a constant distance to the sample. At each point the feedback loop is disabled and a current-voltage curve (I + V curve) is recorded.

Technical Aspects

Demands

- Controlling the tip-sample distance from a few mm down to 0.01E.
- Exact lateral positioning.
- Stabilized tip-sample distance.
- Sharp tip.
- Measuring a current in the range of 0.01nA - 50n A.

Positioning

The large distance range the tip has to be controlled on makes it necessary to use two positioners: a coarse and a fine positioner. The fine positioner is also used as a scanner. Every fine positioner/scanner is made out of a piezocrystal or piezoceramic material.

Piezoelectricity

Piezoelectric Effect: Applying stress on piezo materials causes electric polarisation

Reverse Piezoelectric Effect: An electric field applied to a piezo material causes mechanical strain

Piezocrystals: quartz, $BaTiO_3$ Piezoceramics: $Pb(Ti,Zr)O_3$ (PZT)

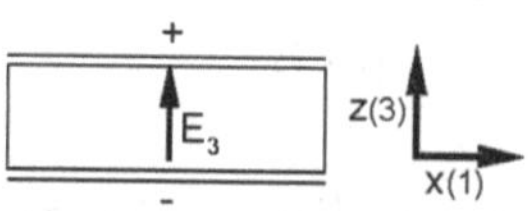

In the following the directions *x*, *y* and *z* are labeled as 1, 2 and 3, respectively. Applying a voltage *V* to a rectangular piece (thickness z) of piezoelectric material as shown above yields an electric field of

$$E_3 = \frac{V}{z}$$

The strain is generated, which can be described by a tensor with xx- and zz-components

$$S_1 \equiv \frac{\delta x}{x}; \quad S_3 \equiv \frac{\delta z}{x};$$

The piezoelectric constants are defined as the ratios of the strain component over the component of the electric field intensity. Thus we get

$$d_{31} \equiv \frac{S_1}{E_3}$$

and

$$d_{33} \equiv \frac{S_3}{E_3}.$$

Typical values for the piezoelectric components of PZT are

d_{31} = – 1 to – 3 E/V

d_{33} =2 to 6 E/V

Example:

EBL#1(Stavely Sensors)

d_{31} = –1.27E/V (tranversal)

d_{33} = 2.95E/V (longitudial)

Fine Positioner (Scanner)

- *Tripod:* The tip is controlled by three orthogonal piezo bars

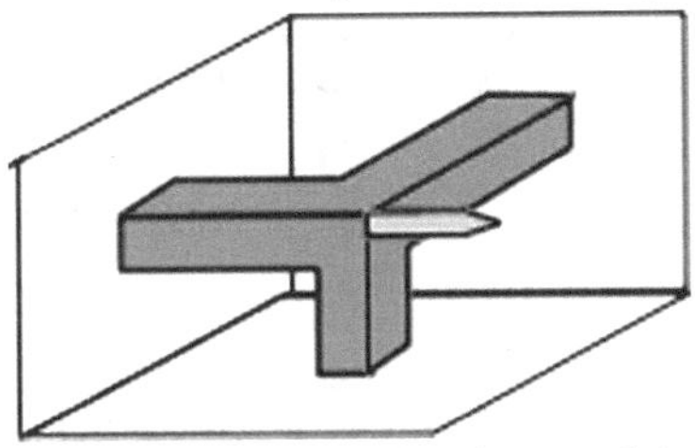

- *Bimorph:* A Bimorph is made out of two thin plates of piezocrystals which are stacked in such a way that a voltage applied on the plates causes stress of opposite sign which generates a deflection (principle is similar to the bimetallic thermometer).
  By combining several bimorphs a three-dimensional scanner can be constructed.
- *Tube Scanner:* A cylinder, made out of piezoceramics, is covered inside and outside with metal, which acts as electrodes. The outer electrode is divided into four sectors.

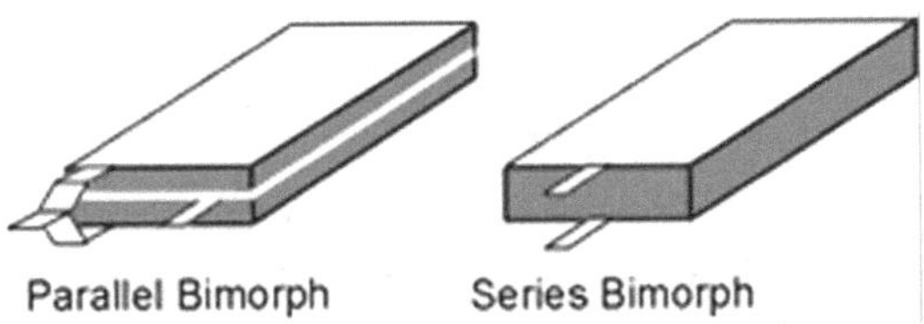

A motion in z-direction (along the longitudinal axis) can be achieved by applying a voltage between the inner and the outer electrode. A deflection in xy-direction is generated by a different bias voltage applied to the two opposite electrodes.

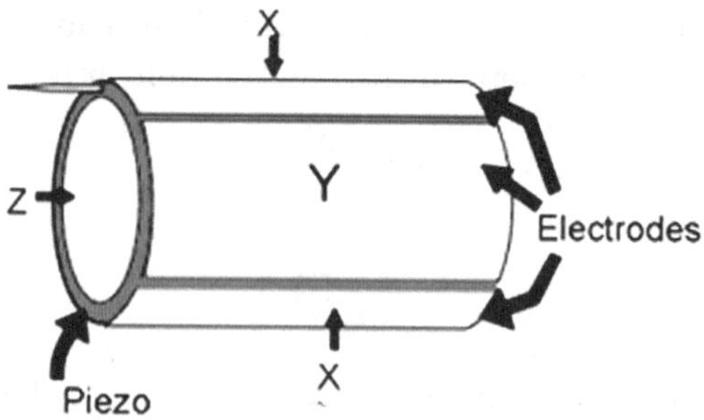

Coarse Positioner

- *Mechanical Positioner:* The tip approaches the sample with the help of springs, demagnifier and microscrews.

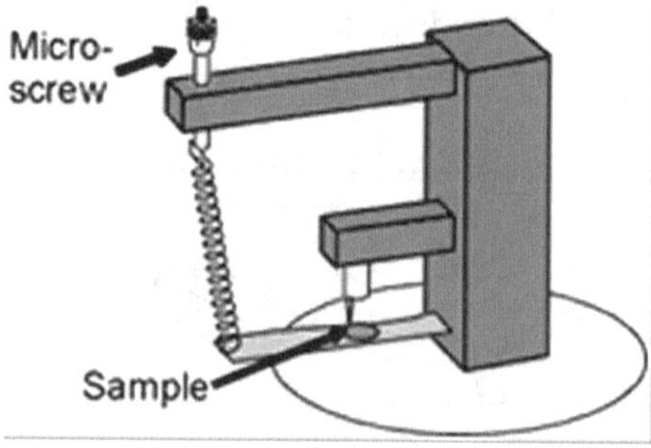

- *Louse:* The sample is mounted on a piezoelectric plate with three electrically isolated feet, which electrostatically can be attached to a ground plate. Thus by expanding and contracting the piezoelctric plate and clamping and detaching the feet the louse can move step by step towards the tip.

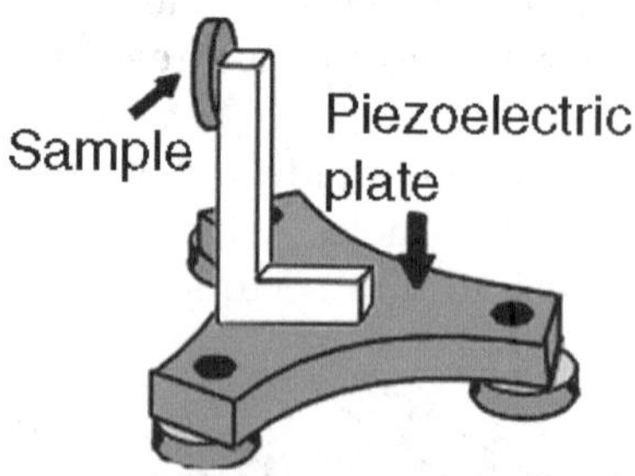

- *Beetle:* The Beetle consists of a plate with three piezoelectric tubes, standing on a cylindric ramp. The sample is mounted in the middle of the ramp whereas the tip with the fine positioner is mounted on the plate. By applying a voltage on the piezoelectric feet, the plate is rotated. Reducing the voltage quickly causes the feet to slip on the ramp due to the inertia of the plate. Thus the tip can be moved towards the sample or away (stick/slip motion).

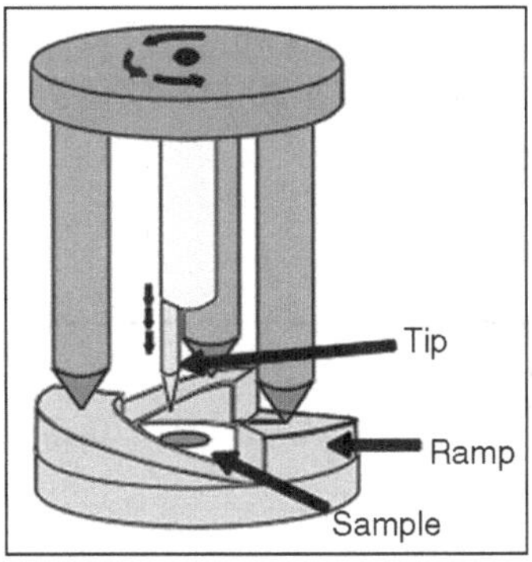

Electronic Circuit

In case of measuring in the constant current mode a feedback circuit has to be build up to control the z-piezo.

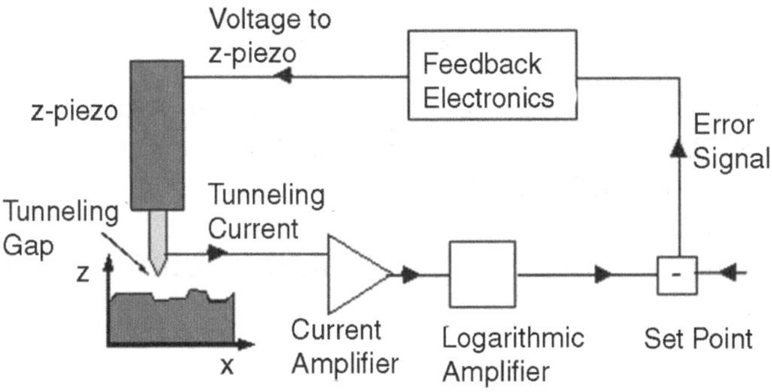

The tunneling current (0.01 mA - 50 mA) is converted into a voltage by a current amplifer.

To get a linear response with respect to the tunneling gap (the current is exponentially dependant on the tip-sample distanance) the signal is processed by a logarithmic amplifier.

The output of the logarthmic amplifier is compared with a predetermined voltage which is used as a reference current. The error signal is passed to feedback electronics, which applies a voltage to the z piezo to keep the difference between the current set point and the tunneling current small.

Care has to be taken to keep the noise signal ratio on a low level. Also the response time of the feedback has to be minimized without loosing accuracy.

Vibration Isolation

The tip-sample distance must be kept constant within 0.01Å to get good atomic resolution. Therefore it is absolutly necessary to reduce inner vibrations and to isolate the system from external vibrations.

Environmental vibations are caused by:

- Vibration of the building 15-20 Hz
- Runnning people 2-4Hz
- Vacuum pumps
- Sound

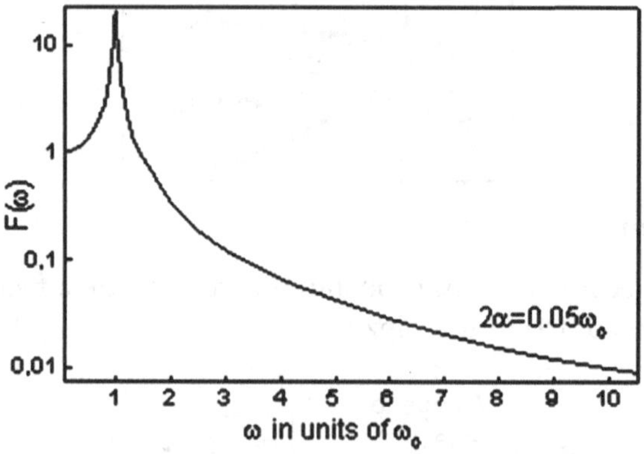

Regarding the transfer function (ratio of amplitudes) of a damped harmonic oscillator with a sinusodial external force, it is clear that the resonant freqency must be much lower than the external frequency to be eliminated. Damping can be done by

- Suspension with springs (including additional eddy current dampers)
- Stacked plate systems
- Pneumatic systems

Tip

The tip is the trickiest part in the STM experiment. It needs a small curvature to resolve coarse structures.

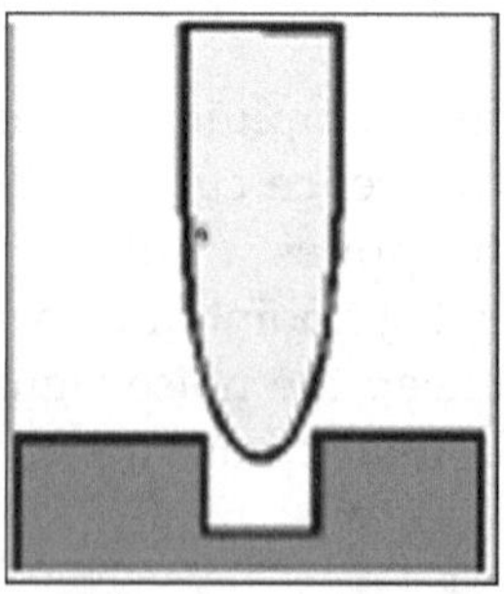

For atomic resoltion a minitip with a one atomic end is necessary. Tips typically are made out of tungsten, platinum or a Pt-Ir wire. A sharp tip can be produced by:

- Cutting and grinding.
- Electrochemical etching.

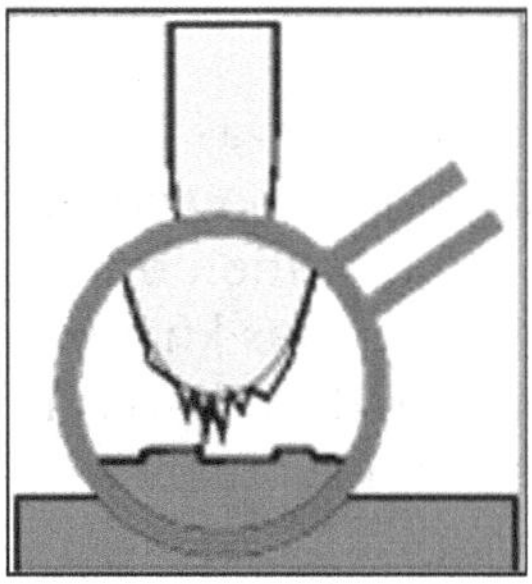

The tungsten wire is put into a solution of NaOH and kept on a positve potential towards a counter electrode. The etching process takes place predominately on the surface of the solution. When the neck is thin enough the wire fractures due to its weight. Thus actually two tips are produced. The tip has to be cleaned with deionized water and pure ethanol or methanol.

Most often the tip is covered with an oxide layer and contaminations from the etchant and is also not sharp enough. Thus other tratments to the tip, like annealing or field evapouration are necessary.

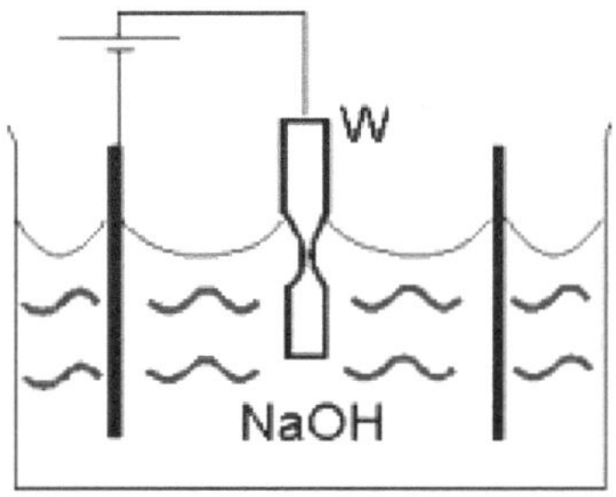

It is also possible to do tip-sharpening during tunneling.

- Sudden rise of the bias voltage to about -7V (at the sample) for 2-4 scan lines.

  By this treatment some W atoms may walk to the tip apex due to the nonuniform electric field and form a nanotip.
- Controlled collision on Si surface.

The tip may pick up a Si-cluster which forms a monoatomic apex with a p_z like dangling bond.

Examples

Determination of surface structures is one of the most important applications of the STM. It is obvious that a lot of surfaces reconstruct to lower their energy. Thus the properties of the surfaces can be completely different from the bulk. Especially due to the size reduction in semiconductor technology surface properties are getting important. A few examples of surface science studies are given below, all illustrated using STM images taken at the Erlangen surface lab.

Si(111)7x7

The imaging of the Si(111)- 7x7 was the first great success of the STM although the structure could not be resolved completely. This has been later achieved with the help of high resolution transmission electron microscopy.

Before the invention of STM it was known from LEED experiments that the surface unit cell is constructed of 49 Si(111) original cells. It was also assumed that the cell has a sixfold symmetry.

STM images show that there are 12 adatoms in each unit cell. Further, in each corner of the cell a hole exists (corner hole). Using the proper tunneling voltage (about -0,35V) it can be seen that the cell consists of two different triangles and thus infact has a threefold symmetry.

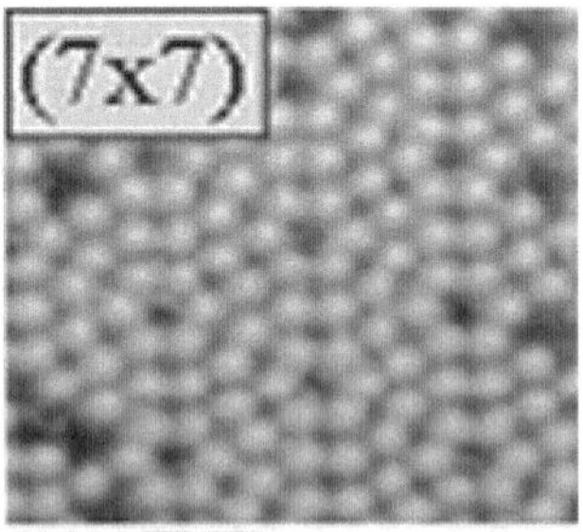

Si 7X7

A complete description of the stucture is given by the DAS model (Dimer Adatom Stacking fault.)

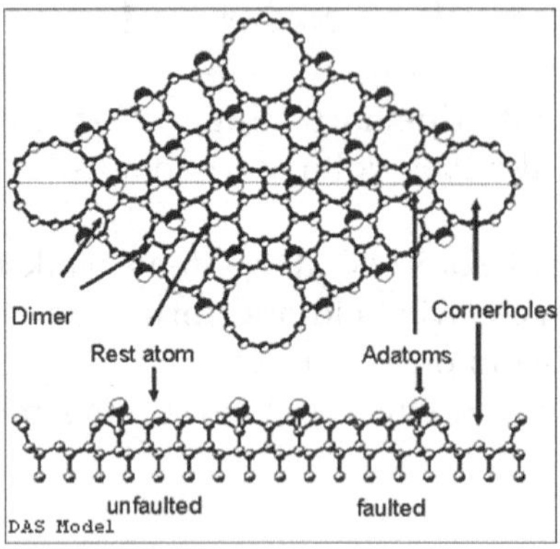

Its features are:

- 12 adatoms per 7 × 7 unit cell
- A hole in each corner.
- Three dimers (two bounded atoms) on each side of the triangular halves.
- Stacking fault in one of the halves.
- 19 dangling bonds in each unit cell (reduced from 49 of the unreconstructed surface).

Au(111)22x$\sqrt{3}$

When the first Au(111) STM images with atomic resolution were taken some features had already been known or assumed by LEED experiments and helium scattering experiments:

- Contraction of the atomic spacing along the [10] direction
- Before reconstruction Au(111) exhibits a fcc stacking
- Reconstruction pushes the Au atoms sideways along the [11] direction towards hcp stacking
- At the domain boundary between hcp and fcp stacking the atoms are squeezed out of their position and form two ridges per unit cell.

Atomic resolution of Au(111) STM images reveal the following results:

- At low resolutions the two ridges can be detected.
- With atomic resolution the atoms can be seen gradual displaced along the [10] direction are (maximum displacement: 0.7E).
- In larger areas the stripe can appear in three directions 120 apart
- In regions where the atoms are raised due to their dislocation the corrugation is larger.

But beside the confirmation of the proposed model of Au(111) reconstruction STM experiments give also information about aperiodic structures, which diffraction experiments are insensitive to.

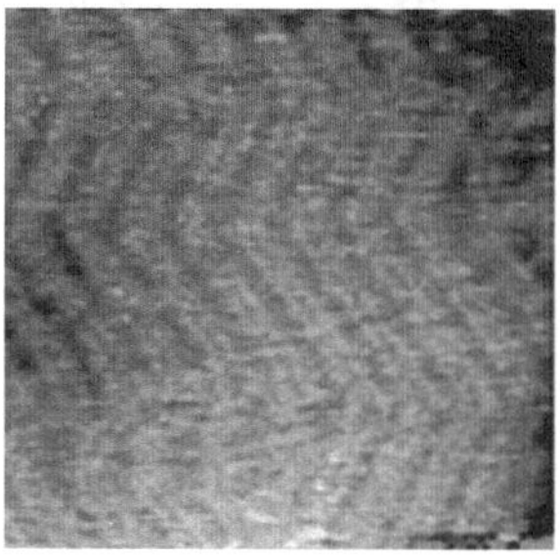

Fig. Herringbone Structure of Au(111)

As gold exhibits a threefold symmetry, the three directions which the ridge can run along are of equal probability.

Thus between regions with different reconstruction, domain boundaries are formed. On large areas the double ridges change orientation about every 280E. This leads to the so called herringbone structure.

On an even larger scale herringbone structures of different orientation form a mosaic structure. It has also become evident that the herringbone structure is quite sensitive to defects like e.g. step edges.

SiC(0001) 3 × 3

With a large band gap, a high break down field and a high saturation drift velocity of the conducting electrons SiC is an interesting material for semiconductor technology. Thus its surface strucure and the surface electronic properties are of a great importance for material growth and device fabrication.

SiC surface show a variety of surface reconstructions. One of those structures is the (3 × 3) phase on SiC(0001) whose actual atomic geometry has long been unresolved.

STM experiments provided some insight in the structure:

- There is a single protrusion per unit cell.
- The corrugation is independant of the tunneling voltage.
- Large cornerholes can not be seen.

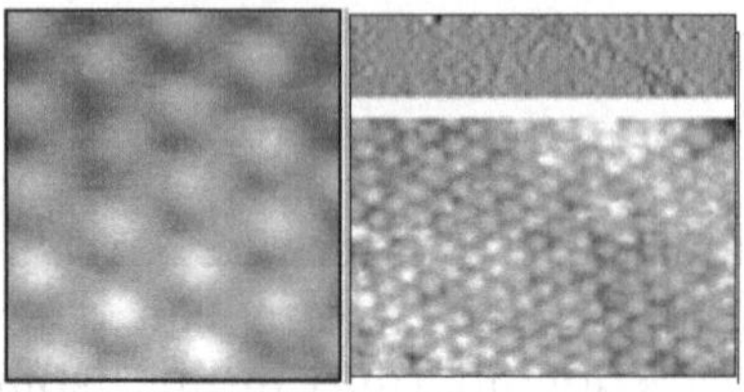

Fig. 3 × 3 4H SiC (0001) SiC(0001)3 × 3

Combining this with results from electron spectroscopy and low-energy electron diffraction (LEED) The surface geometry could be determined in detail.

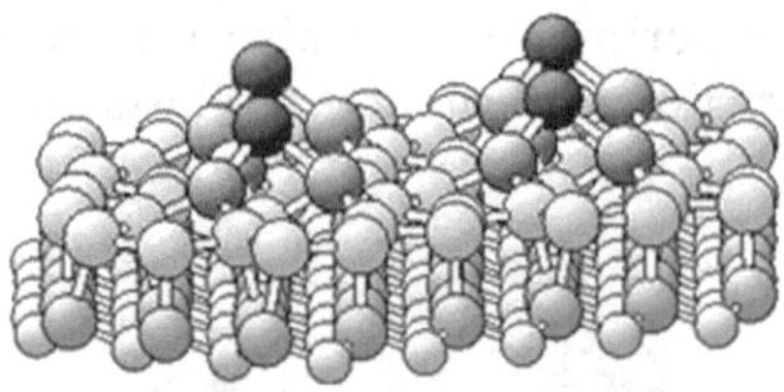

SiC(0001) 3 × 3 Model

The properties of the SiC(0001) 3x3 structure are:

- A compact, planar Si adlayer with sp^2 hybridisation within the layer.
- A twist rotation of the Si-trimer covering the adlayer and its surrounding atoms within the adlayer.
- Single dangling bond per (3x3) unit cell located at the adatom ontop of the trimer.
- Ideal Si-Si bond lengths (2.31-2.39E) and fourfold coordination of all other atoms.

DI AFM NANOSCOPE DIMENSION 3100

Atomic Force Microscope

Title, Ownership & Training: Digital Instruments Nanoscope 3100 atomic force microscope. This tool belongs to the research group of Professor Jeffrey Bokor, and is administered directly by students in his and Professor V. Subramanian's group.

All users must first be cleared by Professor Bokor, pass a written test, and an oral test that certify not only the knowledge but also the proficiency of the operator. It is obligatory that the potential user undergo a significant amount of hands-on training, and gain a solid understanding of the equipment before taking the qualification tests.

Purpose: This manual describes basic operation of tapping mode surface contour measurements.

Scope: Overview/Application: The DI-3100 is an Atomic Force Microscope particularly suited for handling semiconductor wafers and die for device microscopy. It is capable of various Scanning Probe Microscopy techniques including Contact, Tapping, Scanning Tunneling and Lateral Force Modes, although Contact and Tapping Modes are the most prevalently used on the DI-3100.

Both Contact and Tapping Mode are well suited for topographical imaging of surfaces, with vertical resolution ranging from one micron down to sub nanometer scales.

All of these techniques share a common approach where a motor controlling a mechanical tip is placed in a feedback loop as the tip is scanned across a surface. The mechanism upon which the feedback is based is the predominant difference between the modes. Surface imaging and analytical studies of roughness are common uses for this tool.

Sample Preparation/User Policy

The DI-3100 has a vacuum chuck that can hold 4" and 6" wafers. The switch for the vacuum is on the upper right hand side. The vacuum is weak however and will not hold smaller pieces. For smaller individual samples a magnetic sample holder (located on the AFM bench) is available. The magnetic sample holder adds significant height to the sample with respect to the chuck and added caution is required when translating the stage near the SPM. For non-magnetic samples, double-sided adhesive (also located on the AFM bench) tabs can be used to stick samples to the chuck.

Be aware that the adhesive contaminates mounted samples and should only be used for end-of-process samples. Users can also have custom sample holders machined for use with this tool. Minimal force is needed to keep samples in place.

The materials policy for imaging is fairly broad. Most materials related to semiconductor fabrication are allowed in this tool, including gold, photoresist, parylene, copper, etc. If a user feels he/she has a non-standard material, please check with a superuser first.

Samples to be imaged must be relatively planar to avoid having the scan head crash into features on sample. The scan head nominally focuses approximately one millimeter off of the substrate, with the cantilever tilted slightly down. Hence, vertical features or any tilt on the order of one millimeter needs to be handled with caution. Users may use their own judgment if they believe areas of the sample may be safely imaged without endangering the scanner.

However, excessive tall vertical features that cannot be easily navigated around should not be imaged on this device. Once the sample is navigated to a safe area for engaging (landing) the cantilever however, the topographical features to be imaged must have nominal roughness on the order of one micron or less.

Although millimeter tilt or features are tolerable on the sample itself, the area to be imaged must not have any topography higher than approximately one micron. Rough topography on the order of microns can break the cantilever while scanning, but millimeter scaled features may damage the scan head (which is NOT replaceable). *Reread this section after studying Section 9.0 of this manual if necessary. The sample requirements here differentiate between topography encountered while translating the stage to an imaging location (before the scanner is engaged), and topography encountered during an actual scan (after the scanner is engaged).*

A default tip is NOT loaded in the scan head. All users must supply their own tips. All users must use latex gloves while touching any component of the AFM, and must leave the chuck clean. However due to the various materials imaged (e.g. gold, photoresist), wafers imaged by the AFM are no longer VLSI-clean. If there are special needs, please consult staff. The scan head used can image a maximum field of approximately 100 μm × 100 μm. If a larger lateral displacement is required, use other Microlab tools (e.g. ASIQ, Dektak). Both tapping mode and contact mode are commonly used. Other modes are possible on the DI-3100, however users are required to discuss these with superusers.

Applicable Documents: After becoming familiar with the aspects covered in this overview of the AFM, it is recommended that users look over the Scanning Probe Microscopy Training Notebook provided by Digital Instruments. This provides some additional information on some of the data analysis and modification tools described above as well as tips on imaging techniques and artifacts. The section on Imaging Artifacts at the end is particularly recommended. A copy of this training manual is located next to the AFM.

Advance users of the DI-3100 are also encouraged to utilize the Command Reference Manual provided by Digital Instruments. This manual is available upon request.

Definitions and Process Terminology

Tapping Mode: Tapping Mode is the most common imaging technique used with the DI-3100. This mode operates by scanning a tip attached to the end of an oscillating cantilever across the sample surface. The amplitude of oscillation ranges from 20 nm to 100 nm, with the frequency near the resonant peak of the cantilever. The tip lightly 'taps' the surface, altering the oscillatory motion as the scanner moves across the surface. By adjusting the vertical position of the scanner to maintain a constant RMS signal of oscillation, a surface is imaged. The oscillation is measured by a laser positioned by the user to reflect signal into a photodiode detector)

Advantages:

- Higher lateral resolution on most samples (1 nm to 5 nm)
- Lower forces and less damage to soft samples imaged in air
- Lateral forces are virtually eliminated, so there is no scraping

Disadvantages:

- Slightly lower scan speed than contact mode AFM
- For tapping mode, use an etched single crystal silicon probe tip (TESP/RTESP). This probe is a silverish colour and is available in the Microlab for check out. Standard TESP/RTESP tips have a resonant frequency ~300 kHz and have cantilevers 125 μm long.

Safety: Large voltages (upwards of +/- 220V) are applied in order to actuate the piezo. While these voltages nodes are shielded from human contact, care must be exercised, especially if they are accessed through the break-out box.

- *Statistical/Process Data*
- *Available Process, Gases, Process Notes*
- *Operating Procedure*
- Tip Mounting

If you need to load a new tip:

- Open the hood to the AFM and *tighten* (*turn clockwise*) the knob attached to the dovetail. This lifts the SPM slightly so that it can be removed. Unplug the cord and gently slide the SPM out of the dovetail and place it on the workbench.
- Remove the Cantilever Holder from the SPM and place it onto the cylindrical tip mounter.
- Pull the metal spring back, load tip using the etched guides and replace spring. Check that tip is mounted firmly and is secure. Use sharp tweezers when handling the tip, and *only* touch the probe by the side edges.
- Replace the Cantilever Holder. Do Not Apply Twisting Motions when placing the cantilever holder onto the head. Doing so will Damage The Piezo Scanner. The piezo scanner is made of a half-cylinder of piezo crystal, which is extremely fragile and expensive to replace or repair. Place the SPM into dovetail. Gently *loosen* knob to allow the SPM to fall into place. Reconnect the cord.

- You must follow the following procedures exactly. Without starting the nanoscope software first, random voltages (which maybe very large) will be applied to the piezo scanner, causing possible damage.
- Enable the DI-3100 on the wand.
- Start Nanoscope software (icon on desktop).
- Turn on the microscope controller and facilities.
- Under the DI menu, select Microscope Select and choose Quadrexed3100. (A DI Multimode AFM shares this controller and software and may have been setup prior).
- Verify at the bottom Status Bar that TappingAFM Mode is displayed.
- Align the Laser to the Tip
- While looking at the trace of the laser beam, turn the upper knob at the top of the SPM clockwise until the intensity of the trace just becomes very dim (start with the laser trace towards the left). This means the laser has just hit the substrate off of which the 125 μm cantilever is mounted. Turn the upper knob <1/4 turn counter-clockwise to bring the laser just off the substrate.
- By turning the lower knob in either direction (the laser trace should move towards and away from you), find the point at which the laser is just shadowed by the cantilever. It should only require less than 1/8 turn to move on and off the cantilever; if it requires more than this, the laser is most likely hitting the substrate and you need to back off slightly.
- Check the viewport on the side of the SPM to confirm the laser is approximately aligned to the cantilever with the red dot centered.
- Move the laser with the upper knob until the laser is approximately at the end of the cantilever. Adjust either knob in small increments to maximize the Laser Sum Signal, as displayed on the right monitor.
- Adjust the Photodetector.
- Use the two knobs located at the left of the SPM to adjust the photodetector.
- Look at the laser signal on the right monitor and try to centre the red dot at the centre of the cross.
- Try to get ~0V vertical deflection.
- Sum should be ~2 (or slightly less).
- Align the Microscope to the Tip.
- Select Stage! Locate Tip from the top menu. You should do locate tip before focus surface. Doing it the other way risks crashing the tip.
- The tip will move up and down before stopping to help you find it. Zoom out if necessary. There needs to be reflective light to adequately see the tip: Use the chuck if necessary. To move chuck, use the Focus Surface instructions that follow.

- Use the two knobs at the left of the optical objective on the camera to centre the cantilever under the cross hairs.
- Use the trackball to focus on the tip as best as possible (requires holding down the left Focus button on trackball).
- If you cannot find the tip when zoomed in, it may be too far from the surface to reflect light. Use "focus surface" to lower the head slightly and try again.
- Load your sample. Use the chuck and vacuum, magnetic sample holder studs, or provided double sided adhesive to fix your sample onto the chuck. (Note: The travel on the chuck is limited so keep your sample on the upper half of the chuck if possible. The chuck can also be manually rotated if necessary.)
- Select Stage '! Focus Surface from the top menu.
- Check visually to ensure that there is enough clearance such that the sample can move under the SPM with adequate clearance. *There is no interlock or detection mechanism preventing users from crashing their samples into the SPM*, so make conservative estimates. If there is not enough clearance, hold down on the Focus button on the trackball, and push the ball upwards; this lifts the SPM.
- Use the trackball to move the wafer under the SPM (note the top of the trackball orients towards the back of the DI-3100).
- Follow instructions on screen to focus on the surface of your sample. *Watch both the monitor and the SPM itself when focusing; it is possible to crash the head into the sample when trying to focus and this will damage the tip AND SPM. The SPM should be at least 2 mm above the surface (although you will need to be closer than 5 mm) to focus.*
- Tips.
- Use the Lock button in any mode to lock in your fastest movement on the trackball.
- Always Zoom Out before trying rough focusing.
- It is typically easiest to focus on the edge of a wafer (especially if the features on your wafer are small) as this provides a guaranteed 500 μm of relief that is relatively easy to find.
- Focus Surface is one of the few modes that allow you to move the chuck. You can do this step before Locate Tip, as that typically also requires the chuck to be under the SPM.
- Instructions for available functions of the trackball are displayed in the Focus Surface window. This is also true for other modes such as Locate Tip.
- After rough focus, find the features to be imaged, zoom in and refocus. Centre desired feature in view.

Note: Users must always both Locate Tip AND Focus Surface (zoomed fully in) for every new sample that is loaded. Whenever moving to new sections of a sample, the user must also repeat Focus Surface, even if there is no specific topography to be imaged.

The DI-3100 calculates the difference in working distance between the tip and the surface in order to estimate how far it must lower the SPM head.

- Tuning the Cantilever.
- Select View! Cantilever Tune from the top menu (or click the tuning fork icon).
- Check that Auto Tune Controls has a frequency sweep from 100 kHz to 500 kHz, with a Target Amplitude of 2 - 3 V. Rough surfaces may do better with a Target Amplitude set at 0.5 - 1 V.
- Click on the Auto Tune button.
- A classic resonant peak curve should appear with a resonant frequency around 300 kHz. (A frequency outside of 220 - 320 kHz warrants a tip change.) Note the software picks a frequency slightly below the resonant peak to account for damping that occurs when the tip approaches a surface.
- Return to Image mode.
- Set Initial Scan Parameters.
- In the Scan Controls panel, set the initial scan parameters to a conservative setting. A good start is a Scan Size of 1 μm, X and Y Offsets of 0, a Scan Angle of 0, and a Scan Rate of 2 Hz.
- In the Feedback Controls panel, set Integral Gain to 0.5, and Proportional Gain to 0.7.
- Channel 1 is typically set to collect Height data, while Channel 2 collects Phase data. Line Direction should be set to Trace, Real Time Plane Fit should be Line, and Offline Plane Fit to None. Both High and Low Pass Filters should be Off. Other types of data collection can be selected if desired for either channel, although only two channels can be saved at once.
- Engaging the Surface.
- Select Motor! Engage from the top menu (or click on the green down arrow Engage icon) to engage the tip (the motor value located at the bottom of the monitor displays the number of microns the motor has moved down from its original position).
- Wait until the tip engages the surface.
- Once contact with the wafer surface occurs, go to View '! Scope Mode (Or click the Scope Mode icon. This will show the Scope Trace on the right monitor allowing you to view the trace and retrace of the sample.
- Go to Motor/Withdraw 2 - 3 times.
- Using the roller, manually move the SPM 50 - 100 μm more positive.
- Repeat "Engage" until motor no longer fails.

- In the "scope mode", the right-hand side of the screen shows the "Z-Offset Voltage". This is voltage applied to the piezo to maintain constant oscillation amplitude. If it is substantially different from zero, go to Motor! Step Motor and click on "tip up" or "tip down" to bring the Z-Offset to about zero. Doing so will avoid applying a large, sustained voltage to the piezo. Large applied voltages tend to depole the piezoelectric material and cause it to lose its sensitivity permanently.
- Scanning Technique.
- Go to View! Scope Mode (or click on the Scope Mode icon).
- Check to see if Trace and Retrace are tracking each other well (i.e. look similar). If they are tracking, the lines should look the same, but they will not necessarily overlap each other, either horizontally or vertically. It may help here to disable the Slow Scan Axis, forcing the SPM to image the same line over and over.
- *If tracking well (tip is scanning on the sample surface).*
- Click on Setpoint and use right arrow key to gradually increase the Setpoint value, until the tip lifts off the surface (at this point the Trace and Retrace will no longer track each other).
- Next, decrease the Setpoint with the left arrow key until the Trace and Retrace follow each other again.
- Decrease the Setpoint 1 - 2 arrow clicks more to ensure that the tip will continue to track the surface. Occasionally the tip will lift off of the surface during a scan and the setpoint will need to be re-lowered.
- If necessary, increase the Integral Gain until the tracking between Trace and Retrace lines is optimal. Over increasing the Integral Gain will result in noise showing up on top of the signal. Proportional Gain should also be adjusted in the same direction as Integral Gain, remaining 30 - 100% higher in value. Integral Gain is a high frequency parameter, while Proportional Gain is a low frequency parameter. Both modulate the response time of the applied feedback.
- Go to View '! Image Mode (or click on the Image Mode icon) to view the image.
- *If not tracking well.*
- Adjust the Scan Rate, Gains, and/or Setpoint to improve the tracking. Using the arrow keys (left/right) increments each parameter by a recommended amount. Wait 1 - 3 scanlines after each adjustment to allow the SPM to settle.
- Scan Rates and Setpoints depend significantly upon the type of sample being imaged. Recommended scan rates are 2 Hz for scan sizes 1 - 3 μm, 1 Hz for 5 - 10 μm, and 1.0 - 0.1 Hz for larger scans. Samples with larger steps typically need slower scan speeds. Decreasing the Setpoint increases the amplitude of oscillation of the cantilever, although too high a value may lead to distorted images

and damage of soft films. Increasing the Integral and Proportional Gain can also improve tracking by decrease the response time of the feedback loop.

- If there is noise riding on top of image signal (i.e. signal oscillation *not* the same in the trace and retrace directions), decreasing the Gain may help.
- *Tips.*
- Scan Size, Scan Angle, and other parameters can be modified after features are located.
- Decreasing Samples per Line and the Resolution of the scan can speed up rough scans to allow the user to locate small features by producing a quick rough image.
- In Image Mode, Offset and Zoom In/Out can be used to reposition the cantilever within the current scan window without manually typing in x/y offsets (click on execute after placing a cursor or box). Such repositioning however is only accurate within microns of displacement and may not work well with small scans.
- After a large displacement by the scanner (including forcing the SPM to the top or bottom of a scan), there may be significant 'creep' distortion observed in the scanned image. This should fade with time, and image capture can begin after the creep has settled. For large displacements, it may take more than 1 scan for the creep to completely settle out.
- Remember that the scan rate needs to be adjusted when scan sizes are enlarged.
- Typical tapping mode tips are anisotropic in shape and can give higher resolution imaging depending on scan angle. Consult either the DI website or user manuals for more information (this typically only an issue for very thin imaging).
- Further advice and more advance image adjustment techniques are available in DI-3100 manuals that are available upon request from the superuser.

Note: The piezo crystal that is utilized by the SPM is subject to stress and wear, especially with the application of large DC offsets.

After the SPM head is engaged, it relies upon the piezo crystal for lateral displacement (i.e. X Offset and Y Offset); hence large (> ~20 μm) displacements in the scanning field tend to stress the crystal more than average and are to be avoided if possible. Large scans have the same effect, and since large scans often required a slow Scan Rate (and hence long scan time), they tend to particularly stress the piezo. If step heights are the main function for such sweeps, use of either of the Microlab profilometers (asiq/Dektak) is recommended instead.

- Image Capture.

- In Image Mode, image captures consist of raw data from the AFM, yielding all possible information. In Scope Mode the image capture grabs a screen shot of the scope output and is not as useful. *Remember to be in Image Mode when capturing images.*
- Select Capture! Capture Filename to set the file name.
- Make sure the hood is closed while capturing the image to prevent optical noise from disturbing the photodetector.
- Select Capture! Capture (or click on the Eye Icon) to begin capture. At the bottom of the screen, the Capture bar should say either Capture: On or Capture: Next. Capture: On means the capture has begun and will end when the scan reaches the opposite end of the screen. *Changing any essential control parameters will halt the scan and require you to start the scan over.* Capture: Next means the image capture will begin after the scanner reaches the top or bottom of the screen. You can force the scanner to go to the top or bottom by using the respective icon. When the capture is done, Capture:Done will appear in the Capture bar.
- Typically data is saved Offline with no plane fitting. This allows you to save data directly from the AFM, and perform analysis and corrections (leveling, flattening, etc.) offline. If desired, Offline Planefit can be set to Full under the Channel 1/2 Control Panel.
- Clicking on Capture twice in a row enters Capture:Forced mode, where you can force a capture to begin mid-screen.
- By default, captured images go into the directory and can be viewed and analyzed in Offline Mode. *All users are responsible for saving their data to disk immediately after use.* There is a Zip™ and floppy Drive on the computer that can be used to retrieve files. The computer may be formatted at any time as files build.
- Liftoff/Shutdown Procedure.
- When done imaging, lift the SPM by selecting! Withdraw (or clicking the red up arrow icon).
- If a subsequent scan is desired, select Stage '! Focus Surface and move the sample to the new area. *Always be aware of any topography on samples that may be high enough to crash into the head when translating the stage.*
- If ready for unload, check to see if there is enough clearance to withdraw the head. If not, select Stage! Focus Surface and defocus such that the head rises sufficiently.
- Select Stage! Load New Sample. This should further raise the SPM and pull the chuck out to the unload position.
- Turn off the vacuum and unload your sample.
- If double-sided adhesive was used, use methanol or ethanol and Kim-wipes to clean the chuck of all adhesive (to prevent damaging chuck vacuum tubing, do not use acetone).

- Exit software and *then* turn off controllers.
- Disable the AFM on the wand and close the hood.

Note: Exiting the software and then turning off the controllers avoids stressing the Scan Head. However, leaving the controllers ON while turning the computer OFF results in floating voltages driving the piezo in the SPM. This situation is highly stressful for the piezo crystal and will result in rapid deterioration of the Scan Head.

- Data Retrieval/Image Analysis.
- Captured Data can be analyzed by selecting Offline Mode (or clicking on the Offline Mode icon in the upper right hand corner of the left monitor. By default the file browser opens to the Capture Directory, labeled.
- Image analysis can only be performed on one channel of data at a time. Select Image! Select Left <Right> Image to perform analysis on the left <right> image. When an image is selected, typically the left monitor is used to adjust parameters, while the right monitor allows for selection of partial areas, drawing of scales/lines, and execution of desired functions.
- Typically, while in Microscope mode, a Real-time Linefit is used to display collect image data. For most scans, a linefit is required to visually interpret the data. Offline however, it is recommended that the data be captured in its raw form. Hence when first viewing captured data offline, the image will often appear either warped or non-existent. While it is desirable to modify the raw image data as little as possible, often some type of fit or flattening is required. In many cases this fitting process is iterative, with better planefits/ flattening possible as the image becomes clearer. The functions used to fit/flatten/filter data are located under the Modify menu. Commonly used functions include:
- *Planefit Auto/Manual* - Planefits are commonly used to remove bow or tilt from images. Planefit calculates a single polynomial fit for the entire image (or selected areas) and then subtracts the polynomial fit from the image. Manual mode allows for the planefit to be calculated from a manually positioned line.
- *Erase Scan Lines* - Removes scan lines from image due to skips, noise, etc. The selected line is replaced by the average of the two adjacent lines.
- *Flatten* - Flatten may be used to remove image artifacts due to vertical (Z) scanner drift, image bow, skips, etc. It modifies the image on a line-by-line basis, removing the vertical offset between scan lines in the fast scan direction by calculating a least squares fit polynomial for a scan line, and subtracting it from the polynomial fit from the original line. This makes the average Z value of each scan line equal to 0V. *The result is that information in the Y direction is removed.* Hence

when there are regular feature heights, it is best to scan *across* these features in the fast scan direction (i.e. the tip is scanned directly over the step instead of along the step). On smooth surfaces, Flatten has negligible effect on roughness measurements.

- There are many Image Analysis tools available under the Analyse menu. Commonly used tools include:
- *Section* - Depth, height, width and angular measurements can be easily made on cross-sections with Section.
- *Roughness* - Roughness measurements over both an entire image or a selected area, can be calculated. Different parameters can be chosen using the Screen Layout button.
- Different views are also selectable including 3-D isometric views, lit from any angle.

3

The Transmission Electron Microscope

The transmission electron microscope (TEM) operates on the same basic principles as the light microscope but uses electrons instead of light. What you can see with a light microscope is limited by the wavelength of light. TEMs use electrons as "light source" and their much lower wavelength makes it possible to get a resolution a thousand times better than with a light microscope.

You can see objects to the order of a few angstrom (10^{-10} m). For example, you can study small details in the cell or different materials down to near atomic levels. The possibility for high magnifications has made the TEM a valuable tool in both medical, biological and materials research. The Transmission Electron Microscope (TEM) allows the user to determine the internal structure of materials, either of biological or non-biological origin.

POSITRON EMISSION TOMOGRAPHY (PET)

Positron emission tomography (PET) is a test that uses a special type of camera and a *tracer* (radioactive chemical) to look at organs in the body. During the test, the tracer liquid is put into a vein (intravenous, or *IV*) in your arm. The tracer moves through your body, where much of it collects in the specific organ or tissue. The tracer gives off tiny positively charged particles (positrons). The camera records the positrons and turns the recording into pictures on a computer.

PET scan pictures do not show as much detail as *computed tomography (CT) scans* or *magnetic resonance imaging (MRI)* because the pictures show only the location of the tracer. The PET picture may be matched with those from a CT scan to get more detailed information about where the tracer is located. A PET scan is often used to find cancer, to check blood flow, or to see how organs are working.

Why It Is Done

A positron emission tomography (PET) scan is done to:

- Study the brain's blood flow and *metabolic* activity. A PET scan can help a doctor find nervous system problems, such as *Alzheimer's*

disease, Parkinson's disease, multiple sclerosis, transient ischemic attack (TIA), amyotrophic lateral sclerosis (ALS), Huntington's disease, stroke, and *schizophrenia.*
- Find changes in the brain that may cause *epilepsy.*
- Find some cancers, especially *lymphoma* or cancers of the breast, brain, lung, colon, or prostate. In its early stages cancer may show up more clearly on a PET scan than on a CT scan or an MRI.
- See how advanced a cancer is and whether it has spread to another area of the body (metastasized). It is often necessary to do both CT and PET scans to evaluate cancer.
- Help a doctor choose the best treatment for cancer. PET scans may also be done to see whether surgery can be done to remove a tumor.
- Find poor blood flow to the heart, which may mean *coronary artery disease.*
- Find damaged heart tissue, especially after a *heart attack.*
- Help choose the best treatment, such as *coronary artery bypass graft surgery,* for a person with heart disease.

METHODS AND PROTOCOLS OF ELECTRON MICROSCOPY IN MOLECULAR BIOLOGY

The high-resolution electron microscope with its associated equipment can serve as a powerful tool for analyzing molecular structure, interactions, and processes. The last decade not only saw the ever-expanding development of electron microscopy and microanalytical techniques but also witnessed their widespread application in biomedical laboratories and the molecular biology and plant science fields.

Electron microscopy techniques such as transmission electron microscopy (TEM) crystallography, cryo-TEM, and cryo-TEM tomography have been used in the study of membrane proteins, macromolecules, and 3D virus imaging, together with electron energy-loss spectroscopy (EELS) for elemental analysis in biological specimens.

Scanning electron microscopy and X-ray microanalytical attachments have also been applied in molecular biological research for many years. Furthermore, ion mass spectrometry for elemental analysis and mapping used in non-biological fields has also been utilized in biological tissues and cells.

Application of automated crystallography for transmission electron microscopy in the study of grain-boundary segregation

Analytical Electron Microscopy (AEM) has brought significant progress in the study of grain-boundary segregation. Using X-ray energy-dispersive spectrometry (XEDS) in the AEM, elemental segregation information can be related to the crystallographic character of the same boundary via conventional Transmission Electron Microscope (TEM) diffraction techniques.

While significant efforts have been made to improve XEDS analysis of sub-nanometer segregation layers, the methods for crystallographic characterization of grain boundaries have remained the same for several decades and labour-intensive processes. Recently, a method termed Automated Crystallography for TEM (ACT) was developed, which automates crystallographic characterization of grains under TEM observation.

In the present work, we combine ACT and X-ray mapping via EDS in AEM for the study of Sb grain-boundary segregation in a rapidly solidified Cu-0.08 wt% Sb alloy. In contrast with previous reports, a large degree of anisotropy in Sb segregation level between different boundaries is found.

ACT results suggest that one of the several grain boundaries observed with no detectable Sb segregation is very close to a Ó3 coincidence-site lattice structure. The reason for the observed anisotropy in the present alloy is discussed, based upon McLean's theory of segregation.

Development of Crystallographic Methods; Electron Microscopy, X-ray diffraction and Data Mining

The best method to determine an atomic arrangement with a high accuracy and precision is based on diffraction techniques using X-rays and neutrons. Today there is a strong focus of nano-size materials as they exhibit many interesting properties. The small size of these makes it necessary to use structural characterisation methods with high spatial resolution such as high resolution electron microscopy and electron diffraction. This has lead to a strong interest to develop these techniques for structural characterisation, so called electron crystallography.

Electron Crystallography

This project involves the development of new software:

1. Electron microscopy image processing: CRISP
2. Quantification of electron diffraction patterns: ELD
3. Phase and [zone axis] identification from SAED patterns: PhIDO
4. 3D structure determination from HRTEM and SAED data: Triple, Trice

X-ray Energy-dispersive Spectrometry (XEDS)

X-Ray Energy Dispersive Spectroscopy (XEDS) is a well-established technique for characterizing the chemical composition of materials. Electron Backscatter Diffraction (EBSD) is becoming relatively common as well in materials science characterization facilities.

EBSD is helpful for identifying phases and spatial mapping of crystallographic orientation in polycrystals. Combining XEDS with EBSD improves both phase identification as well as the orientation mapping. The combination of XEDS and EBSD for phase identification is well documented.

Combining the chemical composition information provided by XEDS with the crystallographic structure information derived for EBSD patterns enables phases to be identified at SEM resolutions. The limiting factor in the resolution is the interaction volume for X-Rays. The general procedure is to first locate the beam on a specific feature of interest in the microstructure.

Second, obtain an XEDS spectrum and identify the elements present from the spectrum. These elements are then used as a filter on a crystallographic database to select a set of candidate phases. An EBSD pattern is then obtained and crystallographic information extracted from the pattern. This is used as a second level filter to further pare down the list of candidate phases. This is often enough to uniquely identify the phase.

Less utilized is the simultaneous collection of XEDS data during a typical automated EBSD scan. Integrating XEDS data with EBSD data provides a means for studying the effects of local texture on local variations in chemical composition as well as the converse – the role chemistry plays in texture evolution. Combing XEDS with EBSD can also provide a dramatic improvement in the reliability of the orientation mapping results in materials with multiple phases. One of the key elements to orientation mapping of multiphase materials is the ability to differentiate between phases.

This is fairly reliable between phases with dissimilar structure, e.g. the alpha and beta phases in titanium, which are hexagonal and alpha respectively. However, when the phases are similar in structure – as in alpha Titanium, which is hexagonal, and Alumina, which is trigonal –reliably differentiating the phases from individual EBSD patterns is difficult. Without being able to reliably differentiate between the phases, the orientation results are compromised.

However, by using the chemistry as a filter prior to the EBSD analysis enables the phases to be differentiated relatively easily. The general procedure is to simultaneously collect XEDS and EBSD data, extract the critical parameters from these data and record these parameters to file. This procedure is followed for each point in the scan. After the data is collected, the EDS data can be used to identify the potential phase or phases at each point and the indexing of the EBSD pattern performed using the recorded data. This can be done in an automated fashion using cluster analysis.

The importance of diffraction as a structural characterisation technique is clearly revealed by the study of the large and complex proteins. These structures can be a true challenge. Nevertheless the structures of numerous proteins are known today. One way to simplify the study of new protein structures is therefore to develop methods, which uses the knowledge of the amino-sequence and structures of known molecules.

Although more time consuming than many other common analytical tools, the wealth of information available from these experiments is impressive. Not only can you obtain outstanding image resolution, it is also possible to characterize crystallographic phase, crystallographic orientation (both by diffraction mode experiments), produce elemental maps (using EDS), and images that highlight elemental contrast (dark field mode)—all from nm sized areas that can be precisely located. STEM and TEM can be the ultimate failure analysis tools for thin film and IC samples.

Ideal Uses for S/TEM Analysis

- Identification of nm sized defects on integrated circuits, including embedded particles and residues at the bottom of vias.
- Determination of crystallographic phases as a function of distance from an interface.
- Nanoparticle characterization: Core/shell investigations, agglomeration, effects of annealing...
- Catalyst support coverage.
- Ultra small area elemental maps.
- III-V super lattice characterization

Relevant Industries for S/TEM Analysis

- Compound Semiconductors
- Intergrated circuit
- Magnetic media
- Nanomaterials
- MEMS
- Opto-electronics
- Semiconductors

Strengths of S/TEM Analysis

- The ultimate elemental mapping resolution of any analytical technique.
- Sub 0.2 nm image resolution.
- Small area crystallographic information

Limitations of S/TEM Analysis

- Significant sample preparation time
- Samples are often prepared that are <100nm
- Some materials not stable to electron beam

Application NotesTop of Form

- Accurate Thin Film Measurements by High-Resolution Transmission Electron Microscopy (HRTEM) AN451. High-Resolution Transmission Electron Microscopy (HRTEM) is often used to make quantitative measurements of nm scale features due to its outstanding image resolution (< 2 E), its ease of calibration, and high precision and accuracy.

Below are images of a commercially available solid state memory device. The images show both a wide field of view and high resolution image of a region of interest near the gate oxide. Individual columns of silicon atoms are evident in the high magnification image, which are used to calibrate the measurement of the thin film. These columns of atoms are often observed in single crystal materials, while in polycrystalline materials "lattice fringes" are commonly observed and can be used for calibration as well.

When characterizing samples that do not display columns of atoms or lattice fringes, the excellent stability and reproducibility of the TEM allows accurate measurements using an external calibration. Note the excellent strong contrast between the crystalline Si, the gate oxide and the poly Si. TEM is an excellent tool for the characterization of complex structures to determine film thickness, material crystallinity, film integrity, film roughness, coverage and for the observation of the shape of nanoscale structures.

- Electron Energy Loss Spectroscopy (EELS) Characterization of an Ultrathin Multilayer Film AN450
- Transmission Electron Microscopy (TEM) and Scanning Transmission Electron Microscopy (STEM) TN110

BASIC PRINCIPLES OF THE TRANSMISSION ELECTRON MICROSCOPE

The transmission electron microscope (TEM) operates on the same basic principles as the light microscope but uses electrons instead of light. What you can see with a light microscope is limited by the wavelength of light. TEMs use electrons as "light source" and their much lower wavelength makes it possible to get a resolution a thousand times better than with a light microscope. You can see objects to the order of a few angstrom (10^{-10} m). For example, you can study small details in the cell or different materials down to near atomic levels. The possibility for high magnifications has made the TEM a valuable tool in both medical, biological and materials research.

A TEM works much like a slide projector. A projector shines a beam of light through (transmits) the slide, as the light passes through it is affected by the structures and objects on the slide. These effects result in only certain parts of the light beam being transmitted through certain parts of the slide. This transmitted beam is then projected onto the viewing screen, forming an enlarged image of the slide.

TEMs work the same way except that they shine a beam of electrons (like the light) through the specimen(like the slide). Whatever part is transmitted is projected onto a phosphor screen for the user to see. A more technical explanation of a typical TEMs workings is as follows (refer to the diagram below):

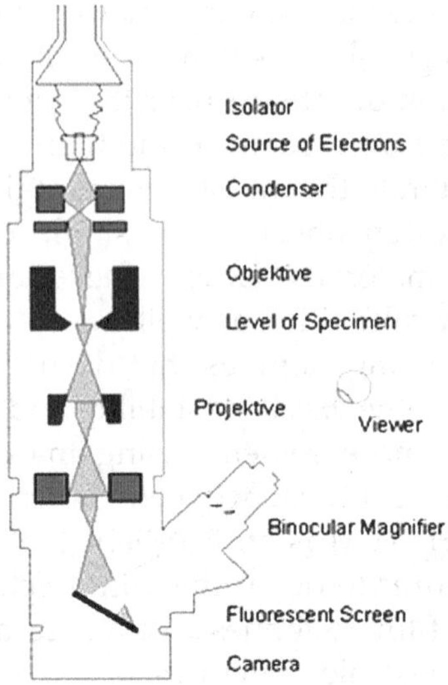

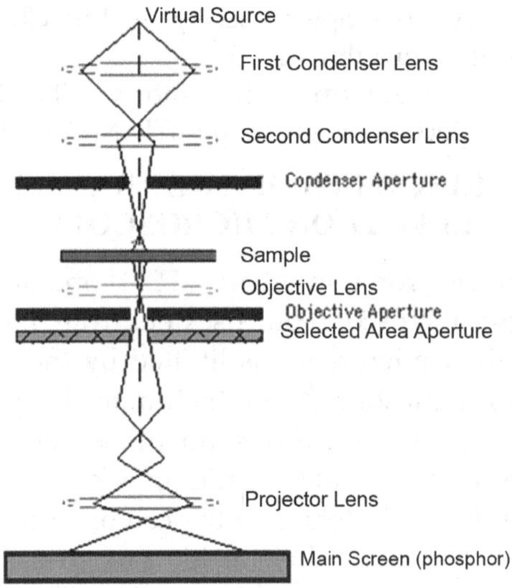

1. The "Virtual Source" at the top represents the electron gun, producing a stream of monochromatic electrons.
2. This stream is focused to a small, thin, coherent beam by the use of condenser lenses 1 and 2. The first lens (usually controlled by the "spot size knob") largely determines the "spot size"; the general size range of the final spot that strikes the sample. The second lens(usually controlled by the "intensity or brightness knob" actually changes the size of the spot on the sample; changing it from a wide dispersed spot to a pinpoint beam.
3. The beam is restricted by the condenser aperture (usually user selectable), knocking out high angle electrons (those far from the optic axis, the dotted line down the centre)

4. The beam strikes the specimen and parts of it are transmitted
5. This transmitted portion is focused by the objective lens into an image
6. Optional Objective and Selected Area metal apertures can restrict the beam; the Objective aperture enhancing contrast by blocking out high-angle diffracted electrons, the Selected Area aperture enabling the user to examine the periodic diffraction of electrons by ordered arrangements of atoms in the sample
7. The image is passed down the column through the intermediate and projector lenses, being enlarged all the way
8. The image strikes the phosphor image screen and light is generated, allowing the user to see the image. The darker areas of the image represent those areas of the sample that fewer electrons were transmitted through (they are thicker or denser). The lighter areas of the image represent those areas of the sample that more electrons were transmitted through (they are thinner or less dense)

The properties of the light microscope as opposed to that of transmission and scanning electron microscopes are collected in a *table.* Transmission Electron Microscope (TEM) versus Transmission Optical Microscope (TOM)

	TEM	**TOM**
Source	100-400 kV electron gun high current densities: 5×10^4 Am^{-2} for tungsten filament 1×10^6 Am^{-2} for field-emission source	Light source
Condenser lens	Electromagnetic lens, focus adjusted by controlling the lens current	Glass lens, focus adjusted by lens position
Specimen stage	Allows for specimen tilt as well as some z-adjustment	Allows for specimen tilt as well as some z-adjustment
Objective lens	Fine focusing of the image by adjusting the lens current	Fine focusing by adjusting the position of the specimen and the objective lens
Final imaging system	Employs electromagnetic lenses to produce image on a fluorescent screen	Eye piece forming image for direct viewing
Recording system	Computer monitor or TV	Normal viewing or photographic films
Experimental set-up in	Vacuum, better than 10^{-6} torr	Air, at atmospheric pressure

Finally some outlines of new and further developments are given:

1. The high voltage electron microscope: it operates with an accelerating voltage of 700 - 3000 kV. Its power of resolution is greater, the preparation can be thicker, the strain on the preparation is smaller. But the enormous technical expenditure is disadvantageous. Only few gadgets exist. New results concerning botany have not been gained.
2. The scanning transmission electron microscope (STEM): In this development of the SEM do the electrons pass through the preparation and the secondary radiation thus generated is used for image formation. Here, too, the expenditure is large, but it is still worthwhile, since large molecules like nucleic acids or proteins or molecular complexes like viruses can be depicted much better and gentler than with the TEM. No news for botany, though.

The interpretation of images gained with electron microscopy is increasingly done with computerized interpretation programs. But they are usually only suitable for the reconstruction of regularly recurring patterns and these, again, are found more often on a molecular level than on a cellular one.

SCANNING TRANSMISSION ELECTRON MICROSCOPY FACILITY STEM OPERATING PARAMETERS

BNL STEM 1

V0	40 keV
Probe	0.25 nm
Beam Intensity	10^3 el/pixel
Dark Field	15 - 40 mR or 40 - 200 mR
Specimens on X,Y Stage	6 Samples in Vacuum at 20° or -180°C
Data Aquisition	Windows PC

Probe The STEM is operated at 40 keV with a probe focussed to 0.25 nm. The sample is maintained at -150°C to eliminate contamination and to reduce mass loss.

Scan The scan raster is 512 × 512 points with a dwell time of 30 μs per pixel (8.5 sec per scan). For most mass measurements, the pixels are separated by 1 nm giving a scan width of 0.512 μm. These scanning conditions provide the best compromise between resolution of the specimen and radiation damage to it.

Detectors Electrons elastically scattered through large angles (40-200 mR), through small angles (15-40 mR), or unscattered (0-12 mR) are collected on separate detectors consisting of scintillators and photomultipliers. Both large and

	small angle signals (normalized by the total beam current) are recorded digitally. The dose is kept below 1000 electrons/nm² to assure that mass loss from radiation damage is less than 2.5% at the -150°C specimen temperature employed.
Image	Digital images are recorded with the STEM from the annular dark field detectors. Areas of the grid to be used for mass measurements are selected for clean background and adequate numbers of both TMV and specimen particles. The background is computed in clear areas in each image, and the intensity minus background is summed over each particle and multiplied by a calibration factor to give a mass value.
Calibration	The STEM calibration factor is checked in each image using TMV which is included as an internal control. The TMV also serves as a qualitative control for the whole process of specimen preparation. If the TMV looks good, the specimen has a chance of being good, but if it looks bad, some step in the specimen preparation has gone wrong. Different kinds of mass measurements can be obtained from this data. The total mass of the individual particles can be used to determine their oligomeric state. For filamentous specimens, the mass per length can be used to determine the spacing of repeating units. Similarly, the mass per area can be a useful measurement on two-dimensional arrays.

NEW MATERIALS AND ICROELECTRONICS

In A Dark, First-floor Room at Georgia Tech's School of Materials Science and Engineering, Dr. Z.L. Wang carefully adjusts the controls of a JEOL 4000EX transmission electron microscope. A black-and-white image appears on the monitor to his right, revealing a jumbled pattern of ridges and valleys that covers the surface of a tiny carbon nanosphere barely a quarter of a micron in diameter. The features are just 34 millionths of a meter — 0.34 nanometers — across.

Photo by Stanley Leary: Equipment in the Electron Microscope Facility, used here by Dr. Z.L. Wang, will benefit research in nanotechnology, novel materials and structures, micro-electronics, new crystals, improved lubricants and living tissues. *(200-dpi JPEG version - 355k)*

This glimpse of the sphere's surface shows researchers that they are on the right track in their efforts to turn the tiny graphite balls into efficient solid lubricants for micromachines. It also gives them clues to the next series of experimental steps they should try.

Research into promising areas of nanotechnology, novel materials and structures, ever smaller microelectronics, new crystals, improved lubricants — and even living tissues — demands the use of electron microscopes able to see features a few hundred atoms in size. To make these costly tools available to researchers from many different disciplines, the Georgia Institute of Technology has formed the Electron Microscopy Facility.

The facility's three transmission electron microscopes and single scanning electron microscope have helped examine the molecular arrangement of gold nanocrystals, the charge distribution on a new type of P/N junction used in microelectronics, the growth processes in experimental quantum-dot semiconductors, lubricant layers for high-density disk drives and the configuration of a new phosphor material for computer displays.

In addition to showing the structure of materials on a very small scale, the equipment can also provide data on the chemical composition of these small objects, map electrical charges and measure magnetic domain.

"We want to play an important role in many research programs going on across Georgia Tech, and to promote projects across schools and across other divisions of campus," explains Wang, who is director of the Electron Microscope Facility and an associate professor in the School of Materials Science and Engineering. "We believe that a facility like this is essential to many exciting areas of research."

Wang sees the facility as a key part of a research process that begins with theory and simulation, proceeds to experimental fabrication, then relies on microscopy for feedback on the results. Rather than simply providing a microscopy service, he wants to help researchers design their projects to gain maximum benefit of the facility's equipment, which also includes sample preparation and digital image analysis equipment.

The cost of developing the Electron Microscopy Facility has been borne by the National Science Foundation, the State of Georgia through the Georgia Research Alliance, and Georgia Tech. "We want to see this facility grow into one of the major microscopy facilities in the nation," Wang says. "This equipment is essential to many key areas of research, and most of the country's largest and best universities now have it."

The equipment includes: Hitachi HF-2000 Field Emission Gun (FEG) Transmission Electron Microscope: This instrument can perform high spatial-resolution chemical microanalysis, high-resolution lattice imaging, and high-coherent beam holographic imaging. It is equipped with a thin window energy dispersive X-ray spectrometer (EDS), which detects not only heavier elements, but also elements as light as carbon. It also has a Gatan parallel-detection electron energy-loss spectrometer, which can be used for quantitative chemical microanalysis of light elements as well as for studying atomic bonding in solid materials. A 180-degree rotational electrostatic biprism suitable for electron holography completes the package.

"The brightness of the highly coherent energy source of the Hitachi HF-2000 allows high resolution lattice imaging at a point-to-point resolution better than 2.3 Angstroms, lattice resolution of one Angstrom, and chemical microanalysis at a spatial resolution of better than 20 Angstroms," Wang notes. "This is an ideal instrument for performing studies of nanostructured materials and interfaces in thin film and composite materials."

The holographic imaging is the only technique that can provide electron phase information, and is therefore suitable for quantitative mapping of electrostatic fields and magnetic fields in materials such as ferroelectric and magnetic recording materials.

To allow in-situ observation of structural evolution, specimens can be cooled to a temperature as low as -160 Celsius using a liquid nitrogen specimen holder, or heated to a temperature of up to 1,300 Celsius using a heated holder. "This capability is vitally important to predicting the lifetime and reliability of advanced materials," Wang notes.

Hitachi S800 Field Emission Gun Scanning Electron Microscope: This equipment can image materials at a resolution of better than 30 Angstroms, and perform chemical microanalysis from bulk specimens.

JEOL 4000EX High Resolution Transmission Electron Microscope: This device routinely provides point-to-point image resolution of 1.8 Angstroms and is best suited for recording high-resolution images of thin foil specimens.

JEOL 100CX II Transmission Electron Microscope: This instrument is primarily used for undergraduate and graduate education, though it also has conventional research uses.

HOW DOES A TRANSMISSION ELECTRON MICROSCOPE WORK?

Transmission electron microscopes use high electric voltages — as much as 400,000 volts — to accelerate a beam of electrons within a vacuum chamber. The beam is then aimed at a thin slice of the material under study. After the powerful beam of electrons passes through the sample, it is focused and projected onto either a monitor or photographic film to provide an image of the structure. Transmission electron microscopes can provide magnification as much as 1.5 million times. Since the electron beam must pass through it, preparation of the sample is critical.

MAGNETIC LENSES GUIDE THE ELECTRONS

A "light source" at the top of the microscope emits the electrons that travel through vacuum in the column of the microscope. Instead of glass lenses focusing the light in the light microscope, the TEM uses electromagnetic lenses to focus the electrons into a very thin beam. The electron beam then travels through the specimen you want to study. Depending on the density of the material present, some of the electrons are scattered and disappear from the beam. At the bottom of the microscope the unscattered electrons hit a fluorescent screen, which gives rise to a "shadow image" of the specimen with its different parts displayed in varied darkness according to their density. The image can be studied directly by the operator or photographed with a camera.

Preparation of Specimen

In a TEM, the specimen you want to look at must be of such a low density that it allows electrons to travel through the tissue. There are different ways to prepare your material for that purpose. You can cut very thin slices of your specimen from a piece of tissue either by fixing it in plastic or working with it as frozen material. Another way to prepare your specimen is to isolate it and study a solution of for example viruses or molecules in the TEM.

You can also stain the specimen in different ways and use markers to locate specific things in the tissue. It can for example, be stained with heavy metals like uranium and lead, which scatters electrons well and improves the contrast in the microscope. Below are two examples described in more detail:

Sections of Embedded Material

Biological material contains large quantities of water. Since the TEM works in vacuum, the water must be removed. To avoid disruption as a result of the loss of water, you preserve the tissue with different fixatives. These cross-link molecules with each other and trap them together as stable structures. The tissue is then dehydrated in alcohol or acetone.

After that, your specimen can be embedded in plastic that polymerize into a solid hard plastic block. The block is cut into thin sections by a diamond knife in an instrument called ultramicrotome. Each section is only 50-100 nm thick.

The thin sections of your sample is placed on a copper grid and stained with heavy metals. The slice of tissue can now be studied under the electron beam.

Negative Staining of Isolated Material

The isolated material (can be a solution with bacteria or a solution with isolated molecules) is spread on a support grid coated with plastic. A solution of heavy metal salt is added.

The metal salt solution does not bind to the material but forms a "shadow" around it on the grid. The specimen will appear as a negative picture when viewing it in the TEM.

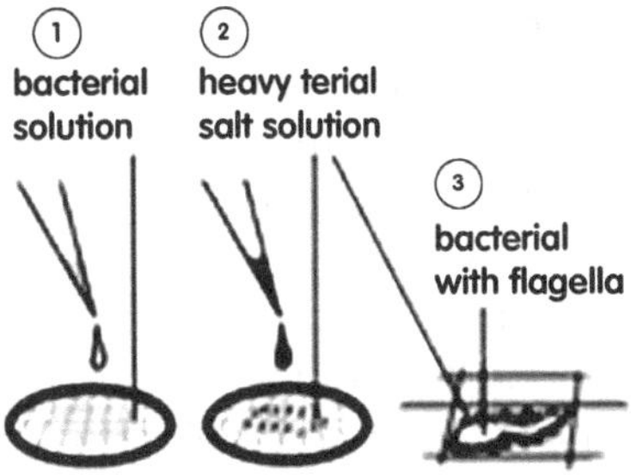

Different Models of Transmission Electron Microscope (TEM)

Materials for TEM must be specially prepared to thicknesses which allow electrons to transmit through the sample, much like light is transmitted through materials in conventional optical microscopy.

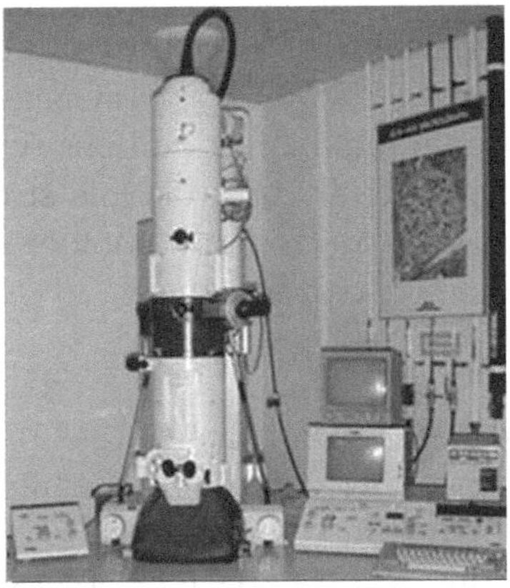

Because the wavelength of electrons is much smaller than that of light, the optimal resolution attainable for TEM images is many orders of magnitude better than that from a light microscope. Thus, TEMs can reveal the finest details of internal structure - in some cases as small as individual atoms. Magnifications of 350,000 times can be routinely obtained for many materials, whilst in special circumstances, atoms can be imaged at magnifications greater than 15 million times.

For biological samples, cell structure and morphology is commonly determined whilst the localisation of antigens or other specific components within cells is readily undertaken using specialised preparative techniques. For non-biological materials, phase determination as well as defect and precipitate orientation are typical outcomes of conventional TEM experiments. Microstructural characterisation of non-biological materials, including unit cell periodicities, can be readily determined using various combinations of imaging and electron diffraction techniques. Images obtained from a TEM are two-dimensional sections of the material under study, but applications which require three-dimensional reconstructions can be accommodate by these techniques.

The energy of the electrons in the TEM determine the relative degree of penetration of electrons in a specific sample, or alternatively, influence the thickness of material from which useful information may be obtained. Thus a 400 kV TEM not only provides the highest resolution available at The University of Queensland but also allows for the observation of relatively thick samples (eg. less than 0.2 micrometers) when compared with the more conventional 100 kV or 200 kV instruments.

Because of the high spatial resolution obtained, TEMs are often employed to determine the detailed crystallography of fine-grained, or rare, materials. Thus, for the physical and biological sciences, TEM is a complementary tool to conventional crystallographic methods such as X-ray diffraction. The CMM has two JEM-1010 Transmission Electron Microscope's which are used predominantly in the high through-put biological discipline. These are compact high performance TEM's with advanced features and functions. The high contrast objective lens polepiece combines the highest possible contrast and brightness with optimum resolution essential for life science studies. The JEOL cool beam gun allows high-brightness, high coherence illumination conditions with low emission current. Other features include user friendly controls, file storage, automatic filament heating, and automatic exposure micrograph photography. The instruments are also equipped with goniometer specimen stages.

JEOL 1011

Similar to the JSM 1010's CMM has a new JSM 1011Transmission Electron Microscope's which is used predominantly in the high through-put biological discipline. This has a compact high performance TEM with advanced features and functions. The high contrast objective lens polepiece combines the highest possible contrast and brightness with optimum resolution essential for life science studies. The JEOL cool beam gun allows high-brightness, high coherence illumination conditions with low emission current. Other features include user friendly controls, file storage, automatic filament heating, and automatic exposure micrograph photography. The instruments are also equipped with goniometer specimen stages.

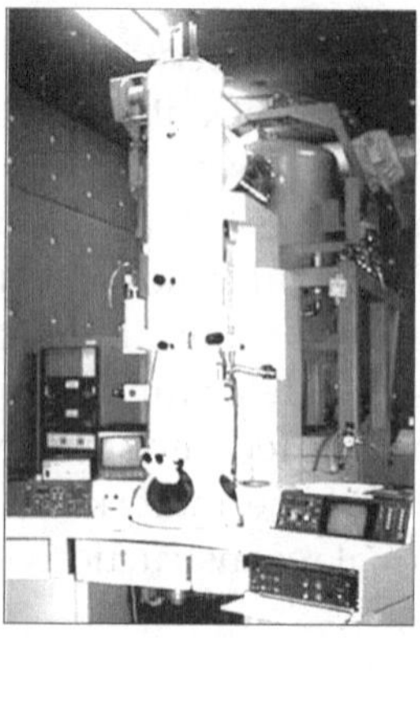

This instrument provides analytical 200 kV TEM capability for detecting elements ranging from carbon to thorium

The JEOL 2010 200kV analytical electron microscope (AEM) is fitted with an Oxford thin-window energy dispersive X-ray (EDS) detector which is capable of detecting elements ranging from carbon upwards.

The 2010 AEM has a modest resolution of 2.5E, has a ±40° stage tilt, has small probe forming capability for convergent beam electron diffraction (CBED), and is set up for maximum sensitivity for elemental detection.

This instrument allows researchers within the materials science disciplines to remain internationally competitive in diffraction and in phase contrast imaging as well as providing elemental analyses from nanometre regions of solid thin-films.

A combination of elemental (EDS) and diffraction (CBED) information provides unambiguous data on the structure and chemistry of the material under investigation. The 2010 and 4010 sample holders are interchangeable it is possible to characterise the elemental composition in the 2010 of phases previously studied by high resolution lattice imaging in the 4010. This unique instrument provides 400 kV TEM capability.

The instrument was purchased and installed as a 4000FX in 1988 and has been upgraded in 1997 to a 4010 TEM, through the combined resources of The University of Queensland, Queensland University of Technology, Griffith University, the Queensland Government and the Australian Research Council.

The University of Queensland is the majority partner in this important regional resource and actively supports research programs in the physical, biological and medical sciences which require these sophisticated analytical methods.

As a TEM, the instrument provides on-screen magnifications up to 2,000,000 times. The optimal imaging resolution is 0.15 nm at 400 kV which allows the direct observation of atoms of, for example, high strength ceramics, high temperature superconductors and light metals.

Both conventional selected area and convergent beam electron diffraction are routine functions. Double-tilt (±25°), single-tilt and liquid nitrogen stages are available for conventional 3 mm diameter samples. The 4010 TEM has the capability to be able to solve atomic structure of surfaces and interfaces in metals, semiconductors and ceramics, allows the study of the structure of internal defects in carbons and light metals, and makes possible the comparison between the atomic structure and macroscopic behaviour in these materials.

Transmission Electron Microscope Philips Technai 120kV

This Philips Technai 120kV electron microscope was installed at the CMM in April 2000.

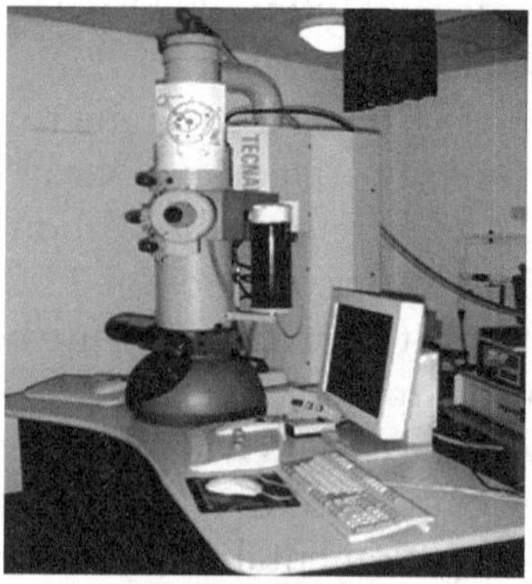

It is a general purpose microscope configured for high contrast with the BioTWIN lens configuration, but together with the ultra-clean high vacuum for contamination free observation and the Compustage for accurate specimen control the Technai 12 also fits the need for cryo-observation.

Another key component to the success of low temperature observation is the installation of a specific cryo-trap effective for the preventing vapour deposition on cryo holders and their specimens while in the column. To facilitate digital capture a CCD camera was integrated above the microscope camera, this CCD camera together with image acquisition software offers the user a wide choice for image manipulation and analysis.

The Technai 12 user interface is comprehensive and provides the multi-user facility the capability to offer each user their individual microscope profile and data storage. The Technai series supports a number of innovative features in the area of the specimen stage, vacuum and imaging systems which we look forward to using in facilitating our investigations.

TECNAI 20

The Philips Tecnai 20 FEG analytical electron microscope (FEG-AEM) has the capability to deliver one nanoamp of current into a probe one nanometre in diameter.

The unique analytical resolution is as a result of the high brightness field emission gun (FEG) which acts as the electron source in this microscope. The instrument can be operated in TEM- microprobe, TEM-nanoprobe or STEM (scanning transmission electron microscopy) modes.

It is fitted with an EDAX thin-window energy dispersive X-ray (EDS) detector which is capable of detecting elements ranging from carbon upwards. The TEM, STEM and EDS operation is fully integrated within a windows NT platform. The instrument has a twin-lens objective system and is capable of ±70 degree stage tilt. This instrument was installed in the year 2000 and is used by researchers in both the materials and the biological sciences.

TECNAI 30

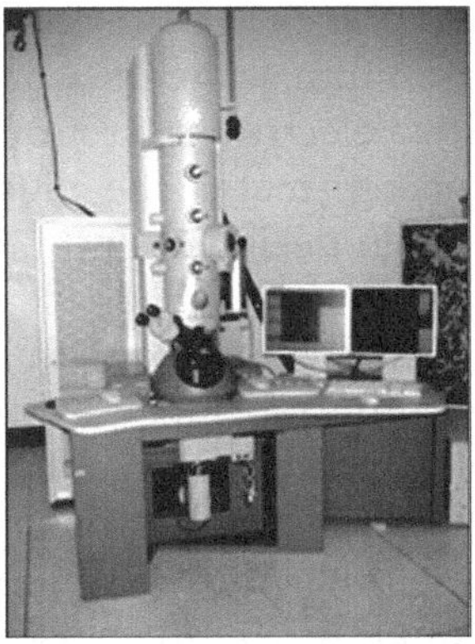

One of the highlights of 2004 was the installation of the new Tecnai F30 cryo-electron microscope within the new cryo-EM facility. This new instrument is the centre piece of the facility and is ideally suited for high-resolution imaging of biological materials including cell sections, 2D crystals of membrane proteins and individual macromolecular assemblies (eg. virus particles and large protein complexes) at liquid Nitrogen temperatures (~ -170°C). Under these conditions high resolution detail is particularly well preserved. Delivery of a 300 kV field emission TEM occurred in early 2004. The FEI Tecnai 30 was the instrument of choice because it is optimised for imaging of biological macromolecules – particularly for cell tomography – with cryogenic cooling of the sample area and reproducible sample tilting to large angles (~ ±70°).

With the addition of the latest digital camera technology to capture ultra high-resolution images combined with an energy filtering spectrometer, this instrument will be unique in Australia and one of a few available in the world. The Tecnai F30 is supported by a second cryo-electron microscope (Tecnai 12) which is also capable of recording low contrast images from cryogenically prepared biological specimens. These microscopes are now supported by a wide array of ancillary equipment much of which was installed in 2004.

Equipment includes:

1. High-pressure freezer. This unit is the first of its kind in Australia and allows the rapid cryogenic freezing of relatively thick tissue and cell samples to preserve high resolution detail.
2. Vitrobot. The Vitrobot is designed to produce thin vitreous ice layers in which single particles or 2D crystals can be suspended for high-resolution cryo-electron microscopy.
3. Carbon evaporation sources. This unit facilitates the rapid production of thin carbon support films on which samples are suspended for electron microscopy.
4. Cryo-microtomes. Cryo-microtomy is used to prepare well preserved sections (including frozen hydrated sections) of biological materials for tomographic 3D image reconstruction.
5. High-resolution digital film scanners. Commercial negative scanners are in place to digitize electron micrographs. In addition a purpose built system designed to match the high-resolution capabilities of the Tecnai F30 has been designed and constructed and is currently being tested.
6. Supercomputing facilities complete with high-end image processing capabilities. Supercomputing facilities and software packages for single particle analysis and tomographic 3D image reconstruction are in place. Early in 2005 the electron cryo-microscopy facility will undergo a further upgrade through the installation of a dedicated local area 32 node, dual CPU Linux cluster (2 x 2.4GHz, 4GB, 80GB).

High-Throughput Cryo-Transmission Electron Microscope (TEM)

Scientific Drivers

Future advances in the biosciences will depend heavily on the ability to link cell biology and structural biology through a comprehensive understanding of the structure and function of individual molecular machinery.

For instance, membrane proteins account for approximately 20-30% of the proteome and form the responsive interface between cellular and sub-cellular compartments and their environment.

Thus, one of the great challenges of cell biology, proteomics and structural biology is atomic-resolution structure determination of membrane protein complexes and dynamic macromolecular assemblies, including whole viruses.

A high-throughput cryo-TEM platform is the only possible tool for elucidating the three-dimension- al structure of entire cells with sufficient resolution to examine the arrangement and interactions of internal macromolecular complexes. Examples of the types of work that this facility will enable include:

- Determining how the 3-D biology of the human islet/beta cell relates to the development, physiology and dysfunction of the human endocrine pancreas and its role in diabetes.
- Characterising caveolae—which have been linked to cancer, cholesterol regulation and muscular dystrophy—by using electron tomography and immunolabelling methods. Revealing the structure of membrane proteins for the design of new and highly specific drugs. With the ability to probe the molecular basis of disease.

Capabilities and the National Research Capacity

The cryo-TEM facility at the University of Queens-land is the only fully established life-sciences facility in Australia or New Zealand capable of collecting and processing atomic-resolution images at low temperatures (-160ºC), as well as undertaking 3-D electron tomography of organelles, cells and tissues at ambient and low temperatures.

As such, it is one of only a handful of such facilities in the world. The NCRIS investment will build on the flagship FEI Tecnai F30 microscope to create a high-throughput platform capable of the high output essential to make genuine inroads into key questions in molecular biology, medicine and biotechnology. The completed system will offer Australian researchers a quantum leap in technology and productivity for the emerging techiques of electron tomography, electron crystallography and single particle analysis.

The high-throughput system will be sited in custom-built laboratories within the Australian Institute for Bioengineering and Nanotechnology (AIBN) and the Queensland Biosciences Precinct (QBP). Practically, this is necessary to build on the previous investments in the FEI Tecnai F30 and associated sample prepara- tion and screening equipment and laboratories.

Scientifically, this builds on the unparalleled experience within the Centre for Microscopy and Microanalysis in cryo-electron microscopy, particularly the recent developments in rapid image-processing pipelines for single particle analysis, electron crystallography and electron tomography.

Moreover, it provides easy access for the approximately 1000 researchers from the Institute for Molecular Bioscience, the new AIBN and the Queensland Brain Institute (QBI).

Nature and Level of Demand

Since commissioning of the Tecnai F30, more than 120 national and international research projects have been undertaken and the instrument is presently near fully usage.

Given the proximity to major existing and new research centres in biosciences and nanotechnology, demand is expected to rapidly increase with the new high-throughput capability. Expected usage is approximately 20 projects (3000 hours) per annum.

Understanding how the SEM Works and how to use it on a College Level

- Electron Source
- Electromagnetic Lens
- Electron Optical Column
- Ray Diagrams
- Electron Beam/Specimen Interactions
- Vacuum
- Specimen Chamber
- Specimen Preparation

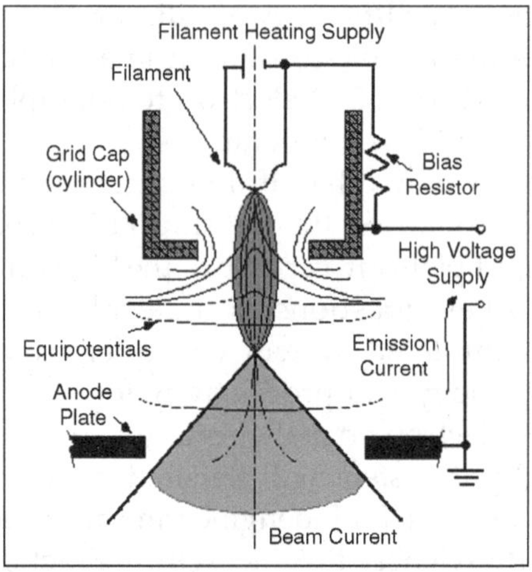

Fig. Forces in a Cylindrical Magnetic Lens

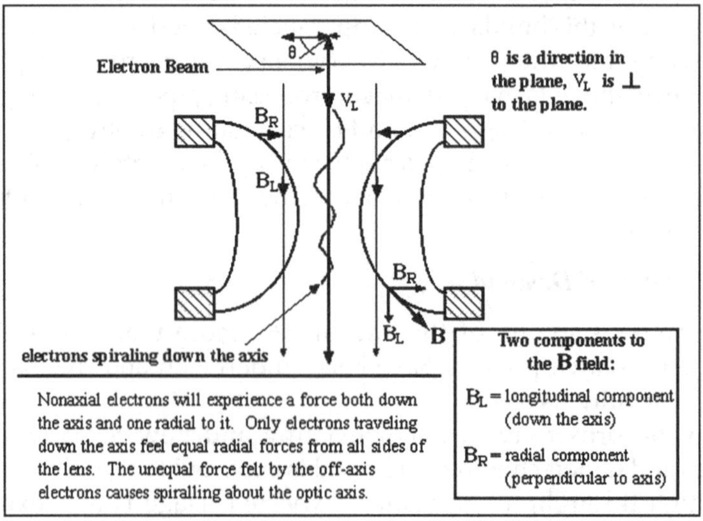

Fig. Beam's Path through the Column

The electron beam comes from a filament, made of various types of materials. The most common is the Tungsten hairpin gun. This filament is a loop of tungsten which functions as the cathode. A voltage is applied to the loop, causing it to heat up.

The anode, which is positive with respect to the filament, forms powerful attractive forces for electrons. This causes electrons to accelerate toward the anode. Some accelerate right by the anode and on down the column, to the sample. Other examples of filaments are Lanthanum Hexaboride filaments and field emission guns.

A beam of electrons is generated in the electron gun, located at the top of the column, which is pictured to the left. This beam is attracted through the anode, condensed by a condenser lens, and focused as a very fine point on the sample by the objective lens.

The scan coils are energized (by varying the voltage produced by the scan generator) and create a magnetic field which deflects the beam back and forth in a controlled pattern.

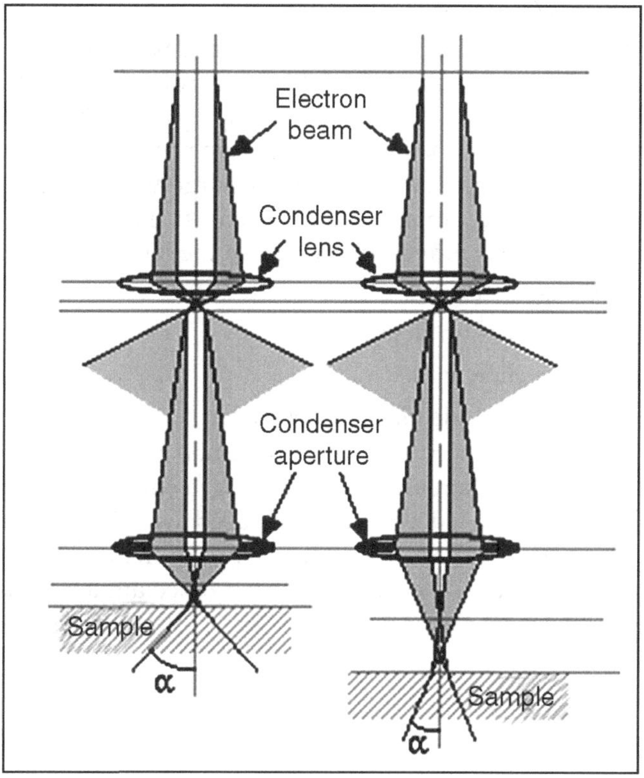

The varying voltage is also applied to the coils around the neck of the Cathode-ray tube (CRT) which produces a pattern of light deflected back and forth on the surface of the CRT. The pattern of deflection of the electron beam is the same as the pattern of deflection of the spot of light on the CRT.

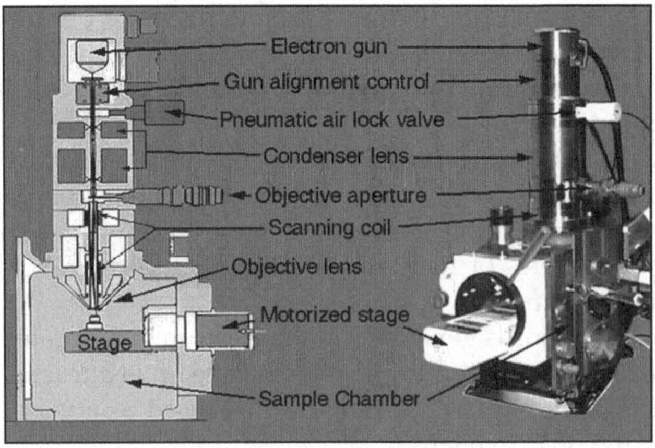

Fig. SEM Ray Diagrams

The electron beam hits the sample, producing secondary electrons from the sample. These electrons are collected by a secondary detector or a backscatter detector, converted to a voltage, and amplified. The amplified voltage is applied to the grid of the CRT and causes the intensity of the spot of light to change. The image consists of thousands of spots of varying intensity on the face of a CRT that correspond to the topography of the sample.

These schematics show the ray traces for two probe-forming lens focusing conditions: small working distance (left) and large working distance (right). Both conditions have the same condenser lens strength and aperture size. However, as the sample is moved further from the lens, the following occurs:

- The working distance *S* is increased.
- The demagnification decreases.
- The spot size increases.
- The divergence angle *alpha* is decreased.

The decrease in demagnification is obtained when the lens current is decreased, which in turn increases the focal length *f* of the lens. The resolution of the specimen is decreased with an increased working distance, because the spot size is increased. Conversely, the depth of field is increased with an increased working distance, because the divergence angle is smaller.

4

Electron Spectroscopic Imaging

The energy filter (Gatan imaging filter, *GIF*), which is installed below a TEM, allows to record electron energy loss spectra (*EELS*) and element specific images (elemental maps) by means of the electron spectroscopic imaging (ESI) technique. Mostly, the so-called three-window method is applied for mappings: an image is taken after a suitable ionisation edge of the corresponding element (post-edge image, $\Delta E3$), two additional images (pre-edge 1, $\Delta E1$, and pre-edge 2, $\Delta E2$) are recorded at energy losses smaller than the ionisation edge.

Only the electrons passing through the selected energy slit contribute to these images. Specimen drift during recording of these images is compensated for by a cross-correlation algorithm implemented in the software of the GIF. The pre-edge images are used for an approximate determination of the unspecific background which is then subtracted from the post-edge image leading to an elemental map with enhanced contrast. Alternatively, the post-edge image can be divided by a pre-edge image (ratio method).

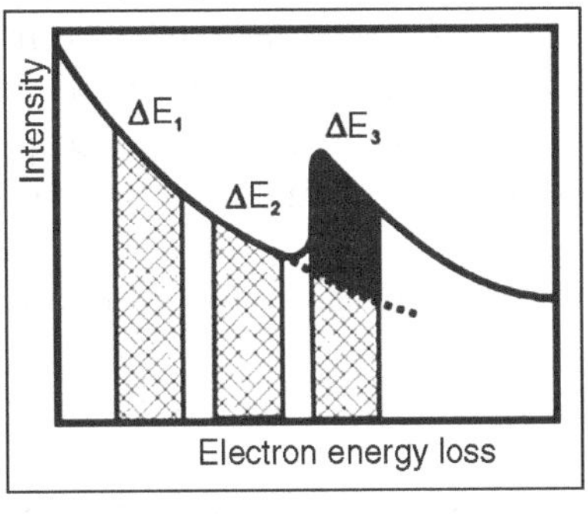

Example:

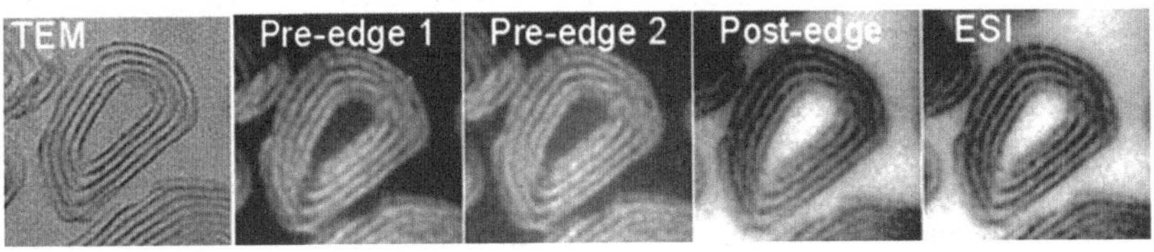

Recording and calculating an elemental map by the three-window method is demonstrated by the corresponding images obtained at the K edge of carbon in a *vanadium oxide nanotube* (TEM investigations reveal that the tubes have an empty core and that a layered builds up the tube walls. Vanadium oxide layers between which the amine molecules are embedded form the walls. Information obtained in reciprocal space by X-ray and electron diffraction led to a decription of the nanotube structure by a cell with a tetragonal metric (a = 0.62 nm and c being equal to the inter-layer distance).

However, one should consider that three-dimensional periodicity is not present in nanotubes due to the bending of the layers. Thus, their structure can only approximately be described in the usual crystallographic terms.). The TEM image shows the cross-section of a single VOx-NT. In the post-edge image, the carbon distribution is clearly visvible (bright contrast at the C-containing sites): C is present around the tube due to the organic resins used during the *cross-sectional preparation* (The observation of the cross-sectional structure of nanotubes is essential for a differentiation between scrolls of layers and concentric layers building up the tube walls. For this purpose, the NTs are deposited on a Si wafer; they lay flat on it, and, after fabricating a sandwich and using common preparation methods, there is a good chance to find NTs in the looked-for orientation.), in the centre of the tubes, and between the VOx layers.

Due to the absence of C there, the VOx layers appear with dark contrast in the ESI and post-edge image, whereas they are bright in the two pre-edge images since these in fact are post-edge images of the V_M edge.

INELASTIC INTERACTIONS OF ELECTRONS WITH MATTER

Ionization

The high-energy electrons of the incident beam can transfer a critical amount of energy to an inner-shell electron of an atom, leading to the ejection of this electron. The ionization energy is provided by the incident electron, reducing its energy. This leads to an ionization edge in the electron energy loss spectrum (*EELS*). Subsequently, the hole in the inner-shell is filled up by an electron with higher energy from an outer shell. This electron gives away a part of its energy, leading to the emission of *characteristic X-rays* or *Auger electrons*.

Secondary Electrons

Electrons in the conduction or valence band do not need much energy (low work function) to be transferred into vacuum. Thus, the energy of secondary electrons (SE) is low (>50 eV). The SEs are mainly exploited in *SEM*.

Phonons

Phonons are lattice vibrations, which are equal to heating the specimen. This effect may lead to a damage of the sample.

Plasmons

Plasmons are longitudinal oscillations of free electrons, which decay either in photons or phonons.

Cathodoluminescence

If semiconductors are hit by high-energy electrons, electron-hole pairs can be formed by promoting an valence electron into the conduction band. Filling this hole with an electron from the conduction band (recombination) leads to the emission of light with a frequency that corresponds to the band gap.

INTERACTIONS OF ELECTRONS WITH MATTER

Two different types of interactions appear:

Elastic Interactions

No energy is transferred from the electron to the sample. The electron either passes without any interaction (direct beam) or is scattered by electrostatic with the positive potential inside the electron cloud. These signals are mainly exploited in *TEM* and *electron diffraction.*

Inelastic Interactions

Energy is transferred from the incident electrons to the sample: secondary electrons, phonons, UV quanta or cathodo- luminescence are produced; shooting out inner shell electrons leads to the emission of X-rays or Auger electrons. These signals are used in *analytical electron microscopy.*

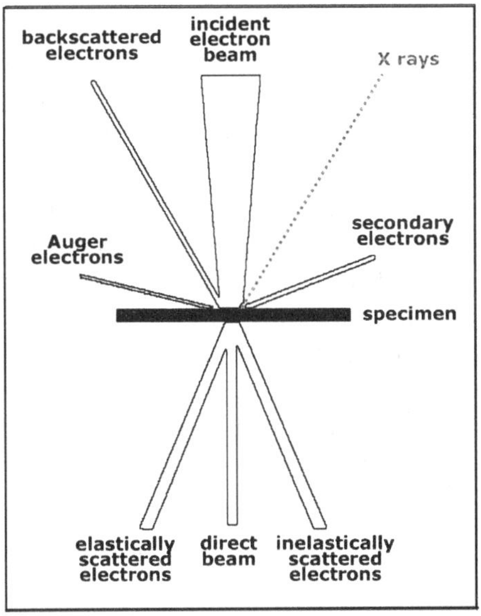

ELASTIC SCATTERING OF ELECTRONS BY ATOMS

An electron penetrating into the electron cloud of an atom is attracted by the positive potential of the nucleus (Coulombic interaction), and its path is deflected towards the core as a result. The Coulombic force F is defined as:

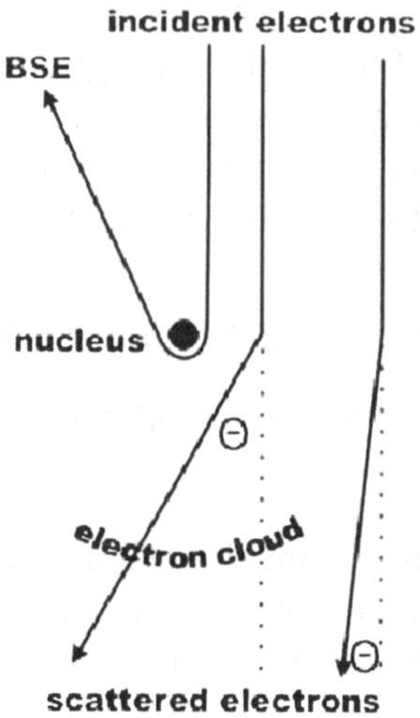

$$F = Q_1Q_2/4\pi\varepsilon_0 r^2$$

with r being the distance between the charges Q_1 and Q_2 and ε_0 the dielectric constant. The closer the electron comes to the nucleus, i.e. the smaller r, the larger is F and consequently the scattering angle. In rare cases, even complete backscattering can occur (back scattered electrons *BSE*). These interactions can be treated as elastic, which means that no energy is transferred from the scattered electron to the atom. This simple model helps to explain basic contrast mechanisms in TEM and STEM. The *mass-thickness* contrast is somehow related to the contrast in optical microscopy, but it is the local scattering power that determines contrast in TEM instead of absorption of light.

The interaction of electrons with heavy atoms having a high charge Q is stronger than with light atoms so that areas in which heavy atoms are localized appear with darker contrast than such with light atoms (mass contrast). In thick areas, more electron scattering events occur of course; thus, these thick areas appear dark er than thin areas (thickness contrast). In particular, this mass-thickness contrast is important in *bright* and *dark* field imaging.

The strong Coulomb interaction of the electrons with the potential of an atom core, which leads to high angle scattering (designated as Rutherford scattering)and even to back-scattering, is employed in *STEM* (Z-contrast imaging) and in *SEM*. By the *HAADF-STEM* method, small clusters (or even single atoms) of heavy atoms (e.g. in *catalysts*) can be imaged in a matrix of light atoms since the contrast is approximately proportional to Z^2 (Z: atomic number).

If a crystalline specimen is transmitted by electrons, then Bragg diffractions happens as well. Each atom in such a regular arrangement acts as a scattering centre.

The scattered electron waves may interact with each other either in a constructive or in a destructive way, which gives rise to a *diffraction pattern*. If a crystal is oriented along a zone axis so that many electrons are strongly scattered to contribute to the reflections in the diffraction pattern, then only a few electrons pass without interactions and therefore this crystal appears with dark contrast in the *BF* image (*diffraction contrast*). In real specimen, all contrast mechanisms, namely mass-thickness and Bragg contrast, occur simultaneously, making interpretation of TEM images sometimes difficult.

SCANNING ELECTRON MICROSCOPY (SEM)

In a scanning electron microscope, a tiny electron beam is scanned across the sample. Simultaneously, the generated *signals* are being recorded, and an image is formed pixel by pixel. In SEM, the signals are observed on the same specimen site as the incoming electron beam (cf. *STEM*, using transmitted electrons). In contrast to TEM methods needing very thin samples, compact samples can thus be investigated by SEM. Valuable information about morphology, surface topology and composition can be obtained. SEM microscopes achieving resolutions below 1 nm are available now.

SEM: Imaging with Secondary Electrons

Imaging with *secondary electrons* provides information about morphology and surface topography. The *LEO 1530 Gemini* has two detectors for secondary electrons (*comparison*).

Examples: The SEM images were obtained with a LEO 1530 Gemini operated at *low voltage* (1 keV).

Signals in Scanning Electron Microscopy

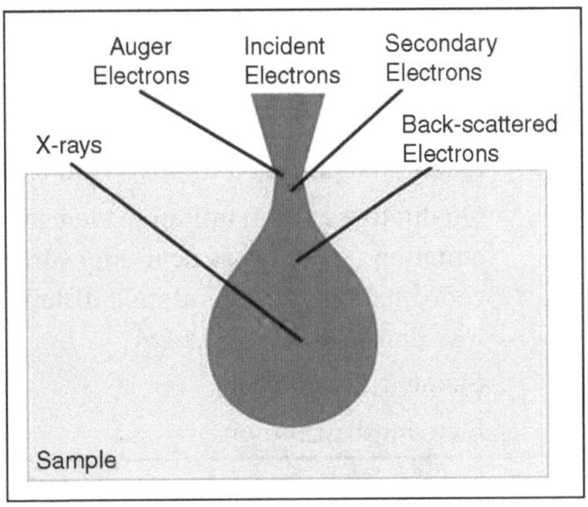

When an electron beam hits matter, it begins to broaden because of strong elastic scattering effects. Simultaneously, inelastic interactions cause an energy loss of the electrons. If the sample is very thick, the energy will completely be transferred to the sample. The overall result is a pear-shaped interaction volume. The penetration depth *t* depends on the electron energy and on the material. Mainly, three signal are measured in the SEM providing different information about the sample:

Secondary Electrons

Since electrons in the conduction or valence band need a small amount of energy (work function) only to be transferred into vacuum, the energy of *secondary electrons* (SE) is low (>50 eV). Because of their low energy, SEs can only escape from the sample if they are generated close to the surface. Therefore, SE images are a means to get topographic images

Back-scattered Electrons

The collision of an electron from the beam with a nucleus leads to the deflection of its path as a result of Coulomb forces (*Rutherford elastic scattering*). Sometimes, the electrons are completely scattered back and leave the surface of the sample. Since heavy atoms with a high atomic number are much stronger scatterers of electrons than light ones, they cause a higher signal. Therefore, images with back-scattered electron contain compositional information (cf., *STEM*).

X-rays

Energy-dispersive (*EDXS*) and wavelength-dispersive (WDXS) spectrometers can be installed at an SEM, providing laterally resolved information about the sample composition.

ANALYTICAL ELECTRON MICROSCOPY (AEM)

Utilized signals: inelastically scattered electrons (energy-loss electrons) and *X-rays*

Methods	*Information*
EDXS	Qualitative and quantitative element analysis
EELS	Qualitative and quantitative element analysis. Bonding situation (energy-loss near-edge strcuture ELNES), coordination and interatomic distances (extended energy-loss fine structure EXELFS)
ESI	Element distribution
STEM + EDXS/EELS	Element distribution

COMPARISON OF EELS AND EDXS

Restrictions: Since electron-matter interactions are mostly elastic, high electron doses are necessary for all methods of analytical electron microscopy. A long recording time may be required for a good signal/noise ratio which is only possible if the sample is stable under the intense electron beam and if no specimen drift occurs. Since ionization edges occur at characteristic energies and have different shapes, not all methods are equally suitable for all elements.

EELS	EDXS
High detection efficiency for low Z elements	High detection efficiency for high Z elements
Elemental, chemical and dielectric information	Elemental information only
Excellent energy resolution (0.3-2 eV) results in few peak overlaps; fine structures of the ionization edges can be analyzed (ELNES, EXELFS)	Bad energy resolution (> 100 eV) causes frequent overlaps
Very efficient and high sensitivity to most elements => fast and efficient mapping technique	Inefficient signal collection and detection => X-ray mapping is time-consuming
Fast technique; complex processing required	Slow technique; only simple data processing required

LIGHT MICROSCOPY

For many biologists, the microscope belongs to the gadgets they use most. The reason is obvious: cells are the basic units of life and their size is, with only few exceptions, below the upper limit of the resolution of the human eye. Numerous statements on the refraction of light by lenses or waterfilled spheres and the resulting magnification proof, that the effects of magnification were already known in antiquity.

In the 13th century, the British natural scientist ROGER BACON used segments of glass spheres as magnifying glasses and recommended them to partially sighted persons as spectacles. The grinding of lenses was perfected in the following centuries.

The physician P. BORCH records, that HANS and ZACHARIAS JENSEN (father and son) from the Dutch town Middelburg invented the compound microscope at the beginning of the 17th century by placing one lens behind the other. They are regarded as the inventors of the telescope, too. Other sources name C.J. DREBBEL (1572 - 1634) from Alkmaar as the inventor of the microscope. Even if this should not be the case, he has helped a lot to make the microscope known. The term microscope was introduced by members of the Italian Accademia de Lincei (Academy of The Lynx), whose most prominent member was G. GALILEI.

Independent of the beginning development of the two-lensed - also called the compound microscope - stronger and stronger lenses became fashionable, too. Again, it was a Dutchman, A. van LEEUWENHOEK, who succeeded in the construction of lenses with 270 fold magnification and a focal length of about one millimetre, with which he worked. His resulting observations constitute the pinnacle of "single-lensed microscopes".

In 1667, the *"Micrographia"* of the British natural scientist ROBERT HOOKE was published. HOOKE examined vegetable tissues with the help of a compound microscope and thus detected their basic elements, the cells. The further development of microscopes and especially the use of microscopes in the 18th century went on rather hesitantly.

This was on one hand caused by the opinion of many natural scientists of reputation, that a lot had still to be discovered even without the use of microscopes.

On the other hand, the interfering chromatic aberrations were a continuing source of misinterpretations. This led to the prejudice, that anything you wanted to see could be seen within a microscope.

As soon as 1695, D. GREGORY considered the combination of several lenses with differing dispersions in order to minimise this handicap. The mathematician L. EULER brought forth a theoretical explanation of achromates (systems of lenses with corrected colors) in 1771 and he suggested the calculation of such achromatic objectives. They were finally built by F. G. BEELDSNYDER (1735 - 1808), who placed a biconcave lens out of flint glass between two biconvex lenses out of another type of glass, that had a different index of refraction.

G. D. AMICI (1786 - 1863) developed achromatic systems of lenses further and invented also immersion objectives. The actual break-through on the way to the construction of modern microscopes was due to the formulation of a theory of image formation within the microscope by E. ABBÉ (1840 - 1905). He proved, that an absolute upper limit of microscopic image formation exists and that it was dependent of the aperture of the objective and the wave length of the light. His theory permitted the development of reproducible high-performance microscopes and their production in series.

The use of newly developed types of glasses by O. SCHOTT (1851 - 1935) led in 1886 to the development of apochromatic objectives (objectives, that remove a rest of colors, the so-called secondary spectrum) by E. ABBÉ and C. ZEISS.

The progress of microscope building enabled a quick series of numerous discoveries in the subjects of histology, cytology and bacteriology. The progress was helped by the development and use of suitable methods of fixation, embedding and cutting, specific dyes and conservatives.

THE THEORY OF LIGHT MICROSCOPY

The closer an object is to the eye, the more details can be distinguished until the limit is reached, below which the eye cannot depict the object clearly any more.

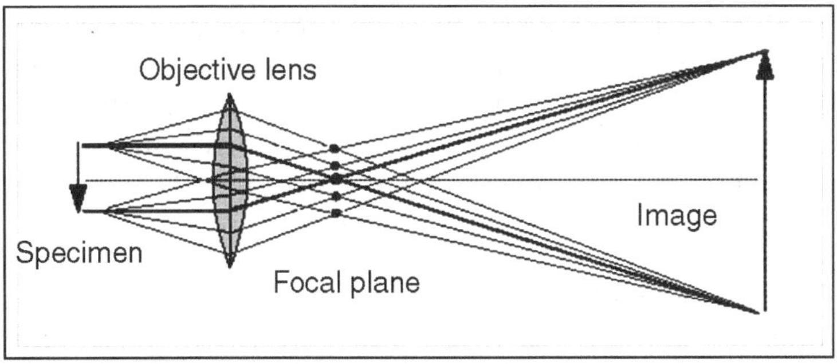

This limit is 250 mm for the average grown-up. The measure for the ability to tell two points apart is called the *power of resolution*. It describes whether two adjoining points can still be perceived as separate. To examine objects below this limit of resolution, a magnifying glass is needed. Its magnification can be calculated in the following way:

$$V = 250\ [mm]/f\ [mm]$$

250 mm is the image distance of the human eye and f is the focal length of the lens. If the magnifying power of the lens is known and the focal length has to be calculated, than the formula has to be used as shown beneath:

$$f\ [mm] = 250\ [mm]/V$$

The picture to the left shows the way the fan of rays of an object that is collected by a single lens is imaged.

If you want to magnify an object even more, two lenses in tandem have to be used (objective and eyepiece or ocular). The resulting construct is a simple microscope. The objective magnifies the object or specimen (O) and a turned up real image (O′) is formed in front of the focal plane of the second lens. The eyepiece than forms an enlarged virtual image (O″), that can be seen as a turned up image at a distance of 250 mm. The magnification of a microscope is thus a product of

$$V_{objective} \times V_{ocular}$$

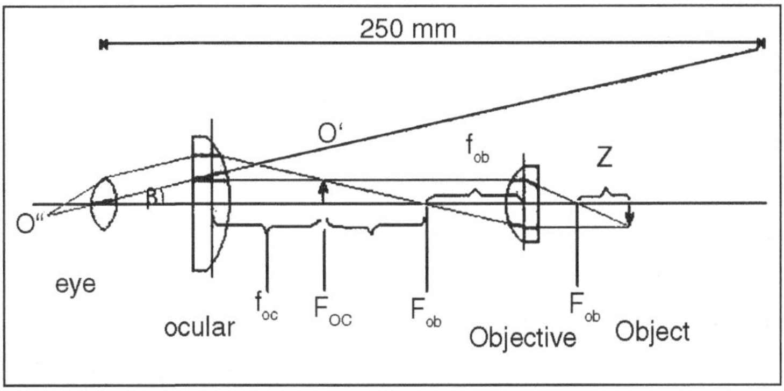

Fig. Light Path in the Microscope: F = Focal plane, O = Object (Specimen), Ob = Objective, Oc = ocular (eyepiece)

Nonetheless the power of resolution is given exclusively by the objective. To describe an objective not only the magnifying power has to be taken in account, but, too, the *numerical aperture*. The numerical aperture is defined as the sine of half the angle of the cone of light from each point of the object, that can be accepted by the objective *(alpha)* multiplied by the index of refraction of the medium in which the object is immersed (n):

$$A = n \sin \textit{alpha}.$$

The medium is usually air with a refraction index of n = 1. *alpha* can never be bigger than 90° and thus the numerical aperture can never outgrow 1. Its largest actual size is 0.95, since the distance between objective and the surface of the cover glass cannot reach zero. The aperture of 0.95 corresponds to an angle *alpha* of roughly 72°.

An increase of the numerical aperture can be achieved by the choice of a medium between objective and object with an index of refraction bigger than that of air. Special oil for immersion with an index of n = 1.515 has proved to be useful. Larger indexes of refraction do not make sense, because the index of refraction of the objective itself (n = 1.525) becomes limiting.

Immersion oil can be used only with specially constructed immersion objectives. If *alpha* has the maximum of 67.5°, the aperture is accordingly 1.515 x 0.92 = 1.40. The degree of resolution (d) is set by the wavelength of light *(lambda)* and the numerical aperture (A_{obj}):

$$d = \textit{lambda}/A_{obj}$$

If *lambda* is 550 nm (green light) the formula runs the following way:

$$d = 550\ [\text{nm}]/2 \times 1.40 = 200\ \text{nm} = 0.2\ \mu\text{m}$$

0.2 μm is the highest theoretical resolution that can be reached with a light microscope. A rough approximation shows that the power of resolution of a light microscope lies at about half the length of a lightwave if a good immersion objective is used. If the limit of resolution of a microscope is known, then the maximal useful magnification can be calculated. A magnification is called useful when two only just clear points are magnified so strongly that they are seen as separate unities by the human eye. At 250mm is the resolution of the human eye about 0.15 - 0.2 mm. The rule of thumb for a useful magnification is thus:

$$500 - 1000 \times A_{obj}$$

An objective with an aperture of 1.4 has accordingly a maximal useful magnification of 1400 fold.

The Wave-Particle Nature of Light

We assume that the dual nature of light as both wave and particle is known and we will in the following deal with some fundamental principles of wave optics in order to be able to interpret microscopic image formation. Our main interest, nevertheless, is the understanding of modern techniques of light microscopy like phase-contrast microscopes, polarising microscopes and interference microscopes. First some terms and definitions:

Amplitude, frequency, wave length, interference, diffraction and phase are the most important parameters to describe a wave. The *wave length* of visible light is between 400 and 800 nm.

Interference

Interference is the mutual influence of two waves on each other, whereby the resulting crests may be either enhanced or flattened (enhancement of amplitude, reduction of amplitude). The extreme case is given by two waves that extinguish each other.

Diffraction

Diffraction is the partial deflection of a ray of light at the corners of opaque objects. For understanding it is useful to cast a look at phenomenons of diffraction at simple gap- or hole-shaped masks. If a screen is placed behind such a mask and light is sent through, the pattern of diffraction can be seen. It consists of a regular pattern of points or lines that fades away towards the margins. The shape of the pattern is dependent on the kind of mask and the object used (for example a microscopic preparation).

But the pattern is always grouped around a central axis (unmagnified picture of the gap or hole of the mask). If a mask with two openings is used, the light is diffracted at the rim of both openings. The diffraction images of the two rays overlie each other and regular enhancements and reductions can be seen in the resulting pattern. To get a magnified image, the first additional maxima of the diffraction image have to be included (E. ABBÉ).

The results in the pictures above have already shown that the distances of the points in a diffraction image are reversed proportional to the corresponding distances in the object. To gain a real image from a diffraction image a collecting lens (the objective of the microscope is one) is needed. This lens has to have a big enough aperture to collect the intensity maxima (– I, + I). The real image is formed by both refracted and non-diffracted light. Diffraction phenomenon at apertures. The closer two openings (or diaphragms) that diffract light are, the larger is the angle of the diffraction's first intensity maxima and the larger has the aperture of the collecting lens to be.

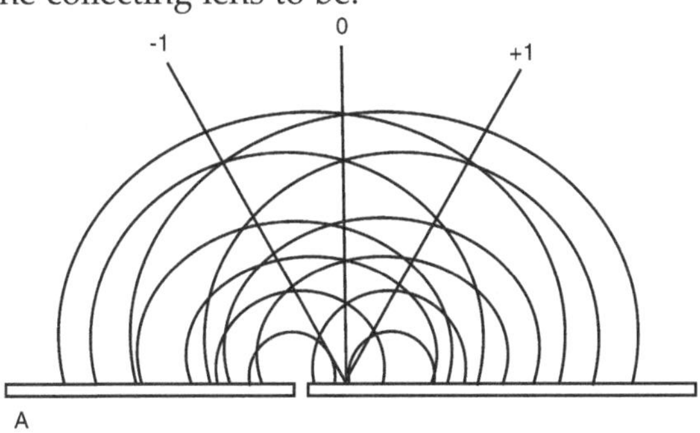

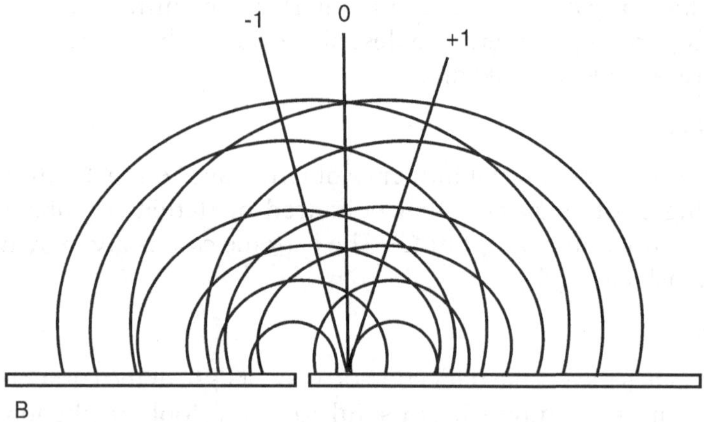

Magnification

Magnifications as those that have been described can, too, be achieved with radiation of other natures. The most important additional device for biologists besides the light microscope is the electron microscope, where the diffraction of a ray of electrons is used.

Another radiation to be mentioned is the short-waved Röntgen-radiation (X ray-radiation), that has proven to be of value for the *elucidation of molecular structures* (M. v. LAUE, L. BRAGG). Since no collecting lenses for Röntgen-radiation exist, the information contained in the diffraction pattern has to be calculated mathematically in order to get to know the structure of the molecule.

5

Aberrations in Optical Systems of Microscopy

ABERRATION IN OPTICAL SYSTEMS

Aberration in optical systems, *i.e.* in lenses or mirrors or a series of them, may be defined as the non-concurrence of rays from the points of an object after transmission through the system; it happens generally that an *image* formed by such a system is irregular, and consequently the correction of optical systems for *aberration* is of fundamental importance to the instrument-maker.

Reference should be made to the articles *Reflexion, Refraction,* and *Caustic* for the general characters of reflected and refracted rays (the article *Lens* considers in detail the properties of this instrument, and should also be consulted); in this article will be discussed the nature, varieties and modes of aberrations mainly from the practical point of view, *i.e.* that of the optical-instrument maker.

OPTICAL MICROSCOPY

The use of "Optical Microscopes" in "Materials Evaluation" is very popular since it enables the researcher to observe the *"internal structure"* of materials. This information can then be used *to improve material processing and properties* which form the foundation of *materials science and engineering (MSE).* The microstructure generally ranges from the atomic scale (0.1nm) to 1mm(1000ìm) with the most widely used scale of 1-1000 ìm. Practically the optical microscopes can be used up to 2000x at which the resolution becomes so poor that objects smaller than 1 ìm cannot be distinguished. But since most materials have grain sizes in the range 1-100ìm optical microscope (OM) is a perfect low cost tool.

Typical microstructural features are grains(single crystal), precipitates, inclusions, pores, whiskers, defects, twin boundaries, etc. Most of the manufacturers use OMs for process control and R&D. There are mainly two types of optical microscopes (OM): *Reflected Light* (RL) and T*ransmitted Light* (TL). In the former the light is reflected from the surface of the specimen while in the latter the sample is so thin (20-50ìm) that light passes through it. A third but less important type may be the *stereo microscopes* that do not require any sample prep and are generally used to observe fracture surfaces, electronic curcuit boards, etc.

Reflected Light Optical Microscopes are the most widely used machines in materials science. Because they involve light reflection the specimen surface must be very well polished to a *"mirror finish"*. In order to have a mirror finish the sample goes through a number of *"sample preparation treatments"*:

1. *Cutting:* Specimen should be cut to <1.25 inch with a hacksaw or a disc saw. Care must be taken not to heat the sample during cutting.
2. *Mounting:* Hot mounting in a press w/phenolic resin or cold mount in polyester. The former is faster and more generally used for metals while the latter is favored for ceramic samples.
3. *Grinding:* Successively sized SiC grinding papers (180-320-600-1200 Grit) are rotated on a wheel (~50-250 rpm) and the sample is pushed face down while watercooling. Small SiC particles are glued to the grinding paper so these are also sometimes called fixed abrasives. These particles slowly remove chips from specimen surface.
4. *Polishing:* Same as grinding except the abrasive particles are loose and no watercooling is performed. Diamond or alumina suspension having particles 0.05 - 30ìm diameter are used.
5. *Etching:* Selective removal of certain microstructural features to reveal the internal structure. Use a hood and do not inhale acid vapour. Wash the specimen thoroughly to avoid damage on objective lenses.

Because OM examination of any material is actually a sampling operation from a larger material, care must be taken not to change the microstructure during sample preparation. It is important that the sample must be *representative*. The sample is examined in OM in two main modes:

1. *Bright Field (BF):* The polished surface appears bright and the surface irregularities such as grain boundaries appear dark. The reason why grain boundaries appear dark can be clearly seen in the figure. Light reflected from the grain boundaries do not go back to the objective. The same thing is true for precipitates that also appear dark. BF mode is more commonly used than DF.
2. *Dark Field (DF):* The incident light is fed from an angle therefore polished smooth surface appears dark and surface irregularities appear bright. DF is less commonly used. Applications are generally in grain size measurements and in judging the quality of polishing.

BF Illumination is used in the following modes:

1. *Plane polarized light (PP):* Most widely used illumination technique in MSE. Grain size measurements, precipitates, inclusions, are all observed in this mode. PP is achieved by inserting a polarizing filter

into the incoming light path. This filter kills the wave's electrical wave component's vibration in all directions except in one direction. This direction corresponds to that of the filter orientation. E.g. light passing through the window through the blinds (jaluzi).

2. *Cross polars or cross-nicols (CP):* When two polarizing filters are placed at 90° to each other i.e.when they form a cross shape. Then no light can pass and we have extintcion. In extinction, the image is totally black when there is no specimen in the microscope or when the specimen is totally isotropic like single phase austenitic steel. Only materials with a measurable level of "birefringence" can show up in CP mode. *Birefringence* is the difference between the refractive indices of a material in different directions ($D = n_x - n_y$). Generally noncubic materials are birefringent. So using this imaging mode one can tell where the cubic and noncubic materials are located throughout the microstructure. This allows the observation of non-cubic phases in the microstructure which may play a major role in some properties like high temperature strength. This information can be very valuable especially in ceramics. CP is less frequently used in RL.
3. *Differential interference contrast (DIC):* New technique. Allows surface relief to be shown much like an SEM-SEI (secondary electron image). A wedge-shaped filter (Nomarski Filter) is inserted in the light path that results in the wavelength of the light to be slightly modified. This produces a retardation of the light wave and consequently a relief effect on the observed image.

TL-OM: Used generally by geologists (petrographic microscope), biologists and ceramists but never by metallurgists because metals are opaque.

SOME SELECTED MACHINES THAT MUST BE PRESENT IN A METALLOGRAPHY LAB

Cutting Machine

A high speed saw that uses 25 cm diameter resin impregnated corundum cutting disc along with watercooling. Disc rotates at around 3000 rpm. Coolant is circulated in the system.

Hot Mounting Press

Samples are hot mounted in a phenolic resin to form a 32 mm diameter pellet containing the specimen.

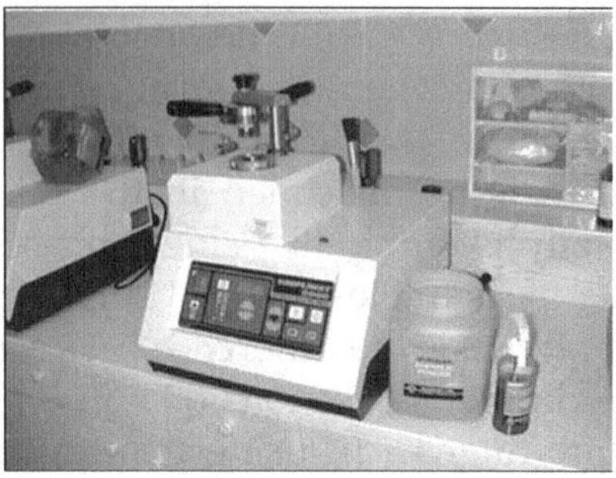

Polishing Machine

Shown in the figure is a fully automatic polishing machine that has a rotating wheel (0-500rpm) and an automatic head.

RL-OM: Nikon Epiphot 300 Inverted Reflected Light Optical Microscope

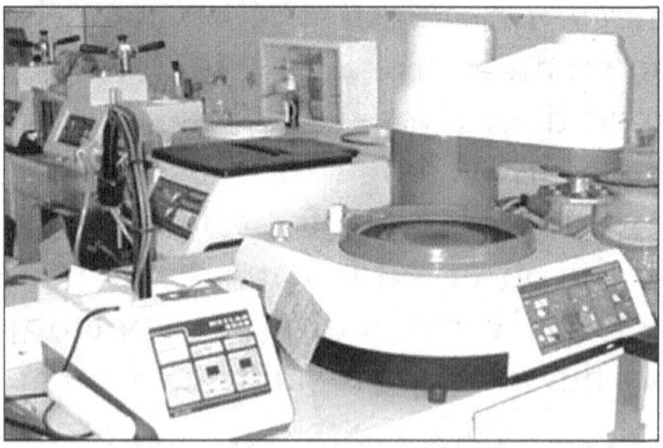

Sample is placed upside down and observed as such. This allows the examination of large samples. Another version of this machine is the upright RL-OM. We have the latter one in our lab.

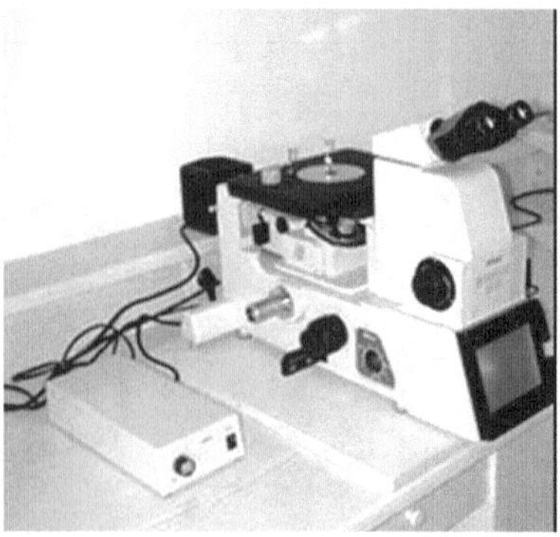

Four Different Image Modes:

1. *SEM BSE:* Scanning electron microscope, Backscattered Electron Image mode.

2. *SEM SEI*: Scanning electron microscope, Secondary Electron Image mode.
3. *RL-OM PP:* Reflected light optical microscope, plane polarized light mode.
4. *RL-OM CP:* Reflected light optical microscope, cross polars polarized light mode.

The sample was taken from a commercial refractory brick. White features in figure are chromite grains while the large feature on top half of the micrograph is a periclase grain.

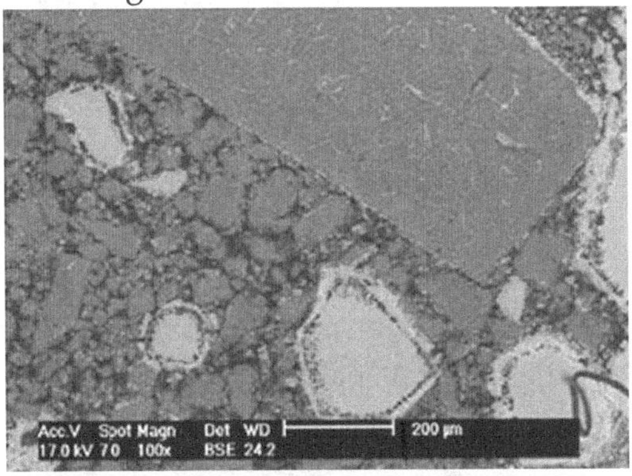

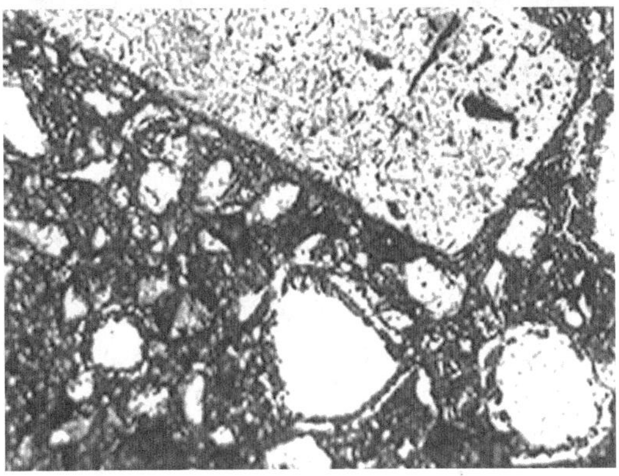

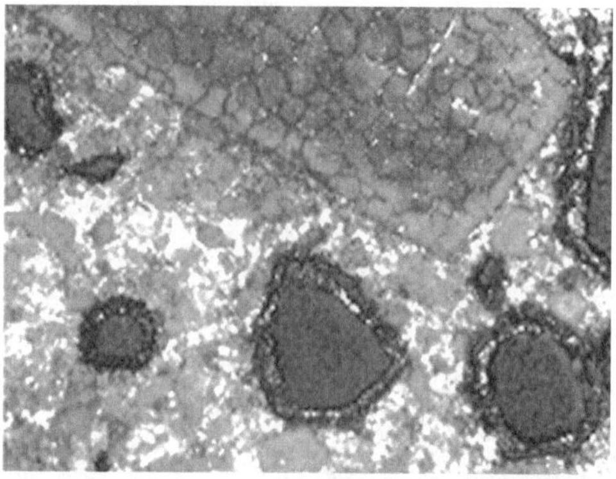

A Ni-Cr Superalloy microstructure:

Straight lines are twin boundaries. Each colour indicates a grain. Occasional

Precipitates are visible. Magnification = ~750x

Cross polars: Each colour is a sign of different crystallographic orientation.

Sample taken from an aircraft part.

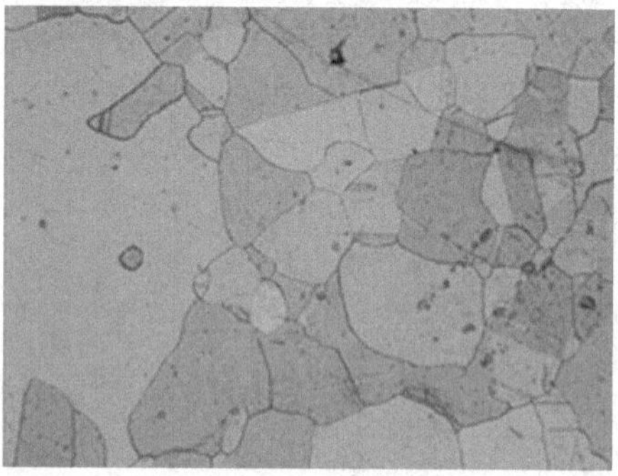

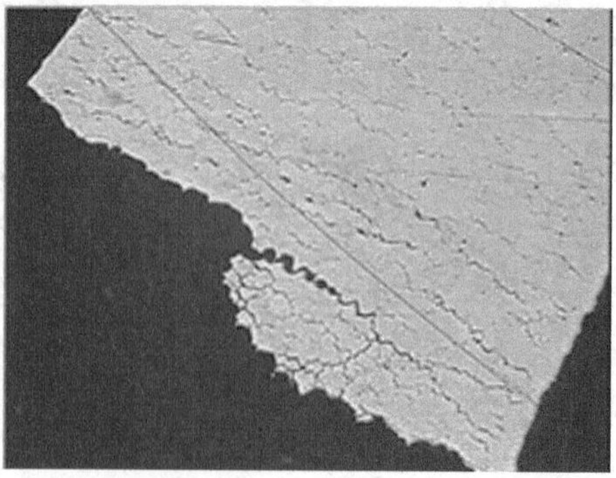

Creep, An aircraft part: (F-110 inner liner) that was used at T>700oC for long periods. This service abuse led to creep cracks in the microstructure. Notice that the grain boundaries are opening up under the stress.

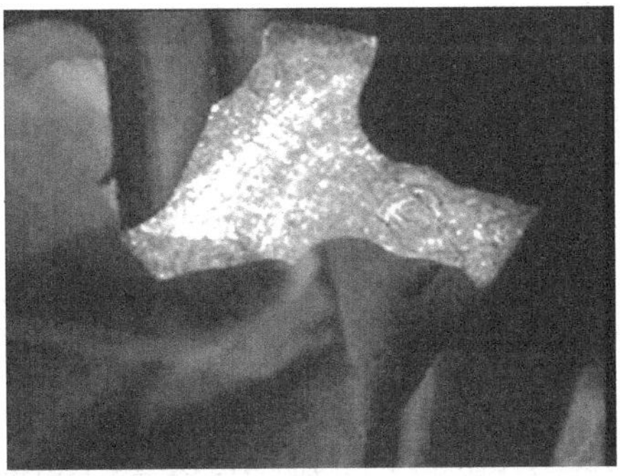

Fatigure crack growth: Notice the progression of crack from bottom right to top left as the fatigue cycle progresses.

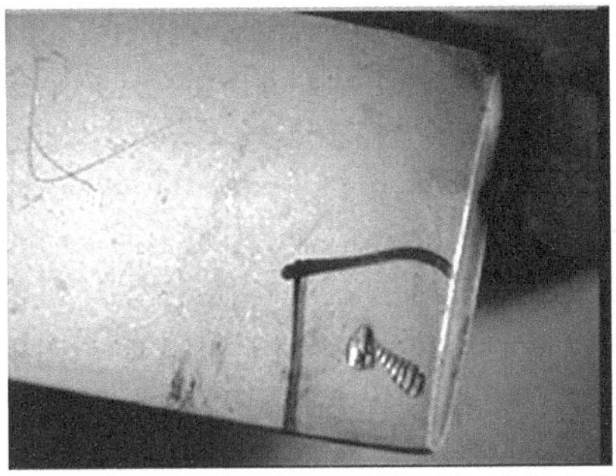

FOD (Foreign object damage) on a J-79 Turbine blade. Notice the bolt-shaped dent created upon impact of the free bolt in the engine.

OPTICAL PROPERTIES OF PONDERABLE BODIES

THEORY OF ELECTRONS

If we want to form an: adequate representation of optical phenomena in ponderable bodies, the conceptions of the molecular and atomistic theories naturally suggest themselves. Already, in the elastic theory, it had been imagined that certain material particles are set vibrating by incident waves of light.

These particles had been supposed to be acted on by an elastic force by which they are drawn back towards their positions of equilibrium, so that they can perform free vibrations of their own, and by a resistance that can be represented by terms proportional to the velocity in the equations of motion, and may be physically understood if the vibrations are supposed to be converted in one way or another into a disorderly heat-motion.

In this way it had been found possible to explain the phenomena of dispersion and (selective) absorption, and the connexion between them (anomalous dispersion). 5 These ideas have, been also embodied into the electromagnetic theory.

In its more recent development the extremely small, electrically charged particles, to which the name of " electrons "has been given, and which are supposed to exist in the interior of all bodies, are considered as forming the connecting links between aether and matter, and as determining by their arrangement and their motion all optical phenomena that are not confined to the free aether.6 It has thus become clear why the relations that had been established between optical and electrical properties have been found to hold only in some simple cases.

However, the general boundary conditions seem to require no alteration. For this reason it has been possible, for example, to establish a satisfactory theory of metallic reflection, though the propagation of light in the interior of a metal is only imperfectly understood. One of the fundamental propositions of the theory of electrons is that an electron becomes a centre of radiation whenever its velocity changes either in direction or in magnitude.

Thus the production of Röntgen rays, regarded as consisting of very short and irregular electromagnetic impulses, is traced to the impacts of the electrons of the cathode-rays against the anticathode, and the lines of an emission spectrum indicate the existence in the radiating body of as many kinds of regular vibrations, the knowledge of which is the ultimate object of our investigations about the structure of the spectra.

The shifting of the lines caused, according to Doppler's law, by a motion of the source of light, may easily be accounted for, as only general principles are involved in the explanation. To a certain extent we can also elucidate the changes in the emission that are observed when the radiating source is exposed to external magnetic foooptics) .

Various Kinds of Light-motion

- If the disturbance is represented by P z = o, P = a cos (*nt - kx+f), 'P Z* = *a'cos (nt* - + *f*), so that the end of the vector P describes an ellipse in a plane perpendicular to the direction of propagation, the light is said to be elliptically, or in special cases circularly, polarized. Light of this kind can be dissolved in many different ways into plane polarized components.
  There are cases in which plane waves must be elliptically or circularly polarized in order to show the simple propagation of phase that is expressed by formulae. Instances of this kind occur in bodies having the property of rotating the plane of polarization, either on account of their constitution, or under the influence of a magnetic field. For a given direction of the wave-front there are in general two kinds of elliptic vibrations, each having a definite form, orientation, and direction of motion, and a determinate velocity of propagation. All that has been said about Huygens's construction applies to these cases.
- In a perfect spectroscope a sharp line would only be observed if an endless regular succession of simple harmonic vibrations were admitted into the instrument. In any other case the light will occupy

a certain extent in the spectrum, and in order to determine its distribution we have to decompose into simple harmonic functions of the time the components of the disturbance, at a point of the slit for instance. This may be done by means of Fourier's theorem.
An extreme case is that of the unpolarized light emitted by incandescent solid bodies, consisting of disturbances whose variations are highly irregular, and giving a continuous spectrum. But even with what is commonly called homogeneous light, no perfectly sharp line will be seen. There is no source of light in which the vibrations of the particles remain for ever undisturbed, and a particle will never emit an endless succession of uninterrupted vibrations, but at best a series of vibrations whose form, phase and intensity are changed at irregular intervals. The result must be a broadening of the spectral line.
In cases of this kind one must distinguish between the velocity of propagation of the phase of regular vibrations and the velocity with which the said changes travel onward.

- In a train of plane waves of definite frequency the disturbance is represented by means of goniometric functions of the time and the coordinates. Since the fundamental equations are linear, there are also solutions in which one or more of the coordinates occur in an exponential function. These solutions are of interest because the motions corresponding to them are widely different from those of which we have thus far spoken. If, for example, the formulae contain the factor e - *zcos* (nt - sy +1) *with the positive constant r, the disturbance is no longer periodic with respect to x, but steadily diminishes as x increases. A state of things of this kind, in which the vibrations rapidly die away as we leave the surface, exists in the air adjacent to the face of a glass prism by which a beam of light is totally reflected. It furnishes us an explanation of Newton's experiment mentioned.*

Velocity of Light

The fact that light is propagated with a definite speed was first brought out by Ole Roemer at Paris, in 1676, through observations of the eclipses of Jupiter's satellites, made in different relative positions of the Earth and Jupiter in their respective orbits. It is possible in this way to determine the time required for light to pass across the orbit of the earth. The dimensions of this orbit, or the distance of the sun, being taken as known, the actual speed of light could be computed.

Since this computation requires a knowledge of the sun's distance, which has not yet been acquired with certainty, the actual speed is now determined by experiments made on the earth's surface. Were it possible by any system of signals to compare with absolute precision the times at two different stations, the speed could be determined by finding how long was required for light to pass from one station to another at the greatest visible distance. But this is impracticable, because no natural agent is under our control by which a signal could be communicated with a greater velocity than that of light. It is therefore necessary to reflect a ray back to the point of observation and to determine the time which the light requires to go and come. Two systems have been devised for this purpose. One is that of Fizeau, in which the vital appliance is a rapidly revolving toothed wheel; the other is that of Foucault, in which the corresponding appliance is a mirror revolving on an axis in, or parallel to, its own plane.

The principle underlying Fizeau's method is shown in the accompanying the course of a ray of light which, emanating from a luminous point L, strikes the plane surface of a plate of glass M at an angle of about 45° A fraction of the light is reflected from the two surfaces of the glass to a distant reflector R, the plane of which is at right angles to the course of the ray. The latter is thus reflected back on its own course and, passing through the glass M on its return, reaches a point E behind the glass. An observer with his eye at E iooking through the glass sees the return ray as a distant luminous point in the reflector R, after the Nam light has passed over the course in both directions trance from M nearly equal to its focal length. The function of this appliance is to render the diverging rays, shown by the dotted lines, nearly parallel, in order that more light may reach R and be thrown back again. But the principle may be conceived without respect to the telescope, all the rays being ignored except the central one, which passes over the course we have described.

The luminous point between two of the teeth at K. Now, conceive that the wheel is set in revolution. The ray is then interrupted as every tooth passes, so that what is sent out is a succession of flashes. Conceive that the speed of the mirror is such that while the flash is going to the distant mirror and returning again, each tooth of the wheel takes the place of an opening between the teeth.

Then each flash sent out will, on its return, be intercepted by the adjacent tooth, and will therefore become invisible. If the speed be now doubled, so that the teeth pass at intervals equal to the time required for the light to go and come, each flash sent through an opening will return through the adjacent opening, and will therefore be seen with full brightness. If the speed be continuously increased the result will be successive disappearances and reappearances of the light, according as a tooth is or is not interposed when the ray reaches the apparatus on its return. The computation of the time of passage and return is then very simple. The speed of the wheel being known, the number of teeth passing in one second can be computed. The order of the disappearance, or the number of teeth which have passed while the light is going and coming, being also determined in each case, the interval of time is computed by a simple formula.

The most elaborate determination yet 'made by Fizeau's method was that of Cornu. The station of observation was at the Paris Observatory. The distant reflector, a telescope with a reflector at its focus, was at Montlhery, distant 22,910 metres from the toothed wheel. Of the wheels most used one had 150 teeth, and was 35 millimetres in diameter; the other had 200 teeth, with a diameter of 45 mm. The highest speed attained was about 900 revolutions per second. At this speed, 135,000 (or 180,000) teeth would pass per second, and about 20 (or 28) would pass while the light was going and coming. But the actual speed attained was generally less than this.

The definitive result derived by Cornu from the entire series of experiments was 300,400 kilometres per second. Further details of this work need not be set forth because the method is in several ways deficient in precision. The eclipses and subsequent reappearances of the light taking place gradually, it is impossible to fix with entire precision upon the moment of complete eclipse. The speed of the wheel is continually varying, and it is impossible to determine with precision what it was at the instant of an eclipse.

The defect would be lessened were the speed of the toothed wheel placed under control of the observer who, by action in one direction or the other, could continually check or accelerate it, so as to keep the return point of light at the required phase of brightness.

If the phase of complete extinction is chosen for this purpose a definite result cannot be reached; but by choosing the moment when the light is of a certain definite brightness, before or after an eclipse, the observer will know at each instant whether the speed should be accelerated or retarded, and can act accordingly. The nearly constant speed through as long a period as is deemed necessary would then be found by dividing the entire number of revolutions of the wheel by the time through which the light was kept constant. But even with these improvements, which were not actually tried by Cornu, the estimate of the brightness on which the whole result depends would necessarily be uncertain.

The outcome is that, although Cornu's discussion of his experiments is a model in the care taken to determine so far as practicable every source of error, his definitive result is shown by other determinations to have been too great by about a t°u part of its whole amount.

In Foucault's determination the measures were not made upon a luminous point, but upon a reticule, the image of which could not be seen unless the reflector was quite near the revolving mirror. Indeed the whole apparatus was contained in his laboratory. The effective distance was increased by using several reflectors; but the entire course of the ray measured only 20 metres. The result reached by Foucault for the velocity of light was 298,000 kilometres per second. The first marked advance on Foucault's determination was made by Albert A. Michelson, then a young officer on duty at the U.S. Naval Academy, Annapolis.

The improvement *Michelson* consisted in using the image of a slit through which the rays of the sun passed after reflection from a heliostat. In this way it was found possible to see the image of the slit reflected from the distant mirror when the latter was nearly 600 metres from the station of observation.

It will be seen that the revolving mirror is here interposed between the lens and its focus. It was driven by an air turbine, the blast of which was under the control of the observer, so that it could be kept at any required speed. The speed was determined by the vibrations of two tuning forks. One of these was an electric fork, making about 120 vibrations per second, with which the mirror was kept in unison by a system of rays reflected from it and the fork. The speed of this fork was determined by comparison with a freely vibrating fork from time to time. The speed of the revolving mirror was generally about 275 turns per second, and the deflection of the image of the slit about 112.5 mm.

In order that the disturbances of the return image due to the passage of the ray through more than 7 km. of air might be reduced to a minimum, an ordinary telescope of the " broken back " form was used to send the ray to the revolving mirror. The speed of the mirror was, as in Michelson's experiments, completely under control of the observer, so that by drawing one or the other of two cords held in the hand the return image could be kept in any required position.

In making each measure the receiving telescope hereafter described was placed in a fixed position and during the " run " the image was kept as nearly as practicable upon a vertical thread passing through its focus. A " run " generally lasted about two minutes, during which time the mirror commonly made between 25,000 and 30,000 revolutions. The speed per second was found by dividing the entire number of revolutions by the number of seconds in the "run"

The extreme deviations between the times of transmission of the light, as derived from any two runs, never approached to the thousandth part of its entire amount. The average deviation from the mean was indeed less than Pa 0 part of the whole. To avoid the injurious effect of the directly reflected flash, as well as to render unnecessary a comparison between the directions of the outgoing and the return ray. a second telescope, turning horizontally on an axis coincident with that of the revolving mirror, was used to receive the return ray after reflection.

This required the use of an elongated mirror of which the upper half of the surface reflected the outgoing ray, and the lower other half received and reflected the ray on its return. On this system it was not necessary to incline the mirror in order to avoid the direct reflection of the return ray. The greatest advantage of this system was that the revolving mirror could be turned in either direction without break of continuity, so that the angular measures were made between the directions of the return ray after reflection when the mirror moved in opposite directions. In this way the speed of the mirror was as good as doubled, and the possible constant errors inherent in the reference to a fixed direction for the sending telescope were eliminated. The revolving mirror was a rectangular prism M of steel, 3 in. high and 12 in. on a side in cross section, which was driven A by a blast of air acting on two fanwheels, one at the top, the other at the bottom of the mirror.

NPO is the object-end of the fixed sending telescope the rays passing through it' being reflected to the mirror by a prism P. The receiving telescope ABO is straight, and has its objective under O. It was attached to a frame which could turn around the same axis as the mirror. The angle through which it moved was measured by a divided arc immediately below its eye-piece. The position AB is that for receiving the ray during a rotation of the mirror in the anti-clockwise direction; the position A'B' that for a clockwise rotation.

In these measures the observing station was at Fort Myer, on a hill above the west bank of the Potomac river. The distant reflector was first placed in the grounds of the Naval Observatory, at a distance of 2551 metres. But the definitive measures were made with the reflector at the base of the Washington monument, 3721 metres distant. The revolving mirror was of nickel-plated steel, polished on all four vertical sides.

Thus four reflections of the ray were received during each turn of the mirror, which would be coincident were the form of the mirror invariable. During the preliminary series of measures it was found that two images of the return ray were sometimes formed, which would result in two different conclusions as to the velocity of light, according as one or the other was observed. The only explanation of this defect which presented itself was a tortional vibration of the revolving mirror, coinciding in period with that of revolution, but it was first thought that the effect was only occasional.

In the summer of 1881 the distant reflector was removed from the Observatory to the Monument station. Six measures made in August and September showed a systematic deviation of +67 km. per second from the result of the Observatory series. This difference led to measures for eliminating the defect from which it was supposed to arise. The pivots of the mirror were reground, and a change made in the arrangement, which would permit of the effect of the vibration being determined and eliminated.

This consisted in making the relative position of the sending and receiving telescopes interchangeable. In this way, if the measured deflection was too great in one position of the telescopes, it would be too small by an equal amount in the reverse position.

As a matter of fact, when the definitive measures were made, it was found that with the improved pivots the mean result was the same in the two positions. But the new result differed systematically from both the former ones. Thirteen measures were made from the Monument in the summer of 1882, the results of which will first be stated in the form of the time required by the ray to go and come.

So far as could be determined from the discordance of the separate measures, the mean error of Newcomb's result would be less than Da km. But making allowance for the various sources of systematic error the actual probable error was estimated at =30 km. It seems remarkable that since these determinations were made, a period during which great improvements have become possible in every part of the apparatus, no complete redetermination of this fundamental physical constant has been carried out.

The experimental measures thus far cited have been primarily those of the velocity of light in air, the reduction to a vacuum being derived from theory alone. The fundamental constant at the basis of the whole theory is the speed of light in a vacuum, such as the celestial spaces. The question of the relation between the velocity in vacuo, and in a transparent medium of any sort, belongs to the domain of physical optics. Referring to the preceding section for the principles at play we shall in the present part of the article confine ourselves to the experimental results. With the theory of the effect of a transparent medium is associated that of the possible differences in the speed of light of different colours.

The question whether the speed of light in vacuo varies with its wave-length seems to be settled with entire certainty by observations of variable stars. These are situated at different distances, some being so far that light must be several centuries in reaching us from them.

Were there any difference in the speed of light of various colours it would be shown by a change in the colour of the star as its light waxed and waned. The light of greatest speed preceding that of lesser speed would, when emanated during the rising phase, impress its own colour on that which it overtook. The slower light would predominate during the falling phase. If there were a difference of Io minutes in the time at which light from the two ends of the visible spectrum arrived, it would be shown by this test. As not the slightest effect of the kind has ever been seen, it seems certain that the difference, if any, cannot imate to T. 0 0.0 0 0 part rt of the entire speed.

The approx case is different when light passes through a refracting medium. It is a theoretical result of the undulatory theory of light that its velocity in such a medium is inversely proportional to the refractive index of the medium. This being different for different colours, we must expects a corresponding difference in the velocity.

Foucault and Michelson have tested these results of the undulatory theory by comparing the time required for a ray of light to pass through a tube filled with a refracting medium, and through air. Foucault thus found, in a general way, that there actually was a retardation; but his observations took account only of the mean retardation of light of all the wavelengths, which he found to correspond with the undulatory theory. Michelson went further by determining the retardation of light of various wave-lengths in carbon bisulphide. The comparison of red and blue light was made differentially.

The colours selected were of wave-length about o 62 for red and 0.49 for blue. Putting Vr and Vb for the speeds of red and blue light respectively in bisulphide of carbon, the mean result compares with theory as follows: Observed value of the ratio V r, Theoretical value (Verdet).. 1.025 This agreement may be regarded as perfect. It shows that the divergence of the speed of yellow light in the medium from theory, as found above, holds through the entirespectrum. The excess of the retardation above that resulting from theory is probably due to a difference between " wave-speed" and "group-speed" pointed out by Rayleigh.

1.0245 But when a flash of light like that measured passes through a refracting medium, the front waves of the flash are continually dying away, and the place of each is taken by the wave following. A familiar case of this sort is seen when a stone is thrown into a pond. The front waves die out one at a time, to be followed by others, each of which goes further than its predecessor, while new waves are formed in the rear.

Larger waves in the middle, moves as a whole more slowly than do the individual waves. When the speed of light is measured the result is not the wave-speed as above defined, but something less, because the result depends on the time of the group passing through the medium. This lower speed is called the groupvelocity of light. In a vacuum there is no dying out of the waves, so that the group-speed and the wave-speed are identical. From Michelson's experiments it would follow that the retardation was about 1/14 of the whole speed. This would indicate that in carbon bisulphide each individual light wave forming the front of a moving ray dies out in a space of about 15 wavelengths.

RAY OPTICS

What are the physics of light? The best way to describe how light travels would be to use *ray optics*. Rays are basically infinitesimally thin slices of waves in the direction that they are traveling.

Reflection

Practically all objects reflect a certain portion of the light falling on them. The light that is reflected off an object is the color that the eyes see. If a ray of light hits a flat, shiny surface, than it reflects off at the same angle. This means the ray that hits the surface, called the *incident ray*, which has an angle of incidence θ_i from the *normal* (an imaginary line perpendicular to the surface at the point where the incident ray encounters it) bounces off as the *reflected ray*, which has an equal angle of reflection θ_r from the normal.

This is leads to the law of reflection which follows:

$$\theta_r = \theta_i$$

The incident ray, the reflected ray, and the normal to the surface all lie in the same plan, and the angle of reflection, , equals the angle of incidence, θ_i.

There are two types of reflection. *Specular reflection* is when parallel light rays strike a smooth, plane surface, and reflect off parallel. This is important in determining the properties of mirrors because it shows how much the mirror reflects without distortion.

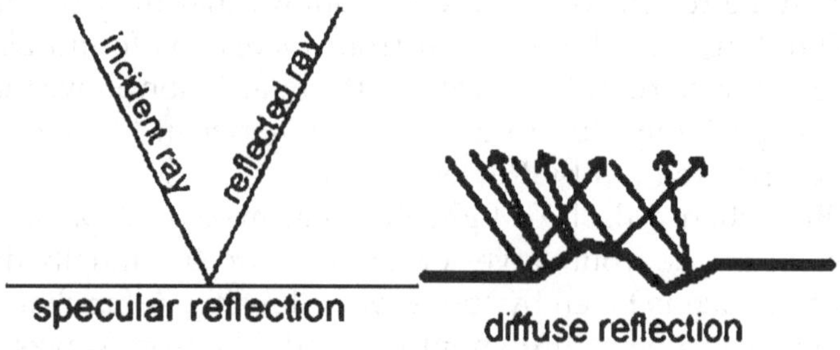

The other type of reflection, *diffuse reflection,* is when parallel rays strike a rough surface, causing them to reflect in all directions, causing distortion.

Images Formed by a Plane Mirror

Looking into a plane mirror, you can see an image of yourself that has four fundamental properties:

- The image is left to upright.
- The image is the same size as you are.
- The image is located as far behind the mirror as you are in front of it
- The image is reversed left to right.

The image being reversed means that when you wave your right hand, the image seems to wave its left hand. The image you see that is behind the mirror is called the *virtual image.*

Since the light rays do not really pass through the image, images formed by a plane mirror are known as virtual images. You can distinguish them from *real images,* which are formed by light that does pass through the image because a white piece of paper placed at the image location will not show anything if the image is virtual, but will show the image if the image is real.

Parabolic Mirrors

A parabolic mirror is a common type of curved mirror. In general shape, the mirror is roughly a section of the surface of a sphere, though as the name suggests, the curve is not perfectly spherical.

It is parabolic. There are two types of parabolic mirrors: concave and convex. Imagine a parabola with the inside surface (the part on the inside of the curve) as a reflective surface. This is a concave parabolic mirror. Conversely, a parabola with the outside as a reflective surface is a convex parabolic mirror.

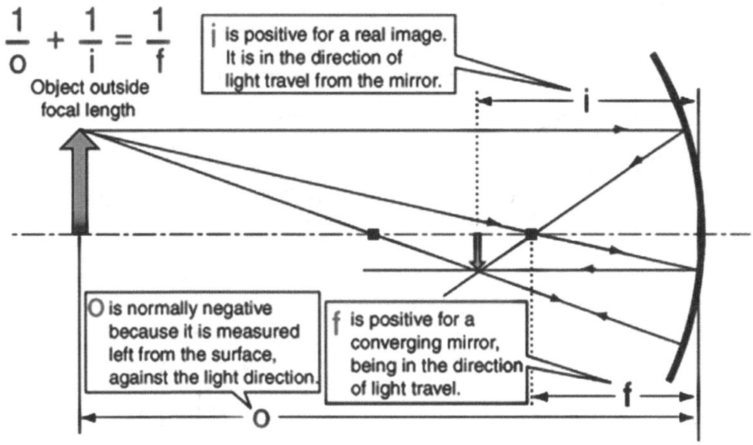

Fig. Concave Mirror

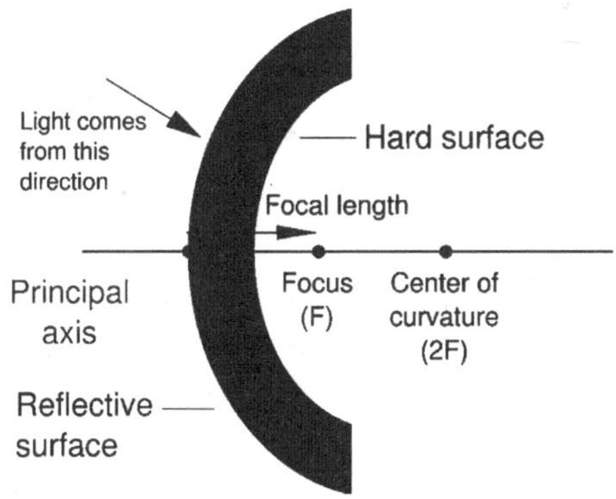

Fig. Convex Mirror

The law of reflection applies two both these types of mirrors as it did with the plane mirror. Only now the normal is drawn perpendicular to the mirror at the point of incidence, which is where the ray strikes the surface of the mirror. Both these types of mirrors have a vertex, shown on the diagram. The line perpendicular to the surface of the mirror at the vertex is called the principal axis.

The ray parallel to the principle axis goes through a point labeled F.

This is the focal point of the mirror. All rays of light parallel to the principle axis reflecting off the surface of the mirror will go through this point. Similarly, a ray going through the focal point will be reflected back parallel to the principal axis.

The focal length of the mirror, *f*, is the distance from the focus to the vertex. The height of the object is otherwise stated *ho*, while the height of the image is *hi*. The distance from the object to the vertex is *do*, and, similarly the distance from the image to the vertex is *di*.

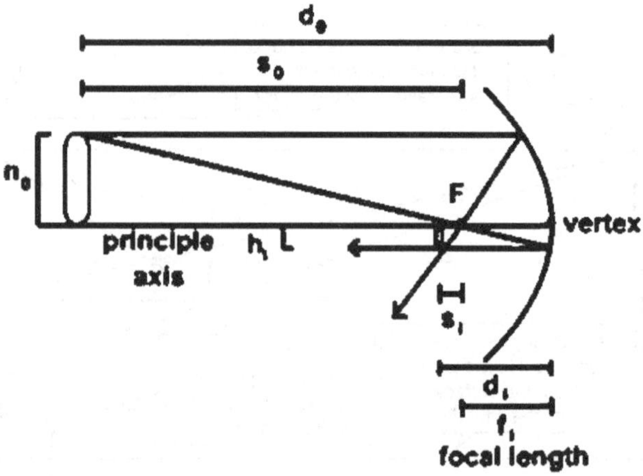

Finally, the distance from the object to the focal point F is *so*, and the distance from the image to the focus is *si*. We label all these distances for ease in use, because there are many useful relationships between them, which follow.

$$\frac{ho}{hi}=\frac{so}{f}=\frac{do}{di}=\frac{f}{si} \text{ as well as } \frac{1}{f}=\frac{1}{di}+\frac{1}{do}.$$

Refraction

Light will always travel in the fastest path from point A to point B. If this path goes through different materials through which light travels at different speeds, then light will not travel in a straight line.

This bending of light when it enters a new medium is known as refraction, and is the reason a stick looks bent when we hold part of it in water. Our eyes are trained to think that light travels to us in a straight line, so when it bends, we think it is the object that is bent, not the light.

Light travels faster in media that are less optically dense and slower in media that are more optically dense. The index of refraction is the ratio of the speed of light in a vacuum to the speed of light in the medium, and is represented by the letter n.

$$n=\frac{\text{Speed of light in a vaccum}}{\text{Speed of light in the medium}}=\frac{c}{v}$$

This index of refraction is quite important, because we can use it to determine at what angle the light bends. We do this using an equation developed by Wilebrord Snell, named, aptly enough, Snell's law. When light hits the boundary between two transparent media, it generally splits into two parts. One part reflects back into the first medium, the other refracts into the second media. The exception to this is when the light hits at such an angle that there is total internal reflection, which means that all the light bounces back into the original medium.

Snell's law of refraction states that when light travels from a medium with a refractive index of n_1 into a medium with a refractive index of n_2, the incident ray, normal ray, and reflected ray all lie in the same plane. Furthermore, θ, (the angle of incidence) and (the angle of refraction are related by the following formula:

$$n_1 \sin = \sin \theta_1 = n_2 \sin \theta_2$$

The speed of light in air is so close to the speed of light in a vacuum, that for most purposes we can use a value of one for the index of refraction of air. Light travels more noticeabley slowly in some other materials, however, and a table of some of these values follows. Of course, light can never travel faster than c, its speed in a vacuum, which is m/s.

Substance (at 20° C)	*Index of refraction, n*
Diamond	2.419
Sodium Chloride	1.544
Water	1.333
Benzene	1.501

EXPLANATION OF REFLECTION AND REFRACTION

Newton supposed the light-corpuscles to be subjected to attractive and repulsive forces exerted at very small distances by the particles of matter. In the interior of a homogeneous body a corpuscle moves in a straight line as it is equally acted on from all sides, but it changes its course at the boundary of two bodies, because, in a thin layer near the surface there is a resultant force in the direction of the normal. In modern language we may say that a corpuscle has at every point a definite potential energy, the value of which is constant throughout the interior of a homogeneous body, and is even equal in all bodies of the same kind, but changes from one substance to another.

If, originally, while moving in air, the corpuscles had a definite velocity vo, their velocity v in the interior of any other substance is quite determinate. It is given by the equation 2mv 2 -1mv 0 2 =A, in which m denotes the mass of a corpuscle, and A the excess of its potential energy in air over that in the substance considered.

A ray of light falling on the surface of separation of two bodies is reflected according to the well-known simple law, if the corpuscles are acted on by a sufficiently large force directed towards the first medium. On the contrary, whenever the field of force near the surface is such that the corpuscles can penetrate into the interior of the second body, the ray is refracted.

In this case the law of Snellius can be deduced from the consideration that the projection w of the velocity on the surface of separation is not altered, either in direction or in magnitude. This obviously requires that the plane passing through the incident and the refracted rays be normal to the surface, and that, if a, and α^2 are the angles of incidence and of refraction, v, and v_2

the velocities of light in the two media, sin α^1 /sin α ^{2} = w/v: w/v ^{2} = v ^{2}/v. The ratio is constant, because, as has already been observed, v_1 and v_2 have definite values. As to the unequal refrangibility of differently coloured light, Newton accounted for it by imagining different kinds of corpuscles. He further carefully examined the phenomenon of total reflection, and described an interesting experiment connected with it. A marked change is observed when a second piece of glass is made to approach the reflecting face, so as to be separated from it only by a very thin layer of air.

The reflection is then found no longer to be total, part of the light finding its way into the second piece of glass. Newton concluded from this that the corpuscles are attracted by the glass even at a certain small measurable distance.

New Hypotheses in the Corpuscular Theory

The preceding explanation of reflection and refraction is open to a very serious objection. If the particles in a beam of light all moved with the same velocity and were acted on by the same forces, they all ought to follow exactly the same path. In order to understand that part of the incident light is reflected and part of it transmitted, Newton imagined that each corpuscle undergoes certain alternating changes; he assumed that in some of its different "phases" it is more apt to be reflected, and in others more apt to be transmitted. The same idea was applied by him to the phenomena presented by very thin layers.

He had observed that a gradual increase of the thickness of a layer produces periodic changes in the intensity of the reflected light, and he very ingeniously explained these by his theory. It is clear that the intensity of the transmitted light will be a minimum if the corpuscles that have traversed the front surface of the layer, having reached that surface while in their phase of easy transmission, have passed to the opposite phase the moment they arrive at the back surface.

As to the nature of the alternating phases, Newton expresses himself as follows: "Nothing more is requisite for putting the Rays of Light into Fits of easy Reflexion and easy Transmission than that they be small Bodies which by their attractive Powers, or some other Force, stir up Vibrations in what they act upon, which Vibrations being swifter than the Rays, overtake them successively, and agitate them so as by turns to increase and decrease their Velocities, and thereby put them into those Fits."

The Corpuscular Theory

Though Newton introduced the notion of periodic changes, which was to play so prominent a part in the later development of the wave-theory, he rejected this theory in the form in which it had been set forth shortly before by Christiaan Huygens in chief objections being:

- That the rectilinear propagation had not been satisfactorily accounted for;

- That the motions of heavenly bodies show no sign of a resistance due to a medium filling all space; and
- That Huygens had not sufficiently explained the peculiar properties of the rays produced by the double refraction in Iceland spar. In Newton's days these objections were of much weight.

Yet his own theory had many weaknesses. It explained the propagation in straight lines, but it could assign no cause for the equality of the speed of propagation of all rays. It adapted itself to a large variety of phenomena, even to that of double refraction but it could do so only at the price of losing much of its original simplicity. In the earlier part of the 19th century, the corpuscular theory broke down under the weight of experimental evidence, and it received the final blow. Foucault proved by direct experiment that the velocity of light in water is not greater than that in air, as it should be according to the formula but less than it, as is required by the wave-theory.

Light as a Wave

The classical description of light as an electromagnetic wave makes some assumptions about its nature, based on what could be observed at the time. Although some of these assumptions have proven to be less than accurate as we learn more about electromagnetic phenomena, they make a good starting point for our discussion about light in particular and electromagnetic waves in general. The two basic assumptions were:

The frequency of the electromagnetic wave can be varied over the entire positive range, but cannot be reduced to zero. This seems intuitively obvious. A frequency of zero would mean no change in the strengths of the electric and magnetic fields. But an electromagnetic wave reqires these fields to be constantly changing in order to exist. And the idea of a negative frequency seems ludicrous.

The energy in the wave is continuously variable, with a minimum energy of any non-zero value, and no maximum value. This also seems intuitive. Looking at sunlight and comparing light intensity on a clear day and a cloudy day, we note that the clouds block some of the sun's energy. The desert sun at Noon is very intense, while the setting sun at extreme northern or southern latitudes is much less noticeable. Yet it is the same sunlight, coming from the same source, so we know that various factors encountered by sunlight must be removing some of the energy from it feel, and scientifically measure the difference.

Of course, other properties of electromagnetic radiation have also been determined, and a number of theories and assumptions have been developed. These have been either confirmed or disproven by experiment. Before we look at the circumstances under which the wave model of light (or any electromagnetic wave) may break down, however, let's look at the wave model itself.Since light was first recognized scientifically as a manifestation of electromagnetic energy, it can be represented as a waveform, like this:

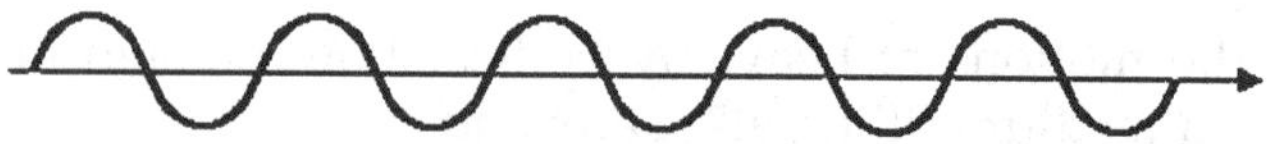

The electrical energy present in the light waveform as it travels in the direction of the arrow, it looks as if the energy level is becoming alternately positive and negative, with momentary crossovers of zero electrical energy. In the basic model of light shown here, this is in fact the case; as with all electromagnetic waves, light energy is constantly changing its form between electrical energy and magnetic energy. The point of maximum magnetic energy coincides with the moment of zero electrical energy.

Beyond that instant, energy shifts again from the magnetic field back into the electrical field, but with a reversal from the previous polarity. This continues as long as that particular ray of light exists.There are other "modes" of propogation which involve more complex interactions between the electric and magnetic fields, but in all cases the Law of Conservation necessarily holds true:

Energy is neither created nor destroyed as it is transformed from one form to another; the total energy in the wave must somehow remain constant throughout the full cycle.

The sine wave shown here represents the strength and polarity of the electrical field associated with the motion of this ray of light. The light itself, assuming no outside influences, travels in the straight line indicated by the blue arrow.

The light energy does *not* "wiggle" back and forth as it moves along its path.As an electromagnetic wave, light has some characteristics in common with all forms of electromagnetic energy. These include *wavelength, frequency,* and *speed of propogation*. These characteristics are actually related to each other, so that any one can be calculated if the other two are known. Let's take a look at each of these characteristics:

Wavelength

Since light is a repeating waveform in motion, it is possible to measure the physical distance between matching points of adjacent cycles of the waveform. This is shown here:

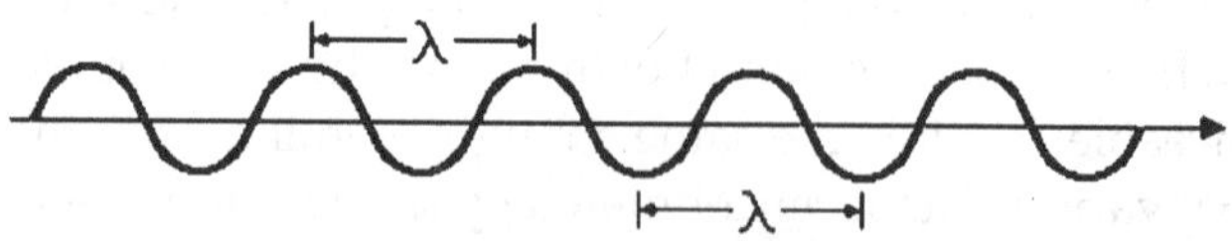

The symbol used to represent this distance is the Greek letter "Lambda" (λ).

The wavelength can actually be measured between any two corresponding points on the waveform. It is convenient to use the most positive point or the most negative point, both of which are shown above.

However, we could have just as easily specified two zero-crossing points, so long as both crossed the zero line in the same direction.Remember that the light itself does not wave back and forth along its path of travel. What we are actually measuring here is the distance traveled through space by this ray of light, while its electrical field goes from its maximum positive value, through zero to its maximum negative value, and then through zero again to once more reach its maximum positive value.

This distance is normally measured in meters (m) or some decimal fraction of a meter, such as centimeters (cm). The correct units of measurement are meters per cycle (m/cycle) or some appropriate derivation. In the case of light, the wavelength is so short that a specific distance, called the *angstrom* (Å), has been defined.

One ångstrom = 10^{-10} m or 10^{-8} cm.

Visible light has a characteristic wavelength in the range of approximately 3900 Å to 7700 Å. Electromagnetic energy outside this range is no longer visible to the human eye.

Speed of Propogation

The speed at which light travels through any medium is determined by the density of that medium. The presence of matter, even transparent matter, will slow the light down. Even air will have some effect, and glass has a more significant effect on the speed at which light will travel through it.Ever more sophisticated experiments have determined the speed of light quite accurately. According to current knowledge:

Speed of light in a vacuum = $2.997925 \pm 0.000002 \times 10^{10}$ cm/sec.

As made famous in Einstein's equation, the letter c is used as a general symbol for the speed of light.

Frequency and Period

In any electromagnetic wave, it takes time for the energy in the wave to change from electrical format to magnetic and then back again. The amount of time required to do this twice, covering one complete cycle, or wavelength of the signal is known as the *period* of the wave. Thus, the period of any wave, measured as some amount of time per cycle, is in fact the time interval that corresponds to the physical wavelength of the signal.

The *frequency* of the wave is the inverse or reciprocal of the period. That is, the frequency is the number of cycles of the waveform that occur in one second of time. For many years this was simply measured in units of cycles per second. Recently, however, the specific name *hertz* (abbreviated Hz) has been designated as the appropriate unit to indicate cycles per second.In general equations, the letter f is used to indicate frequency in hertz.The basic mathmatical formula that relates wavelength, frequency, and the speed of light is:

$$c = f\lambda$$

The wave theory of light was happily adopted and accepted until it was found to fail to explain some observed and measured phenomena, in consistent and repeatable experiments. The two phenomena that upset this model are the photoelectric effect and blackbody radiation. These two effects could only be explained by assuming that light energy propogates as a series of independent "corpuscles," or bundles. This gave rise to the more recent particle theory of light.

THIN LENS FORMULA AND OPTICAL INSTRUMENTS

Given the laws of reflection and refraction, one can see in principle how the passage of light through an optical instrument could be traced. For each of a number of initial rays, the change in the direction of the ray at each mirror surface or refractive index interface can be calculated. Between these points, the ray traces out a straight line.

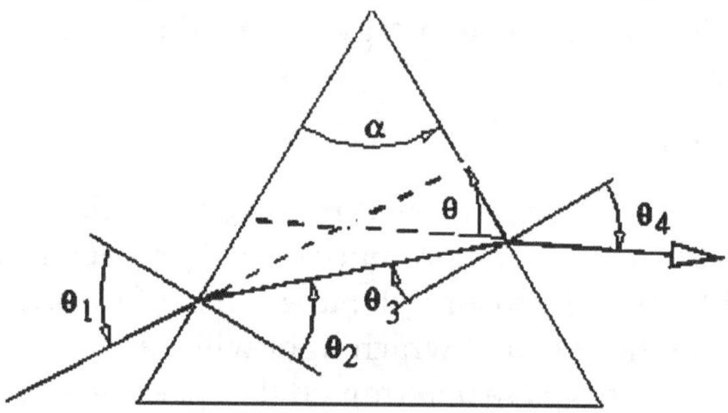

Fig. Bending of a Ray of Light as it Passes Through a Prism.

Though simple in conception, this procedure can be quite complex in practice. However, the procedure simplifies if a number of approximations, collectively called the *thin lens approximation,* are valid. We begin with the calculation of the bending of a ray of light as it passes through a prism.

The pieces of information needed to find θ, the angle through which the ray is deflected are as follows: The geometry of the triangle defined by the entry and exit points of the ray and the upper vertex of the prism leads to

$$\alpha + (\pi/2 - \theta_2) + (\pi/2 - \theta_3) = \pi,$$

which simplifies to

$$\alpha = \theta_2 + \theta_3.$$

Snell's law at the entrance and exit points of the ray tell us that

$$n = \frac{\sin(\theta_1)}{\sin(\theta_2)} \quad n = \frac{\sin(\theta_4)}{\sin(\theta_3)},$$

where n is the index of refraction of the prism. One can also infer that

$$\theta = \theta_1 + \theta_4 - \alpha.$$

This comes from the fact that the the sum of the internal angles of the shaded quadrangle is $(\pi/2 - \theta_1) + \alpha + (\pi/2 - \theta_4) + (\pi/2 - \theta_4) + (\pi + \theta) = 2\pi$.

Combining equations allows the ray deflection θ to be determined in terms of θ_1 and α, but the resulting expression is very messy. However, great simplification occurs if the following conditions are met:

- The angle $\alpha << 1$.
- The angles $\theta_1 << 1$ and $\theta_4 << 1$.

With these approximations it is easy to show that

$$\theta = \alpha(n-1) \text{ (small angles).}$$

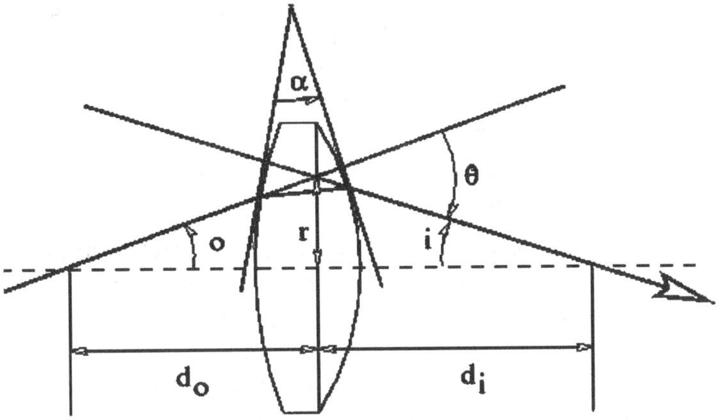

Fig. Light Ray Undergoing Deflection Through an Angle Q by a Lens. The Angle a is the Angle between the Tangents to the Entry and Exit Points of the Ray on the Lens.

Generally speaking, lenses and mirrors in optical instruments have curved rather than flat surfaces. However, we can still use the laws for reflection and refraction by plane surfaces as long as the segment of the surface on which the wave packet impinges is not curved very much on the scale of the wave packet dimensions. This condition is easy to satisfy with light impinging on ordinary optical instruments. In this case, the deflection of a ray of light is given by equation if α is defined as the intersection of the tangent lines to the entry and exit points of the ray.

A *positive lens* is thicker in the center than at the edges. The angle α between the tangent lines to the two surfaces of the lens at a distance τ from the central axis takes the form $\alpha = C_{\tau}$, where C is a constant. The deflection angle of a beam hitting the lens a distance τ from the center is therefore $\theta = C\tau$ $(n-1)$. The angles o and i sum to the deflection angle: $o + i = \theta = C\tau\ (n-1)$. However, to the extent that the small angle approximation holds, $o = \tau/d_o$ and $i = \tau/d_i$ where d_o is the distance to the object and d_i is the distance to the image of the object. Putting these equations together and cancelling the τ results in the *thin lens formula*:

$$\frac{1}{d_o} + \frac{1}{d_i} = C(n-1) \equiv \frac{1}{f}.$$

The quantity f is called the *focal length* of the lens. Notice that $f = d_i$ if the object is very far from the lens, i. e., if d_o is extremely large.

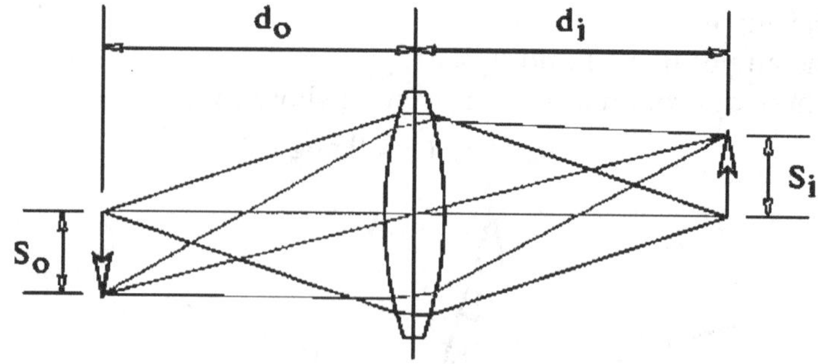

Fig. A Positive Lens Producing an Image on the Right of the Arrow on the Left.

The image is produced by all of the light from each point on the object falling on a corresponding point in the image. If the arrow on the left is an illuminated object, an *image* of the arrow will appear at right if the light coming from the lens is allowed to fall on a piece of paper or a ground glass screen. The size of the object S_o and the size of the image S_i are related by simple geometry to the distances of the object and the image from the lens:

$$\frac{S_i}{S_o} = \frac{d_i}{d_o}.$$

Notice that a positive lens inverts the image.

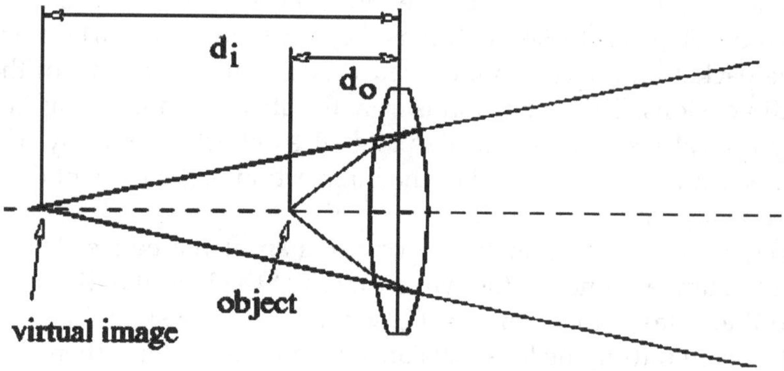

Fig. Production of a Virtual Image by a Positive Lens.

An image will be produced to the right of the lens only if $d_o > f$. If $d_o < f$, the lens is unable to converge the rays from the image to a point. However, in this case the backward extension of the rays converge at a point called a *virtual image,* which in the case of a positive lens is always farther away from the lens than the object. The thin lens formula still applies if the distance from the lens to the image is taken to be negative. The image is called virtual because it does not appear on a ground glass screen placed at this point.

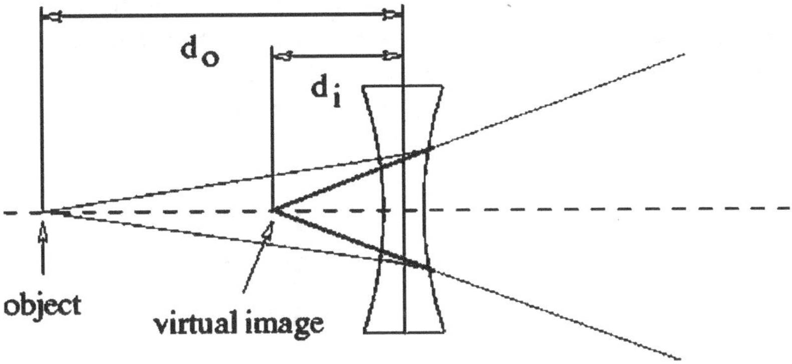

Fig. Production of a Virtual Image by a Negative Lens.

A *negative lens* is thinner in the center than at the edges and produces only virtual images. The virtual image produced by a negative lens is closer to the lens than is the object. Again, the thin lens formula is still valid, but both the distance from the image to the lens and the focal length must be taken as negative. Only the distance to the object remains positive.

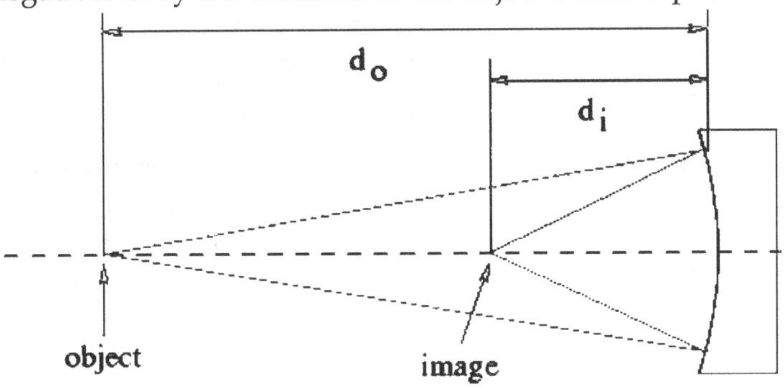

Fig. Production of a Real Image by a Concave Mirror.

Curved mirrors also produce images in a manner similar to a lens. A concave mirror works in analogy to a positive lens, producing a real or a virtual image depending on whether the object is farther from or closer to the mirror than the mirror's focal length.

A convex mirror acts like a negative lens, always producing a virtual image. The thin lens formula works in both cases as long as the angles are small.

FERMAT'S PRINCIPLE

An alternate approach to geometrical optics can be developed from Fermat's principle. This principle states that light waves of a given frequency traverse the path between two points which takes the least time. The most obvious example of this is the passage of light through a homogeneous medium in which the speed of light doesn't change with position.

In this case shortest time is equivalent to the shortest distance between the points, which, as we all know, is a straight line. Thus, Fermat's principle is consistent with light traveling in a straight line in a homogeneous medium.

Fermat's principle can also be used to derive the laws of reflection and refraction.

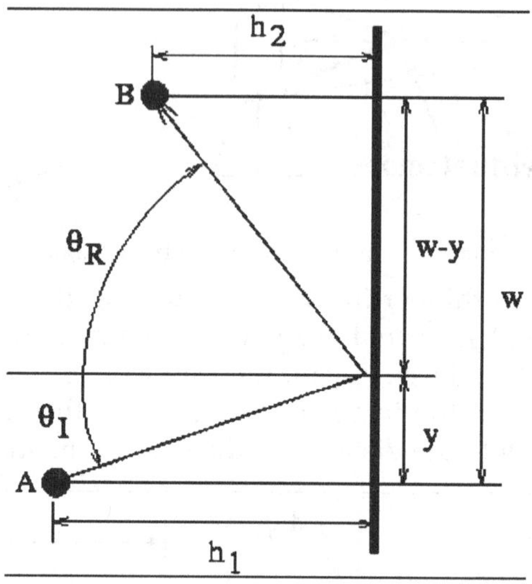

Fig. Definition Sketch for Deriving the Law of Reflection from Fermat's Principle. θ_I is the Angle of Incidence and θ_R the Angle of Reflection.

A candidate ray for reflection in which the angles of incidence and reflection are not equal. The time required for the light to go from point A to point B is

$$t = \left[\left\{h_1^2 + y^2\right\}^{1/2} + \left\{h_2^2 + (w - y)^2\right\}^{1/2}\right] / c$$

where c is the speed of light. We find the minimum time by differentiating t with respect to y and setting the result to zero, with the result that

$$\frac{y}{\left\{h_1^2 + y^2\right\}^{1/2}} = \frac{w - y}{\left\{h_2^2 + (w - y)^2\right\}^{1/2}}.$$

However, we note that the left side of this equation is simply sin θ_I, while the right side is sin θ_R, so that the minimum time condition reduces to sin θ_I = sin θ_R or $\theta_I = \theta_R$, which is the law of reflection.

A similar analysis may be done to derive Snell's law of refraction. The speed of light in a medium with refractive index n is c/n, where is its speed in a vacuum. Thus, the time required for light to go some distance in such a medium is times the time light takes to go the same distance in a vacuum the time required for light to go from A to B becomes

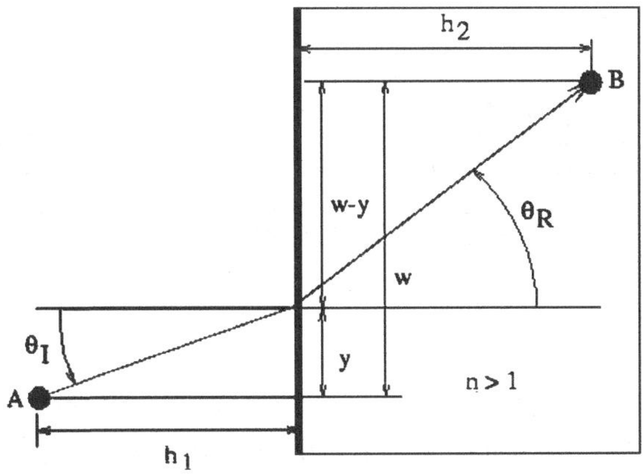

Fig. Definition Sketch for Deriving Snell's Law of Refraction from Fermat's Principle. The Shaded Area has Index of Refraction $n > 1$.

$$t = \left[\left\{h_1^2 + y^2\right\}^{1/2} + n\left\{h_2^2 + (w-y)^2\right\}^{1/2}\right]/c.$$

This results in the condition

$$\sin\theta_I = n\sin\theta_R$$

where θ_R is now the refracted angle. We recognize this result as Snell's law.

The minimum time may actually be a *local* rather than a global minimum the global minimum distance from A to B is still just a straight line between the two points! In fact, light starting from point A will reach point B by *both* routes — the direct route and the reflected route.

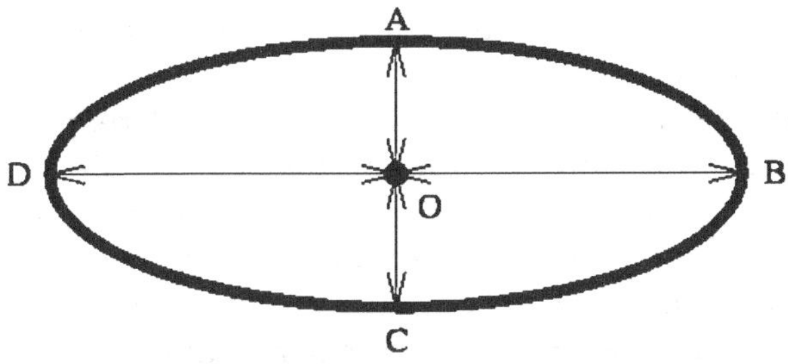

Fig. Ellipsoidal Mirror Showing Minimum and Maximum Time Rays from the Center of the Ellipsoid to the Mirror Surface and Back again.

It turns out that trajectories allowed by Fermat's principle don't strictly have to be minimum time trajectories. They can also be *maximum* time trajectories In this case light emitted at point O can be reflected back to point O from four points on the mirror, A, B, C, and D. The trajectories O-A-O and O-C-O are minimum time trajectories while O-B-O and O-D-O are maximum time trajectories.

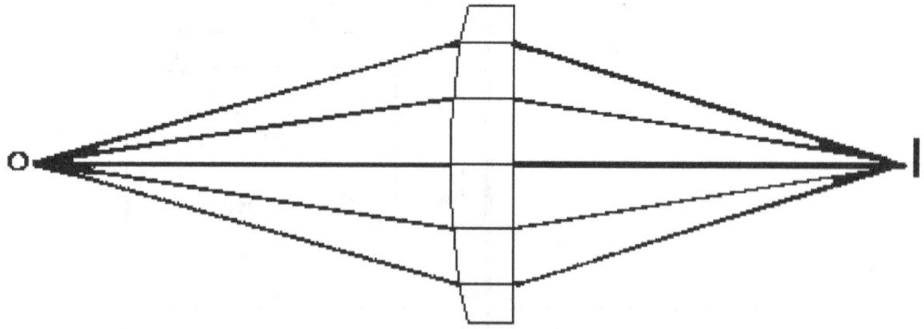

Fig. Ray Trajectories from a Point O being Focused to Another Point I by a Lens.

All the rays from point O which intercept the lens end up at point I. This would seem to contradict Fermat's principle, in that only the minimum (or maximum) time trajectories should occur. However, a calculation trajectories in this particular case take the same time. Thus, the light cannot choose one trajectory over another using Fermat's principle and all of the trajectories are equally favored. Note that this inference applies not to just any set of trajectories, but only those going from one focal point to another.

GEOMETRICAL OPTICS AND NEWTON'S LAWS

The question that motivates us to study physics is "What makes things go?" The answers we conceive to this question constitute the subject of dynamics. This is in contrast to the question we have primarily addressed so far, namely "How do things go?" As noted earlier, the latter question is about kinematics. Extensive preparation in the kinematics of waves and particles in relativistic spacetime is needed to intelligently address dynamics. This preparation is now complete.

Three different dynamical principles based respectively on pre-Newtonian, Newtonian, and quantum mechanical thinking. The last two are still in common use, and we first show that they are consistent with each other in the realm in which Newtonian and quantum mechanics overlap, i. e., in the geometrical optics limit of quantum mechanics. For simplicity, this relationship is first developed in one dimension in the non-relativistic limit. Higher dimensions require the introduction of partial derivatives, and the relativistic case will be considered later.

Only forces which are "conservative" in Newtonian mechanics, i. e., exhibit a potential energy, have a correspondence in the quantum mechanical world. Forces of this type receive our primary attention. Certain ancillary concepts in Newtonian mechanics such as work and power are introduced at this stage.

MULTICOLOR CONFOCAL MICROSCOPY

Refinements in design have simplified confocal microscopy to the extent that it has become a standard research tool in cell biology. However, as confocal microscopes have become more powerful, they have also become more demanding of their optical components. In fact, optical aberrations that cause subtle defects in image quality in widefield microscopy can have devastating effects in confocal microscopy.

Unfortunately, the exacting optical requirements of confocal microscopy are often hidden by the optical system that guarantees a sharp image, even when the microscope is performing poorly. Optics manufacturers provide a wide range of microscope objectives, each designed for specific applications. This report demonstrates how the trade-offs involved in objective design can affect confocal microscopy.

Chromatic Aberration with Two Objectives

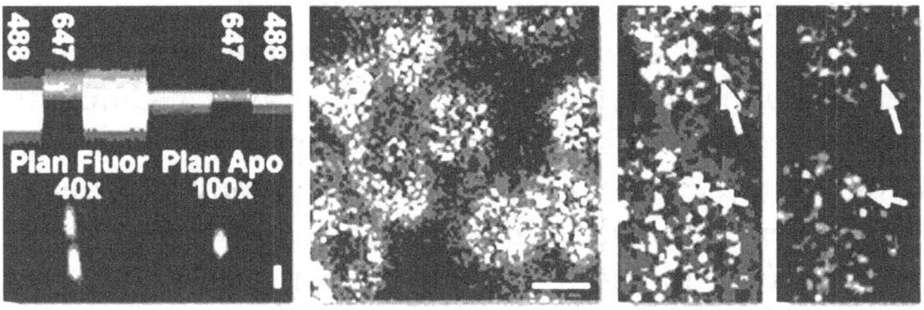

(a) (b) (c) (d)

Over the past ten years, confocal microscopy has developed from a technique limited to specialists in microscopy into a standard research tool. Its proliferation of applications results as much from the rapid technological developments in confocal microscopy as from the maturing user interface of commercial confocal microscope systems.

The latest systems are nearly turnkey systems with which even novice microscopists can rapidly collect high-quality images. Ironically, the same technical developments that have stimulated the spread of confocal microscopy in experimental biology have also pushed the limits of the optics of confocal microscopes in a way that makes understanding the optical properties of a confocal microscope more important than ever. The most common application of confocal microscopy is to compare the distributions or behaviors of multiple probes in the same cells.

Such studies have been made possible by the development of confocal microscopes that are capable of efficiently collecting multiple colors of fluorescence and developing new dyes that have extended the useful spectrum of fluorescence microscopy. Depending on the configuration of the microscope, such studies may require optics that use wavelengths of light ranging from the ultraviolet through the infrared.

The requirements for accurate colour imaging have been further increased by the development of colorimetric methods of quantitative microscopy, such as fluorescence ratio measurements of ion concentrations. Freedom from chromatic aberration is only part of the equation of optical design, which also considers monochromatic aberrations and parameters such as high photon efficiency, field size, field flatness, working distance, and the ability to image deep into aqueous biological tissues.

Since the design of microscope optics reflects a compromise of these various parameters, manufacturers typically design a variety of different microscope objectives, with each representing a particular set of design trade-offs and each suited to a particular application. The studies discussed here demonstrate that the choice of microscope objective can have profound effects on the results of confocal microscopy experiments. They emphasize the importance of carefully choosing a microscope objective that is appropriate for the experimental application.

An ideal lens would focus all colors of light to the same point. In reality, all lenses have chromatic aberration, a property in which different colors of light are focused to different points. When observing a sample through the microscope eyepiece, this defect makes objects appear to have colored edges. When imaging a sample in colour confocal microscopy, this defect results in different colors of excitation illumination being focused to different points in the sample and different colors of emission being collected from different points in the sample. Horizontal displacements in the image plane, termed lateral chromatic aberration, result in different magnification of different colors.

This problem can be minimized by restricting the analysis to the centre of the microscope field. However, vertical colour displacements along the focal axis, referred to as axial chromatic aberration, are present throughout the microscope field. This is clearly a problem for any researcher trying to use colour confocal microscopy to determine the relative distribution of multiple probes. The images presented in figures show how the results of colour confocal imaging critically depend on the nature of the microscope objective. The figures provide a comparison of the performance of a plan fluor 40x objective, which is designed for maximum UV light transmission, with that of a plan apochromat 100x objective, designed with minimal chromatic aberration.

For one aspect of the comparison, a vertical series of images of the reflection from the face of a glass coverslip was collected, using either 488 or 647-nanometer light. As the different colors of light are reflected by a single surface, a lens free of chromatic aberration would focus the different colors to the same focal plane. When reproduced as a vertical cross-section, images collected with an ideal lens would show a single horizontal line in which the two colors overlap completely. The images in the top half of figure illustrate *xz*-sections of the glass reflection, with the focal (*z*) axis oriented vertically, and demonstrate that the plan apochromat 100x objective does a credible job of approaching this ideal, resolving the two colors of light to a depth within 0.1 micrometer of one another.

In contrast, the minimally corrected 40x plan fluor objective detects the reflection of 647-nanometer light approximately 1.2 micrometers above the 488-nanometer light. In the images of the glass reflection, 647-nanometer light is shown in red, and 488-nanometer light is shown in blue. The scale bar indicates a distance of 1 micrometer.

Correction of Axial Chromatic Aberration

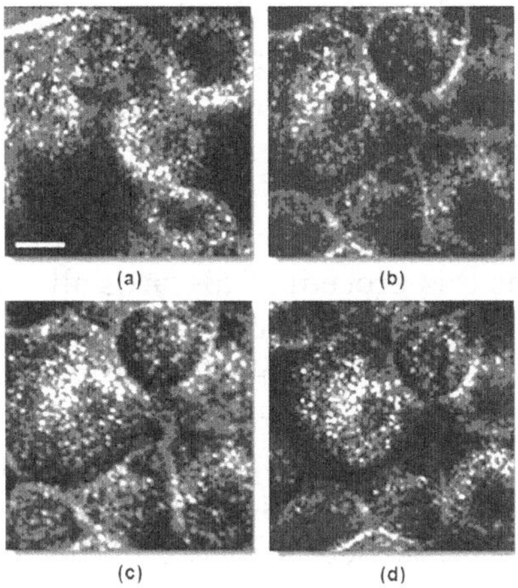

The effect of this vertical discrepancy is apparent in fluorescence. The bottom half of this panel shows vertical cross-sections of image volumes of beads labeled with three fluorophores. When imaged with the 100x plan apochromat objective, the three colors are coincident, resulting in a white, reasonably round image. In contrast, the 40x objective generates an image in which the far-red fluorescence (680-nanometer emissions, depicted in blue) is distinctly vertically displaced from that of either the red (600 nanometer) or the green (520 nanometer) fluorescence. In addition, a slight horizontal displacement of the far-red fluorescence reflects lateral chromatic aberration. To provide a biological test specimen, advantage was taken of the fact that endosomes can be labeled by incubating cells with fluorescently tagged endocytic ligands. The large number of molecules included in each endosome ensures that each will contain the same ratio of fluorescent probes.

Thus, endosomes provide an excellent test of colour confocal imaging as they present subresolution objects in biological samples with colour that will reflect the relative fluorescent contributions of each fluorophore. Figure shows a colour image of cells in which the endosomes have been labeled with both fluoresceintransferrin (F-Tf, which fluoresces green) and Cy5-transferrin (Cy5-Tf, which fluoresces far-red), collected using the 100x plan apochromat objective. The co-localization of the two probes is apparent in the constant yellow-orange colors of the individual endosomes and is more apparent in comparisons of the high magnification images of F-Tf and Cy5-Tf. In these images, the scale bar represents a length of 10 micrometers.

When switching to the 40x plan fluor objective, however, slight differences in the focal plane are apparent even between the red and green fluorescence, as shown in cells labeled with Tf conjugated to both fluorescein and rhodamine. The differences in focal planes between far-red and green fluorescence result in startling differences in the apparent distributions of F-Tf and Cy5-Tf, which now appear to be completely distinct. For a cell biologist who is assessing the relative distributions of various probes, these displacements would have disastrous consequences. The discrepancy in focal plane can be circumvented by summing the entire vertical series of the image volume into a single projection, which removes the effect of differences in the focal plane of the two colors. The consistent yellow colour of every endosome in the projection demonstrates the constant ratio of the two probes in each endosome. However, as this procedure discards all vertical information, it is seldom an appropriate way of presenting confocal images.

The effects of axial chromatic aberration can also be minimized by measuring the axial offset of different colors and combining the images collected from the focal planes appropriate for each colour. An example of this procedure is shown in figure, which shows the combined green and far-red images collected 1.2 micrometers apart with the 40x plan fluor objective. The scale bar in the images of figure represents a 10-micrometer length. The correction achieved by combining images from different focal planes is apparent when comparing the images of the individual probes, shown in figure. Whereas figure and figure show poor agreement in distribution between F-Tf and Cy5-Tf in images collected at the same focal plane, a comparison of figures shows that the far-red image can be superimposed with the green fluorescence image collected 1.2 micrometers deeper. In these images, the scale bar corresponds to a length of 5 micrometers..

Color Focal Plane Correction

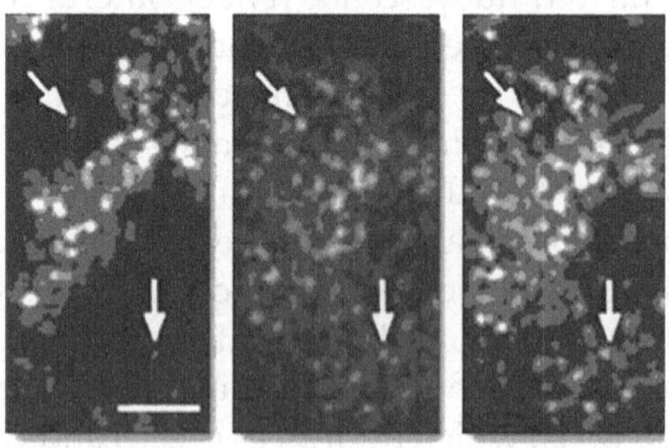

(a) (b) (c)

While chromatic aberration results in different colors of light being focused to different points in the image volume, spherical aberration can profoundly decrease the signal in a confocal microscope. A spherically aberrated lens focuses axial and peripheral rays to different points, thus blurring the image of a point source of light. In the same way that the confocal pinhole so effectively improves image contrast by rejecting out-of-focus light, it effectively eliminates much of the fluorescence of an object imaged with spherical aberration

For many samples, the primary source of spherical aberration derives from the difference between the refractive index of the immersion medium and the mounting medium. Until recently, the highest-resolution, best-corrected microscope objectives were designed for use with oil as an immersion fluid. For these objectives, spherical aberration is minimized only when the entire light path has the refractive index of immersion oil (which is the same as that of glass) and accumulates with distance into a medium with a different refractive index.

Since most samples — particularly living samples — are mounted in media with a refractive index that is significantly lower than immersion oil, spherical aberration has thus limited the depth of image volumes using oil immersion objectives. Figure shows the effects of spherical aberration in confocal microscopy, where the oil immersion 100x plan apochromat objective was used to collect an image of cells labeled with F-Tf and mounted at a depth of either 0 micrometers (the surface of the coverslip) or 35 micrometers into an aqueous medium. In both cases, the endosomes appear sharply defined, but the accumulated spherical aberration of a light path 35 micrometers into an aqueous medium has profoundly compromised the fluorescent signal in figure. The scale bar represents a length of 10 micrometers in each of the images presented in figure.

Recently, optical manufacturers have addressed this problem by designing objectives that use water as the objective immersion medium. For aqueous samples, the matched refractive index of immersion and sample media makes spherical aberration independent of imaging depth. This enables these objectives to collect images as deeply as the working distance permits, frequently hundreds of micrometers into a sample. The success of this design is shown in figure, which shows images of cells labeled with F-Tf, collected either zero or 66 micrometers into an aqueous medium using a water immersion 60x plan apochromat objective. Here, the fluorescent signal is unaffected by the light path through the aqueous sample medium. This new generation of high numerical aperture, plan apochromat water immersion objectives has significantly helped confocal microscopy realise its potential in three-dimensional biological imaging.

The colour performance of this water immersion objective falls between that of the oil immersion plan fluor and the plan apochromat objectives discussed previously. The cross-sectional images of the reflection of a glass surface in the top half of Figure show that 647-nanometer light is focused approximately 0.6 micrometer above 488-nanometer light. Similar patterns for volumes collected either 0 or 63 micrometers into an aqueous medium show that this discrepancy is independent of imaging depth. The bottom-left image of those in figure shows that this colour discrepancy results in the far-red image of a triple-labeled bead being displaced from that of either the red or green fluorescence images.

In the fluorescent bead images, 520-nanometer emissions are shown in green, 600-nanometer emissions in red, and 680-nanometer (far-red) emissions in blue. The bottom-right portion of this panel demonstrates the critical importance of coverslip thickness correction in minimizing spherical aberration in this kind of objective. Since the correction for spherical aberration depends on the length of the optical path through the coverslip, it is adjustable by a collar that is set according to the thickness of the coverslip. While the bead cross-section on the left was collected with the collar set to the measured coverslip thickness (174 micrometers), the image on the right was collected with the collar misadjusted to 150 micrometers.

The spherical aberration that results from an improper collar setting severely attenuates the fluorescence signal and compromises vertical resolution. Although this misadjustment was chosen to dramatize the point, we emphasize that such an error is easily encountered in practice because the actual thickness of coverslips can vary by more than 40 micrometers around their nominal value. Proper collar setting can only be ensured through measurements of individual coverslips. In all images in figure, the focal axis is oriented vertically, and the scale bar indicates a distance of 1 micrometer.

Spherical Aberration

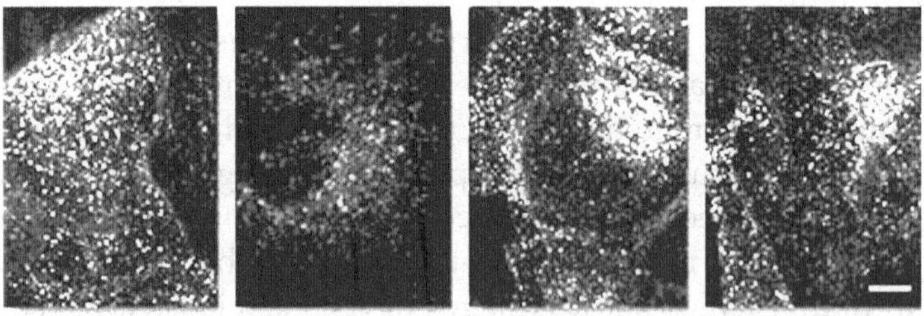

(a) (b) (c) (d)

The images shown in figures are of a field of cells labeled with both F-Tf and Cy5-Tf, and were collected at a depth of 63 micrometers into an aqueous buffer. The scale bar represents 10 micrometers in these two images. Although the images appear sharp, the axial chromatic aberration of this objective has the effect of making the distributions of F-Tf and Cy5-Tf appear discrete.

Nonetheless, the projection of the vertical series of images of these cells shows that the two probes equally label all endosomes. In agreement with the reflection images shown in figure, axial chromatic aberration is similarly apparent in images of endosomes collected at a depth of zero micrometers (at the coverslip surface). Little difference is observed in the focal plane of red and green fluorescence in images of endosomes labeled with Tf conjugated to both fluorescein and rhodamine, but F-Tf and Cy5-Tf again appear to label discrete populations of endosomes.

The distribution of the two probes can be better compared when allowing for the difference in focal plane and combining the Cy5-Tf image with a fluorescein image collected 0.6 micrometer deeper. The co-localization of the two probes is now apparent in the constant yellow colour of the endosomes shown in figure. Apparent differences in the distribution of the two probes shown in images collected at a single focal plane disappear when the Cy5-Tf image is compared with the F-Tf image collected 0.6 micrometer deeper. The scale bar represents 10 micrometers length in figures, and corresponds to 5 micrometers in figures. The effects of chromatic aberration can also be quantified by measuring fluorescence ratios in images of endosomes labeled with multiple probes. Semi-logarithmic plots of ratio histograms are presented in figure for cells labeled with F-R-Tf, with images collected using the 100x plan apochromat (red curve), the 40x plan fluor (blue curve), and the 60x plan apochromat water immersion objective (green curve). Figure displays ratio histograms of cells labeled with F-Tf and Cy5-Tf using the same three objectives. Figure shows that the ratio of rhodamine to fluorescein (red to green) emissions is reasonably constant for all three objectives.

In contrast, figure shows that while the 100x plan apochromat still shows minimal variation in the ratio of Cy5 to fluorescein (far-red to green), the water immersion 60x plan apochromat shows more variation, and the 40x plan fluor shows even more. The effects of chromatic aberration in the 60x plan apochromat and 40x plan fluor are more apparent when the distributions for a single focal plane are compared to those measured in the projections of the vertical image series for each objective. In each case, the narrow distribution of fluorescence ratios in the projected images reports the nearly constant ratio of the two probes in the endosomes, which is misrepresented in any single focal plane image.

Chromatic and Spherical Aberration in Water Immersion

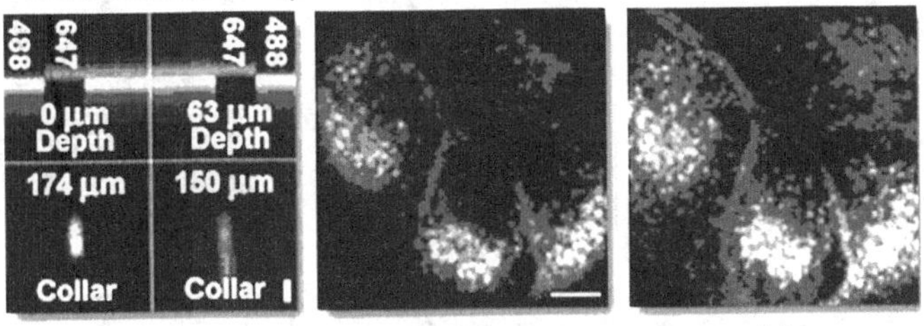

(a) (b) (c)

The studies presented here demonstrate vividly how chromatic and spherical aberrations compromise the performance of confocal imaging. At the same time, they emphasize the critical importance of objective choice in confocal microscopy. Poor colour correction will lead to erroneous interpretations of the relative distributions of multiple probes, one of the major applications of confocal microscopy. The ratio quantifications also demonstrate how chromatic aberrations compromise microscope quantitation. Spherical aberration resulting from mismatched immersion and sample media worsens vertical resolution and can obliterate fluorescence detection altogether. Although particularly apparent in the high-contrast images provided by confocal microscopy, these errors also compromise images collected by conventional epifluorescence and transillumination techniques.

Chromatic aberration has been characterized primarily with respect to UV fluorophores. Other results demonstrate significant axial chromatic aberration in the far-red as well, a range increasingly used by microscopists with the development of new far-red fluorescent probes and lasers capable of exciting them.

Although the studies presented here involve characterization of point sources, these samples readily display problems that would be manifest (but not necessarily obvious) with more extensive probes. Such probes include the cytosolic dyes that are frequently used in fluorescence ratio measurements of ionic concentrations.

For example, fluorescence of the cytosolic pH indicator SNARF-1 is excited by 488-nanometer light, and pH is measured from the ratio of fluorescence at 580 nanometers to that at 640 nanometers. For measurements of the cytosolic pH of thin cells, it is easy to imagine how the fluorescence ratio can be influenced by the position of the cell relative to the separate focal planes of the green excitation light, the red emissions and the far-red emissions.

Although these errors will frequently simply add variation to measurements (doubling or tripling standard deviations in the presented quantifications), they can also systematically affect ratio measurements, for example, when comparing ratios in different parts of cells of varying thickness.

Axial Chromatic Aberration in Water Immersion Objectives

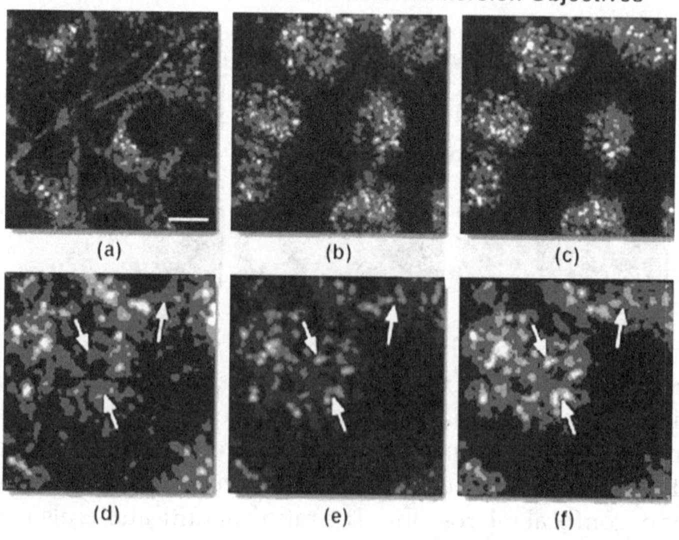

A final consequence of chromatic aberration, although not discussed here, is how it affects fluorescence detection. In a confocal system with chromatic aberration, the difference between the excitation and emission wavelengths results in the excitation of one volume and imaging of a different volume, attenuating the fluorescence signal. For beam-scanning systems (most confocal microscopes), this loss of signal worsens with distance off axis as the imaged spot is focused farther and farther from the confocal pinhole.

It must be emphasized that these observations are not unique to the particular objectives used in the data presented, or to a particular manufacturer. The characterizations of the colour correction were identically reproduced in three examples of the 60x water immersion objective and two examples of the 100x oil immersion objective on both Quantum and Diaphot microscope stands. Significant axial chromatic aberration has been identified in objectives from every major manufacturer. Indeed, the ubiquity of chromatic aberration has led to the development of alternative microscope designs that address chromatic aberration either by imposing special auxiliary correcting lenses or avoiding refraction altogether and turning to reflecting objectives.

Nonetheless, microscope optics is likely to continue to evolve, and preliminary experience with the new generation of CFI60 optics from Nikon, Inc. indicates that axial chromatic correction is much improved. Superficially, it may appear that some objectives are simply better than others. However, as discussed previously, objective design represents a compromise of design parameters.

As such, each objective reflects a different set of design compromises according to its intended application. Nonetheless, it is possible that researchers will require more than what is currently available in optical design, forcing them to design experiments that do not stretch the limits of objective design. If an oil immersion objective must be used to image a sample in an aqueous medium — for example, in studies of living cells — spherical aberration can be minimized either by minimizing the optical path through the aqueous medium or by using an oil with a refractive index tailored to the spherical aberration induced by the aqueous optical path.

The problem of chromatic aberration can be minimized in several ways. If a particular experiment requires the use of an objective with significant chromatic aberration, the most obvious solution is to simply avoid the use of far-red fluorescing dyes such as Cy5 altogether. The results discussed in this review demonstrate satisfactory agreement between green and red fluorescence with all the tested objectives. In general, it is best to use dyes with maximum excitation and emission near the specific wavelengths for which a particular objective has been corrected. A second solution, discussed above, is to collect images of each fluorophore in a vertical series of focal planes and combine the different colour images according to the difference in focal plane between the two.

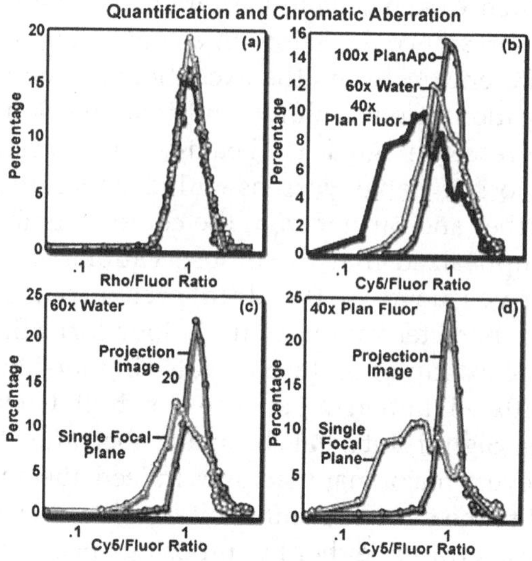

This is a simple solution, but one that will not be appropriate when rapid image acquisition is necessary, as with live cells. Analyses not shown here also indicate that this solution does not have sufficient accuracy to correct images intended for ratio quantification. A more expensive (or less widely accessible) solution is to use two-photon microscopy, which is inherently unaffected by chromatic aberration as long as no detector pinhole is used. However, its multicolor capability is only now being explored.

6

Electron Microscopy and Surface Structure

HIGH RESOLUTION ELECTRON MICROSCOPY AND SURFACE STRUCTURE FACILITY

This facility provides unique equipment to investigate the atomic scale structure of both the surface and subsurface region of a sample combined with in-situ growth and chemical characterization.

Equipment

1. *Spear:* The Specimen Preparation Evaluation Analysis and Reaction System is a system of interconnected Ultra high Vacuum (UHV) chambers for the preparation and study of surfaces and interfaces. The Analytical Chamber is equipped with a PHI Electronics Duoplasmatron Ion Gun, a PHI Electronics Dual-Anode X-ray Source, an FEI Thermally Assisted Schottky Field Emission Electron Gun, and the PHI Electronics Spherical Electron Energy Analyzer. The ion gun can produce oxygen, argon, or xenon ions which, when coupled with a secondary electron detector, generate useful Secondary Electron Microscopy (SEM) images during sputtering.
   The electron energy analyzer can acquire XPS data from either Al K-alpha or Mg K-alpha x-rays from the X-ray source, as well as highly spatially resolved Auger electron data generated by the electron gun. The electron gun can also be used for the direct heating of samples. The Transfer Chamber allows samples to be moved between the Load-Lock, the Analytical chamber, and various other connected systems (UHV-HREM, MIBE, SINBAD) while at all times remaining in a UHV environment.
2. *Uhv-hrem:* The Hitachi UHV-H9000 High Resolution Electron Microscope is the first conventional transmission instrument to achieve routine mid 10×10^{-11} torr operation. It is equipped with a Gatan Parallel EELS and CCD Camera, both interfaced to computers. Routinely the instrument will resolve at about 0.18nm for bulk spacings and atomic structure of surfaces at about the 0.25nm level.

3. *Mibe:* The main goal of the Magnetron and Ion Beam Epitaxy System is to combine deposition processes with the analytical capabilities of SPEAR and an UHV TEM. Samples can be grown in MIBE and then transferred to SPEAR where analytical techniques can be performed. It has two metal ion sources currently fitted with boron and carbon targets, two d.c. magnetron sputtering sources used to form ZrN and CNx multilayer materials, and one gas ion source.
4. *Sinbad:* The Stabilizing Ion and Neutral Beam Assisted Deposition system is a specially designed system for the in-situ investigation of thin solid films. It is designed to handle the deposition of thin films onto thin 3mm TEM ready samples and investigate the effect of energetic particles on materials. The sample manipulation stage can be used for d.c. biasing of the sample as well as resistive heating during deposition and the unit is equipped with a single position electron-beam evaporator, a 4 keV ion-gun, and a compact electron cyclotron resonance (ECR) plasma source. It is also connected to the SPEAR system.
5. *Computers:* Two Linux clusters, JosephSmith and HyrumSmith, are used for running DFT calculations and compiling code. Additionally, computers are used for image simulation and diffraction calculations. The primary software is NUMIS, a combination of Semper 6 and multislice/imaging programs.

   Originally these calculations were run on UNIX machines, but are gradually being moved to PC's running an X-windows environment. Additionally, the EDM software package, which is continually being updated, is used to combine various aspects of image processing and manipulation of high resolution images and diffraction patterns as well as direct methods.

ELECTRON MCIROSCOPY TECHNIQUES IN ELECTRON DIFFRACTION

Electron diffraction is one of major electron mciroscopy techniques. Traditionally, It has been used for phase identification, structure and symmetry determination, foil thickness measurement, lattice parameter measurement, disorder and defect identification.

Recent development has significantly improved the quantitative analysis of electron diffraction intensities and has brought new types of highly accuracte electron diffraction techniques for structure refinement and structure factor measurement.

The recent development in the new generation of field-emission electron microscopes and energy-filter promise further development in single atomic column scattering and diffraction from nanometer-sized molecules, clusters, wires and other low-dimensional objects.

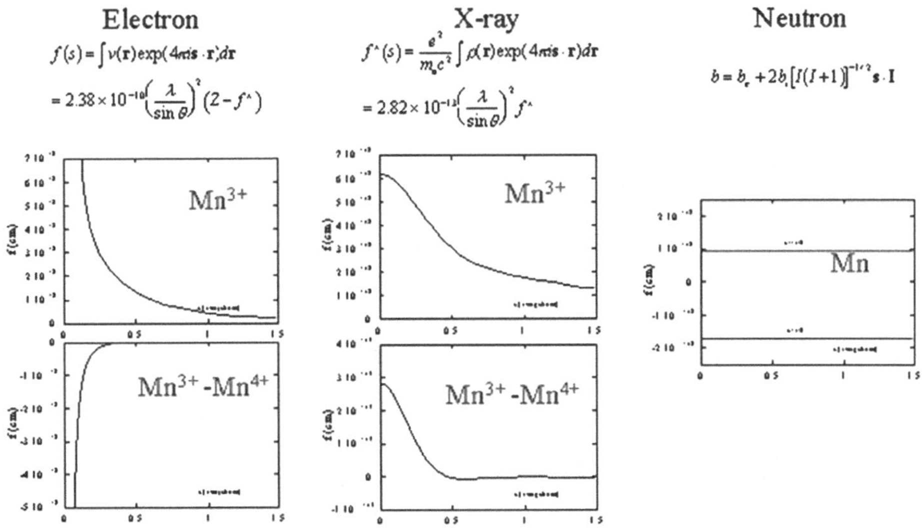

There are three types of electron diffraction: Low Energy Electron Diffraction (LEED), Transmission High Energy Electron Diffraction (THEED) and Reflection High Energy Electron Diffraction (RHEED). Both LEED and RHEED are major surface characterization techniques.

THEED is commonly associated with transmission electron microscopy and molecular scattering. Treatment of LEED is very different from THEED and RHEED, where surface reflection and the quantum mechanical effect of exchange and correlation are dominant. For high energy electrons, diffraction can be described based on individual atomic scattering in the same way as neutron and x-ray diffraction.

Electron diffraction differs from X-ray and neutron scattering from its ability to form images and small probes (angstroms). Equally important, electrons interact with matter through the strong Coulomb potential.

As results, electrons interact very strongly and are very sensitive to matter's valence electrons or bonding electrons. The figure below shows a comparison between electron, X-ray and neutron scattering factors:

LOW ENERGY ELECTRON DIFFRACTION (LEED)

Most suitable for long-range order determination (does not work on disorderedor amorphous materials). Can provide short range information using detailed theoretical analysis.

General Diffraction Ideas

5000 eV photon (x-ray):

$$E = \frac{h.c}{\lambda}$$

$$\lambda = \frac{6.6 \times 10^{-34}\, Js \times 3.00 \times 10^{8}\ \text{ms}^{-1}}{5 \times 10^{3}\ \text{V}(=\text{JC}^{-1}) \times 1.60 \times 10^{-19}\ \text{C}}$$

$= 2.479 \times 10^{-10}$ m

$= 2.5$ Å

20 eV electron:

$$\lambda = \frac{h}{p} \text{ de Broglie wavelength}$$

$$p = \text{m. v} = \sqrt{2\text{m(KE)}} = 2.42 \times 10^{-24}\ \text{kg.m.s}^{-1}$$

$$1 = \frac{6.63 \times 10^{-24}\ \text{J.s}}{2.42 \times^{-24}\ \text{kg.m.s}^{-1}}$$

= 2.74 5 10–10 n = 2.7 Å

Electrons of KE 20-200 eV can be diffracted by lattice of atomic dimensions

Low energy electrons are *strongly* back-scattered by the electrons of the surface atoms (unlike x-ray diffraction) - electrons of 20-200 eV penetrate only ~10-50 E into surface

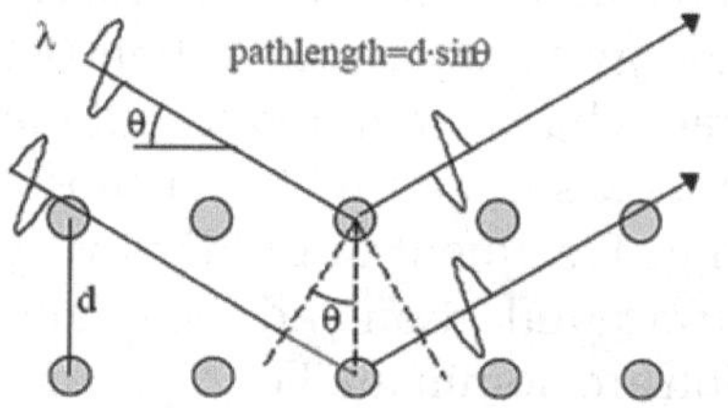

$$n\lambda = 2d\sin\theta \qquad \textbf{Bragg Equation}$$

$$n\left(\frac{h}{\sqrt{2meV}}\right) = 2d\sin\theta$$

Instrumentation for LEED

- Electron gun produces focussed e-beam 10 nA-10 μA E 20-200 eV.

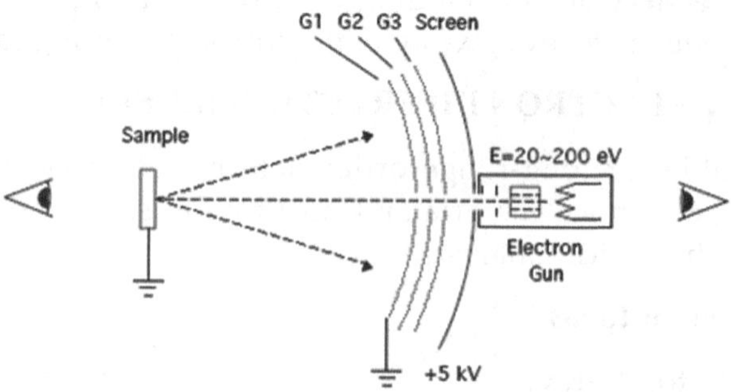

- Magnetic shield expels residual magnetic fields.
- Sample positioned at "focus" of hemispherical grids.

- Diffracted electrons (elastically scattered) and secondary electrons (inelastically scattered) back-scattered towards LEED optics in field free region.
- Diffracted electrons – spots
- Secondary electrons - diffuse background

After passing through G1 (ground) accelerated towards phosphor screen
Negative potentials applied to G2 and G3 to repel secondary electrons

Electrons strike phosphor → photons
Front-view LEED versus rear-view LEED
Image captured on photographic film or video camera

Rear-view LEED system. Width about 10", height about 12 ". The grids, G1,G2 and G3, are visible at the top of the picture and the view-port is at the bottom. The magnetic shield has been removed.

Dynamical LEED

Kinematic LEED considers incident electron scattered once (top layer of atoms) - works fine for symmetry/size of adsorbate or unit cell.

Dynamical LEED considers incident electron scattered multiple times (1st, 2nd, 3rd... layers of atoms) - necessary to account for intensity of spots.

Essential ingredients of dynamical LEED theory:

Calculation of amplitude (A) and phase (d) due to -

(a) ion core scattering
(b) multiple scattering
(c) inelastic events
(d) surface vibration (effect of temperature)

Dynamical LEED Methodology

Intensity Tells us About Interlayer Spacing

When combined with multiple scattering theory, gives information about "height" of adsorbate atoms and relaxation phenomena

Can measure intensity in two ways:

1. vary incidence angle - I(q) curves
2. vary incidence energy - I(V) curves

Methodology

1. Initial guess for structure.
2. Calculate I(V) or I(q) curves using dynamical LEED theory.
3. Compare theory and experimental data.
4. Refine guessed structure until best agreement is obtained.

REFLECTION HIGH ENERGY ELECTRON DIFFRACTION (RHEED)

In LEED, low energy electrons were used to:

- Provide large elastic scattering cross-section for *back*-scattered electrons.
- Keep the penetration depth of the electrons short.

In RHEED, another solution is used:

- Provide large elastic scattering cross-section for *forward*-scattered electrons.
- Keep penetration depth small by using grazing incidence.

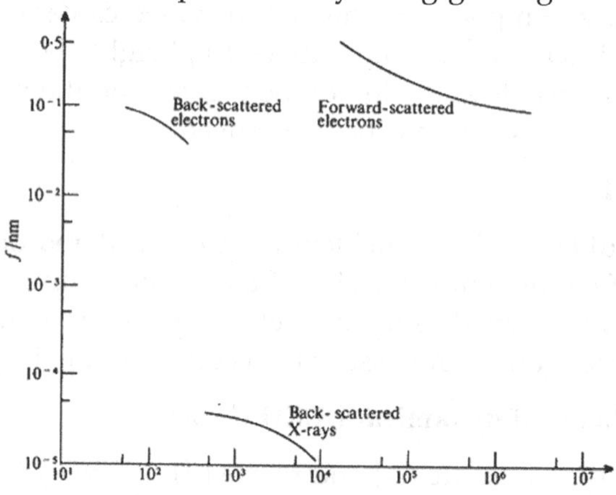

Fig. Approximate Atomic Scattering Amplitudes for Aluminium as a Function of Energy of Electrons and x-rays.

- Electrons 5-100 keV directed towards sample at very near qi 90° - keeps even long-pathlength (high-energy) electrons in near-surface region.
- Forward-diffracted electrons strike phosphor screen.
- Grid used to repel secondary and inelastically scattered electrons.

Applications of RHEED

RHEED is very sensitive to surface roughness on 100-1000 E scale.

Since pathlength of RHEED electron is high, electron may pass through a small surface feature and cause true bulk-like diffraction.

Such bulk diffraction:

- Reduces intensity in RHEED streaks.
- Causes sharp spots in RHEED pattern.

Can use RHEED to follow growth of one material on another (epitaxy) - As, Ga molecular beams on GaAs(001) surface.

Extended X-ray Absorption Fine Structure (EXAFS)

Effect is due to *interference* between outgoing electron wave (photoelectron) and part of outgoing electron wave scattered by neighboring atoms

- Oscillations in x-ray intensity 50~200 eV from absorption edge.
- Does not occur for isolated atoms.
- Outgoing photoelectron (KE > 50 eV) behaves like free electron.
- Most interference from nearest neighbors.
- Sensitive to *short-range* order (unlike diffraction) - works for amorphous, polycrystalline, glassy materials, liquids.
- Can select absorption edge for one particular species (C, O, N...).

If we assume that outgoing and back-scattered electrons can be treated as plane waves, constructive interference occurs when

$$n.\lambda = 2.r$$

Destructive interference will occur when

$$\frac{n.\lambda}{2} = 2.r$$

At intermediate values of l, intermediate interference results that affects the intensity of the outgoing wave.

Energy of outgoing photoelectron (at absorption edge threshold photoelectron has zero KE)

$$E = hv_{photon} - hv_{threshold}$$

Since,

$$E = \frac{1}{2}m.c^2 = \frac{p^2}{2m}$$

and

$$p = \frac{h}{\lambda}$$

we can write

$$E = hv_{photon} - hv_{threshold}$$

$$l = \frac{h}{(2m[hv_{photon} - hv_{threshold}])^{1/2}}$$

Constructive interference occurs whenever,

$$n\lambda = 2.r$$

$$n = \frac{h}{2.r(2m[hv_{photon} - hv_{threshold}])^{1/2}}$$

where *n* is really a measure of the "degree of interference"

HIGH RESOLUTION SURFACE INVESTIGATIONS

There are a wide range of analytical techniques which may be used for materials characterization depending on the type of information needed. For high resolution surface investigations, two commonly used techniques are Atomic Force Microscopy (AFM) and Scanning Electron Microscopy (SEM)– respectively. Each of these techniques resolves surface structure down to the nanometer scale. However, the image formation mechanisms are quite different, resulting in different types of information about the surface structure.

The occurrence of the SEM and AFM side-by-side is becoming more common in today's analytical laboratories. This article will compare and contrast the two techniques with respect to specific types of surface measurements, and demonstrate how these analytical techniques provide information which is complementary in nature.

History

The first SEM was constructed in 1938 by von Ardenne by rastering the electron beam of a Transmission Electron Microscope (TEM) to essentially form a Scanning Transmission Electron Microscope (STEM). In 1942, Zworkin et. al. developed the first SEM for bulk samples. This configuration contains many of the basic principles of today's SEMs. Cambridge Scientific Instruments produced the first commercial instrument in 1965. A number of improvements have occurred since this time, resulting in an increase in resolution from 50nm in 1942 to ~0.7nm today. Besides the development of morphological imaging, the SEM has been developed to detect signals which are used to determine compositional information, such as X-rays, backscattered electrons, cathodoluminesce, Auger electrons, and specimen current.

The development of the AFM was preceded by the development of the Scanning Tunneling Microscope (STM) in 1981 at IBM Zurich Research Laboratory by Binnig and Rohrer.

Its ability to view the atomic lattice of a sample surface earned the inventors the Nobel Prize in Physics in 1986. Although the STM provides subangstrom resolution in all three dimensions, it is limited to conductive and semiconductive samples. To image insulators as well as conductors, the Atomic Force Microscope (AFM) was developed in 1986, and the first commercial AFMs were produced in 1989 by Digital Instruments, now Veeco Instruments.

AFM provides three-dimensional surface topography at nanometer lateral and subangstrom vertical resolution on insulators and conductors. From this beginning, the field of Scanning Probe Microscopy (SPM) was born which consists of a family of techniques that involves scanning a sharp tip across the sample surface while monitoring the tip-sample interaction to form a high resolution image.

Although the AFM has become the most commonly used form of SPM, many other SPM techniques have been developed which provide information on differences in friction, adhesion, elasticity, hardness, electric fields, magnetic fields, carrier concentration, temperature distribution, spreading resistance, and conductivity.

Imaging Mechanisms

Scanning Electron Microscopy

The operation of the SEM consists of applying a voltage between a conductive sample and filament, resulting in electron emission from the filament to the sample. This occurs in a vacuum environment ranging from 10^{-4} to 10^{-10} Torr.

The electrons are guided to the sample by a series of electromagnetic lenses in the electron column. The resolution and depth of field of the image are determined by the beam current and the final spot size, which are adjusted with one or more condenser lenses and the final, probe-forming objective lenses.

The lenses are also used to shape the beam to minimize the effects of spherical aberration, chromatic aberration, diffraction, and astigmatism. The electrons interact with the sample within a few nanometers to several microns of the surface, depending on beam parameters and sample type. Electrons are emitted from the sample primarily as either backscattered electrons or secondary electrons. Secondary electrons are the most common signal used for investigations of surface morphology. They are produced as a result of interactions between the beam electrons and weakly bound electrons in the conduction band of the sample.

Some energy from the beam electrons is transferred to the conduction band electrons in the sample, providing enough energy for their escape from the sample surface as secondary electrons. Secondary electrons are low energy electrons (<50eV), so only those formed within the first few nanometers of the sample surface have enough energy to escape and be detected.

High energy beam electrons which are scattered back out of the sample (backscattered electrons) can also form secondary electrons when they leave the surface. Since these electrons travel farther into the sample than the secondary electrons, they can emerge from the sample at a much larger distance away from the impact of the incident beam which makes their spatial distribution larger.

Once these electrons escape from the sample surface, they are typically detected by an Everhart-Thornley scintillator photomultiplier detector. The SEM image formed is the result of the intensity of the secondary electron emission from the sample at each x,y data point during the rastering of the electron beam across the surface.

Atomic Force Microscopy

AFM consists of scanning a sharp tip on the end of a flexible cantilever across a sample surface while maintaining a small, constant force. The tips typically have an end radius of 2nm to 20nm, depending on tip type. The scanning motion is conducted by a piezoelectric scanner which scans the tip in a raster pattern with respect to the sample (or scans the sample with respect to the tip).

The tip-sample interaction is monitored by reflecting a laser beam off the back of the cantilever into a split photodiode detector. By detecting the difference in the photodetector output voltages, changes in the cantilever deflection or oscillation amplitude are determined.

The two most commonly used modes of operation are contact mode AFM and TappingMode AFM, which are conducted in air or liquid environments. Contact mode AFM consists of raster-scanning the probe (or sample) while monitoring the change in cantilever deflection with the split photodiode detector.

A feedback loop maintains a constant cantilever deflection by vertically moving the scanner to maintain a constant photodetector difference signal. The distance the scanner moves vertically at each x,y data point is stored by the computer to form the topographic image of the sample surface. This feedback loop maintains a constant force during imaging.

TappingMode AFM consists of oscillating the cantilever at its resonance frequency (typically ~300kHz) and lightly "tapping" the tip on the surface during scanning. The laser deflection method is used to detect the root-mean-square (RMS) amplitude of cantilever oscillation. A feedback loop maintains a constant oscillation amplitude by moving the scanner vertically at every x,y data point. Recording this movement forms the topographical image.

The advantage of TappingMode over contact mode is that it eliminates the lateral, shear forces present in contact mode. This enables TappingMode to image soft, fragile, and adhesive surfaces without damaging them, which can be a drawback of contact mode AFM.

Comparison of Techniques

There are a number of different ways to compare and contrast these two techniques. Although investigations that use both SEM and AFM to characterize a material are common, there are just a few studies that directly discuss the complementary nature of the techniques.

A comparison of these techniques will be conducted with respect to 3 factors:

1. Surface Structure,
2. Composition, and
3. Environment.

The comparisons are presented for typical equipment configurations and operating procedures.

Surface Structure

Although both SEM and AFM are similar in lateral resolution, there are situations in which one technique can provide a more complete representation of the sample surface, depending on the information desired. One principle difference is in how the two techniques process vertical changes in topography. Below we will discuss measurements of different vertical scales of topography, beginning with very smooth surfaces and working up to very rough surfaces to determine how the surface topography affects the ability of each technique to perform the measurement.

Atomically Smooth Surfaces

Atomically smooth surfaces can occur either naturally, such as on mineral surfaces, or by processing, such as polishing and epitaxial growth on semiconductor, data storage, and optical surfaces. Note that, unlike SEM, the AFM can measure in all three dimensions (x, y, and z) with a single scan. Since the AFM has a vertical resolution of <0.5E, it can resolve the 1.4E monoatomic silicon steps on the surface as well as calculate an RMS roughness of 0.7E. On a sample this smooth, the SEM has difficulty resolving these features due to the subtle variations in height.

Thin Films

On most thin films, the SEM and AFM produce a similar representation of the sample surface. A common application of surface investigations of thin films consists of determining changes in morphology with variations of deposition parameters, such as temperature, pressure, time, etc.

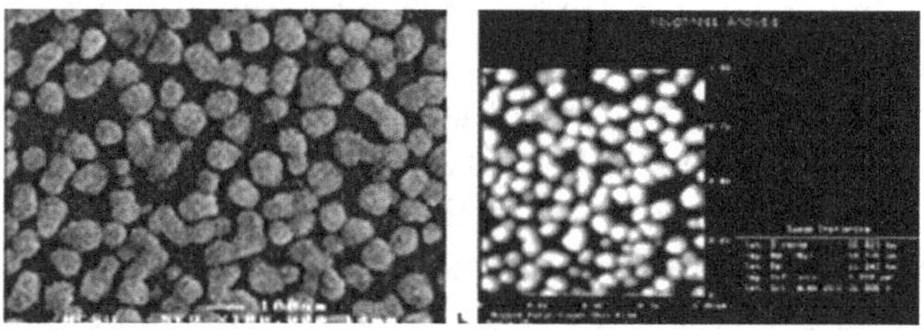

The above figure shows SEM and AFM images of a polysilicon thin film at approximately the same lateral magnification. The two images show similar surface structure, however, they differ in the other types of information that can be acquired on this sample.

The three dimensional nature of the AFM can be used to calculate changes in roughness and surface area variations due to differences in deposition parameters. For the SEM, a large area view of the variations in surface structure can be acquired all at once (such as several mm's), whereas a 100¼m x 100¼m area is typically the largest area viewed by an AFM. These images are an example of "rugged" polysilicon films which are used as capacitors in memory devices.

By making these films rough, the surface area is increased which makes it possible to hold more charge without increasing the lateral dimensions of the capacitors on the chips. By adjusting the deposition parameters and using the AFM to analyse the surface area of the films, the deposition parameters needed to produce a film with the maximum surface area were determined. Another example of the difference between the two techniques is in interpreting subtle differences in height. In the SEM image, changes in slope can result in an increase in electron emission from the sample surface, producing a higher intensity in the image.

However, it can sometimes be difficult to determine whether the feature is sloping up or down. The only other option would be to cleave the sample through one of these features and look at the sample in crosssection, which would be tedious and time consuming. Since the AFM data contains the height information, determining whether a feature is a bump or pit is straightforward. This information was used in the study of the growth mechanisms of GaP on Si during chemical beam epitaxy deposition. Determination of whether these features were small bumps or depressions would have changed how the deposition process was altered to produce an epitaxial GaP film.

High Aspect Ratio Structures

Semiconductor processing commonly requires measurements of high aspect ratio structures such as trenches and via holes. In a SEM, these structures are typically measured in cross section by cleaving the wafer and imaging the sample on end to obtain the dimensions of the structure. In contrast, the AFM image of a trench or via is made by scanning over the sample surface. The ability of the AFM to measure these structures nondestructively makes it possible for the wafer to be returned to the production line after the measurement is acquired.

Although the SEM measurement is destructive to the sample, the ability to image the undercuts of these lines is a useful application that AFMs are not typically designed to perform with the exception of the Dimension X3D Automated AFM (please see Veeco Instruments Application Notes on AFM Metrology of undercuts and nearvertical side walls with model Dimension X3D).

Rough Surfaces

One of the key advantages of the SEM with respect to other types of microscopy is its large depth of field. This ability makes it possible to image very rough surfaces with millimeters of vertical information within a single image. The depth of field and small beam size makes it possible to image the fibers far below the top layer. This ability also makes it possible to measure very rough surfaces over larger lateral areas as well.

Although the AFM can measure vertical surface variations below 0.5E, its ability to measure a tall structure comes from how far the scanner can move vertically. Standard scanners typically have 5 to 6¼m of vertical range, however, in some configurations the vertical range approaches 10¼m or larger.

For scanning areas that have heights of greater than 5 to 10¼m's of variation, the SEM would be better suited for the analysis. The convoluted three dimensional Y_2O_3 oxide crystal shown growing out of a relatively flat Y_2O_3 thin film on a Si substrate is easily imaged in the SEM. Although the AFM would have problems imaging the obtuse angles and enclosed areas of this surface, the roughness of the Y_2O_3 film can be measured whereas in the SEM image the surface roughness is not evident. Therefore, the two techniques together give a more complete "picture" of the sample.

Composition

SEM is the only one of the two techniques which provides elemental analysis, however, both SEM and AFM are associated with techniques which can provide compositional information through analyzing materials and physical properties of the sample. Some of the most common of these methods are described next.

SEM

Along with the secondary electron emission which is used to form a morphological image of the surface in the SEM, a number of other signals are emitted as a result of the electron beam impinging on the surface. Each of these signals carries information about the sample which provide clues to its composition. Two of the most commonly used signals for investigating composition are x-rays and backscattered electrons. X-ray signals are commonly used to provide elemental analysis by the attachment of an Energy-Dispersive Spectrometer (EDS) or Wavelength-Dispersive Spectrometer (WDS) to the SEM system.

X-ray emission results from inelastic scattering between the beam electrons and the electrons of the sample atoms. This interaction results in the ejection of an inner shell electron from the atom, creating a vacancy that is filled by an outer shell electron.

This jump from an outer to inner shell results in a change in energy that produces either a x-ray or Auger electron. The emitted x-ray has energy equal to this change. The x-rays are then detected by either a lithium-drifted silicon detector for an EDS system, or a gas proportional counter detector for a WDS system.

Backscattered electrons are the result of beam electrons being scattered back out of the sample. In this case, the incident beam electrons undergo a number of scattering events within the specimen in which very little energy is lost, allowing these electrons to go much deeper into the sample than secondary electrons and still emerge from the sample surface to be detected. The percentage of beam electrons that become backscattered electrons has been found to be dependent on the atomic number of the material, which makes it a useful signal for analyzing the material composition. Once these electrons escape from the surface they are detected by either the Everhart-Thornley detector or a solid state detector.

AFM

Although an AFM does not provide elemental analysis, it can supply compositional information by differentiating materials based on physical properties, such as stiffness, elasticity, compliance, friction, adhesion, magnetic and electrostatic fields, carrier concentration, temperature distribution, spreading resistance, and conductivity. Many of these techniques consist of looking simultaneously at another signal while performing standard AFM imaging. One of the most common techniques for mapping differences in materials properties is PhaseImaging.

PhaseImaging is conducted during TappingMode AFM operation by monitoring the phase lag between the oscillating drive signal used to drive the cantilever and the oscillating detection signal from the photodiode detector. This signal will indicate differences in viscoelasticity and/or adhesion across the imaged area.

This technique is commonly applied to mapping the distribution of polymers in a heterogeneous system, or mapping the distribution of filler, such as silica or carbon black, in a polymer matrix. Other ways to get similar information include Force Modulation AFM, which maps differences in elasticity across the sample surface, and Lateral Force Microscopy (LFM), which maps differences in friction across the sample surface.

There are also techniques that can be used to investigate long range forces across the imaged area. Magnetic Force Microscopy (MFM) and Electric Force Microscopy (EFM) map the magnetic and electrostatic field gradients, respectively, which extend from the sample surface.

These techniques are performed by using either a magnetic or conductive probe to map the attractive and repulsive forces between the tip and the sample. MFM is commonly used to detect the domain structure of magnetic bits written on magnetic media, to evaluate the performance of magnetic heads, and to investigate the magnetic structure of experimental materials.

This is conducted by a routine called LiftMode in which a TappingMode topographic image and a magnetic image are acquired over the same area. LiftMode consists of first collecting a line scan in TappingMode of the surface topography. The tip is then lifted above the surface and a second scan is made over the same line using the saved topographic scan to maintain a constant tip-sample separation.

The long-range magnetic forces shift the resonance frequency of the oscillating cantilever, which is detected to produce the magnetic image. Although the AFM is applied so that it is nondestructive to the sample surface, it can be used to study differences in mechanical properties by performing nanoindention to investigate hardness differences between materials. This hydrated samples. These two techniques compensate for this need by different means: an environmental chamber for a SEM, and a fluid cell for the AFM.

Second, the SEM is required to work in a vacuum environment due to the nature of the technique which brings up the issues of vacuum compatibility of the sample, the conductivity of the surface. To image poorly conductive surfaces without sample charging may require conductive coatings or staining, which may alter or obscure the features of interest; or it may require low-voltage operation or an environmental chamber, which may sacrifice resolution.

For SEM, hydrated samples are addressed by placing a specimen in an environmental chamber with either an electron transparent window or a small aperture for the beam to enter the chamber. The chamber is typically flushed with an inert gas saturated with water vapour. Common applications are to either investigate hydrated surfaces to preserve their surface structure when hydrated, or to reduce charging on insulating samples.

For the electron technique uses a diamond tip mounted on a stiff, stainless steel cantilever. A TappingMode AFM image is collected with the probe to determine the area of interest for indentation, the nanoindention is then made at a specified force, and an image is then collected of the indented area.

In this example, the two films demonstrate a difference in hardness from indentations made at the same forces, producing different sized indents. Scratching and wear testing may also be conducted with this configuration to investigate adhesion and delamination of films under a small applied force.

Environment

One of the primary differences between these two types of microscopy is the environment in which they are performed, i.e., SEM is only conducted in a vacuum environment. In addition to vacuum, AFM is conducted in an ambient, gas or liquid environment. There are several issues which make environment an important issue. First, there is a frequent need in fields such as biology and biomaterials to study beam to interact with the surface in this configuration, it must go though an environment of gas and water vapour.

One drawback of this configuration is that it will result in an increase in scattering of the electron beam on the way to and from the surface, which may result in the sacrifice of image quality and resolution. One of the primary attractions to the AFM is its ability to image insulating surfaces at high resolution in liquid. Imaging samples in a hydrated state with an AFM is commonly performed by enclosing the sample and probe in a liquid environment. Since AFM does not rely on conductivity, the image and scanning mechanism are not disturbed by the presence of the liquid. Common applications for AFM investigations in liquid are in the biological sciences, biomaterials, crystal growth, force interaction studies, and for investigating processes *in-situ*.

The resolution of the image will be determined by the radius of the tip, the applied force, and the noise floor of the instrument. Because of these factors, this configuration allows the study of hydrated specimens at a lateral resolution of 1 to 5nm and a vertical resolution down to 0.5E without sample damage, as seen in the image of the GroES chaperon. With the appropriate accessories, AFM can also be used in varied gaseous environments and at elevated temperatures. The latter is particularly important for research and development of polymers.

7

Electron Optics and Electron Microscopy

Electron optics was born in 1927, when Hans Busch showed that the elementary lens equation is applicable to electron image formation. Interest was soon awakened in France and a book on the subject by Louis de Broglie appeared in 1950. We draw attention to the highlights of electron optics and microscopy over the decades, with particular reference to aberration correction, and to some little known French contributions to the story of the scanning electron microscope and to Fourier optics.

The natural starting-point for this occasion is the book *Optique électronique et corpusculaire* by Louis de Broglie, published in 1950 and based on lectures delivered in the Institut Henri Poincaré during the academic years 1946–47 and 1947–48. This is a remarkably complete work, all the more impressive in that de Broglie had not contributed to geometrical electron optics and, according to his preface, had not had access to any recent books on electron optics and electron microscopy; the only works on electron optics cited are the early text of Brüche and Scherzer (1934) and the dissertation of Paul Chanson (1947).

He appears not to have been aware of the important collection of articles published in book form by Busch and Brüche in 1937 (which had first appeared in the *Zeitschrift für Technische Physik*) nor of the book by von Ardenne (1940), which was reasonably widely available and is among the references in Chanson's thesis. It is less surprising that de Broglie had not seen the excellent book published in England in 1939 by Myers (though this too is listed by Chanson) but surely copies of Zworykin *et al.* (1945) would have reached post-war France. We know from the bibliography of Grivet's lecture at the Réunions d'Etudes (1946) that he had seen several of the major German publications, though he too cites neither Myers nor Zworykin *et al.*

Electron optics has its beginnings more than 20 years earlier. Surprisingly, wave electron optics preceded geometrical electron optics: the association of a frequency and hence a wavelength with charged particles by de Broglie (1925) antedates the recognition that a magnetic coil has the same effect on a beam of electrons as a glass lens on light rays. It was not until 1927 that Hans

Busch (1884–1973) showed that the elementary lens relation is applicable in both cases. With hindsight, it seems odd that the discovery was not made much sooner.

The seed of such an idea is to be found in the work of Hamilton, nearly a century earlier, and at any time since then, and especially since the demonstration by Thomson in 1897 and 1899 that cathode rays are light charged particles, someone might have asked himself the question: 'If the same basic principle governs mechanics and optics, what is the mechanical analogue of a lens?' But apparently no-one did.

The thirty years that separated Thomson's first paper on the electron (a word that he himself eschewed) and Busch's lens equation saw repeated attempts to improve the concentration of electrons in the spot of the cathoderay tube, which Braun had invented in 1897, before the nature of the rays had been elucidated! We note in passing that Dennis Gabor (who was later to invent electron holography) worked on this CRT problem and in fact constructed a real electron lens, a coil enclosed in an iron yoke, to concentrate the beam but, as he ruefully admitted later, he did not realise at the time how his 'lens' worked.

In the research unit of Professor Matthias in the Berlin Technical University, Ernst Ruska (1906–1979) was given the task of verifying Busch's lens formula experimentally by his research supervisor, Max Knoll (1897–1969). In 1931, he formed a magnified image by combining two primitive lenses and the electron microscope came into being. That story has been told in detail by Ruska himself (1979, 1980) and we therefore move on to a result obtained in 1936 by Otto Scherzer (1909–1982), which has had an immense influence on electron optical research ever since.

The resolution of the electron microscope is limited by the geometric and chromatic aberrations of the objective lens of the instrument (once all parasitic aberrations caused by mechanical imperfections have been eliminated).

Only the aberrations that do not vanish close to the optic axis are of importance in a magnifying system because only the region of the specimen close to the axis is observed. The relevant aberrations are the *spherical* and *chromatic* aberrations. The spherical aberration blurs the image since rays that travel far from the optic axis are focused more strongly than those close to the axis; the chromatic aberration again blurs the image since electrons with slightly different wavelengths are focused more or less strongly.

This latter aberration is so large that highly monochromatic beams must be used (today, the energy spread of the electron beam in a microscope does not exceed an eV or a few eV for a beam energy of 200–300 keV).

These aberrations are characterized by two coefficients, the spherical aberration coefficient *Cs* and the chromatic aberration coefficient, *Cc*. The unwelcome result established by Scherzer (during the long night hours when he was on Home Guard duty, next to a telephone that never rang) is expressed by the following formulae:

$$C_s = \frac{1}{16}\int_{z_0}^{z_i} \{b^4h^4 + 2(hb' + h'b)^2h^2 + 2b^2h^2h'^2\}dz$$

$$C_c = \frac{1}{4}\int_{z_0}^{z_i} b^2h^2dz$$

in which *b(z)* is a multiple of *B(z)*, the magnetic induction on the optic axis, and *h(z)* is a solution of the paraxial equation of motion. It is immediately obvious that the integrands are positive definite (or strictly, non-negative definite) and hence that these aberrations can never be eliminated by clever lens design. Moreover, their values for real lenses are so huge that electron lenses must be operated at very small numerical apertures.

Loopholes in Scherzer's derivation were immediately sought, notably by the great electron optician Walter Glaser (1906–1960), who observed that the *Cs* integral can be transformed by partial integration into the form

$$CS = \frac{1}{48}\int_{z_0}^{z_i} (5b^2 - bb'' + 4b^4)h^4dz = \int_{z_0}^{z_i} f(B)b^4dz$$

The function *f(B)* contains only the field distribution *B(z)* and its derivatives. Glaser argued that the field obtained by setting the integrand equal to zero and solving the resulting differential equation for *B(z)* would yield a field distribution free of spherical aberration.

Moreover, he did indeed succeed in finding a solution, subsequently exploited for beta-ray spectrometers but useless for microscopy since it does not admit the existence of a pair of real conjugates. In passing, we might wonder whether a referee today would recommend publication of Glaser's paper, as Scherzer's formula (above) shows that such an attempt could never succeed for all the terms in the integrand must vanish if the integral is to be equal to zero, but Glaser was undeterred by such negative thinking. Many years later, Werner Tretner, a pupil of Scherzer's, established the lower limits on *Cs* and *Cc* in practical lenses. Just after the war, a series of Réunions d'Etudes et de Mises au Point was presided over by de Broglie.

The theme was 'L'Optique Electronique' and the lectures give a vivid snapshot of the state of electron optics in France in 1945. They open with an account of 'Mécanique ondulatoire en optique électronique' by de Broglie himself (22 May 1945), in which we are reminded that "L'optique électronique aurait pu avoir en France son développement initial. DQs 1927–28, je signalais à l'un de mes premiers élQves l'intérêt qu'il y aurait à développer l'optique géométrique des électrons.

Malheureusement, il n'a pas poursuivi son travail dans cette direction et moi-même, absorbé par des recherches plus générales sur la mécanique ondulatoire, je n'ai pas approfondi ces questions." Contributions by C. Magnan on the electrostatic microscope of the CollQge de France, destined to be converted into an ion or proton microscope and by André Lallemand on his 'electron telescope' bring us to a noteworthy review by Emmanuel Fauré–Frémiet (29 May 1945) of the 'Applications du microscope électronique à l'étude des problQmes de la biologie cellulaire et bactérienne'. The bibliography is astonishingly complete, so far as work in the USA and Germany is concerned and it is worth noting that Fauré–Frémiet had apparently seen the American facsimile reproduction of von Ardenne's book of 1940 and the collection *Das Übermikroskop als Forschungsmittel*.

Paul Chanson and André Ertaud then describe the properties of electrostatic lenses and their measurement with the aid of an electrolytic tank. The following two contributions, by Pierre Grivet (1911–1992) and Gaston Dupouy (1900–1986), are landmarks in the history of the electron microscope in France. The authors describe the construction of an electrostatic and a magnetic electron microscope during the wartime years and the results obtained with them.

Grivet's chapter contains a good bibliography, revealing that he too had had access to some of the German and American publications of the previous years and in particular, to the special electron microscope issue of the *Jahrbuch der AEG Forschung* (1940) and to the book describing the achievements of AEI in this area published three years later. Dupouy's lecture, a splendid example of his style, is followed by a 'Discussion', opened by de Broglie thus: "Je remercie Mr Dupouy de sa trQs belle conférence. Il nous a montré les beautés, les finesses de l'optique électronique magnétique, ainsi que les résultats intéressants qu'il a obtenus. Si la France arrive à ecrire quel ques pages nouvelles dans les livres de l'optique électronique, ce sera en partie à Mr Dupouy qu'elle le devra". The collection concludes with an essay by L. [*sic*] Léauté on 'Les applications du microscope électronique à la métallographie'.

For the examination of metals, Léauté (who was in fact André Léauté, 1882–1966, Professeur at the Ecole Polytechnique) mentions first the emission microscope and then reproduces an image of a replica obtained by Mahl, inventor of this technique. The third method mentioned is by far the most unexpected: Léauté, following the example of von Ardenne, developed a scanning electron microscope, now completely forgotten. The Notice sur la Vie et l'OEuvre de André Léauté (1882–1966) by Jacques Pomey (1969) tells us a little more about this instrument and we therefore quote it at length: "Pour un esprit comme le sien, l'enseignement ne peut pas se concevoir sans un laboratoire, qu'il crée à l'Ecole polytechnique, et où il initie ses élQves à la méthode expérimentale.

Ceux-ci étudient et réalisent de nombreux appareils originaux d'électronique, conçus par leur maître. Parmi ces chercheurs il convient de citer MM. Boisot, R. Brachet, Cl. Brachet, F. Davoine, Fouretier, G. Mayer, Taillade et surtout L. Cartan, qui fut victime de la cruauté nazie, ce dont A. Léauté n'a pas pu se consoler! A ceux-là il convient aussi d'ajouter sa fille aînée, Madame J. Guigan, qui avait reçu une formation scientifique complQte et qui a travaillé pendant dix ans auprQs de son pQre, dans ce laboratoire de l'Ecole Polytechnique. Le principe du microscope électronique à balayage et à émission électronique secondaire est dû à Von Ardenne, mais celui-ci ne parvient pas à le réaliser et abandonne [in fact, von Ardenne did build such an instrument, which was destroyed in the wartime bombing of Berlin].

A. Léauté reprend cette idée, car il en mesure tout l'intérêt pour la métallurgie, à une époque où les métallographes ne disposaient ni du procédé d'amincissement des échantillons par polissage électrolytique de Jacquet, ni de faisceaux électroniques de trQs haute énergie; à cette époque la technique des répliques débutait à peine. Ce microscope à émission secondaire devait donc jouer par rapport au microscope électronique classique par transparence le même rôle que le microscope de H. Le Chatelier par réflexion vis-à-vis du microscope classique histologique utilisant les coupes minces transparentes. Il présente aussi pour autre intérêt de pouvoir en principe avoir un pouvoir résoluteur et une profondeur de champ comparables à ceux du microscope électronique classique, ce qui permettrait de l'utiliser directement en micro fractographie. Dans son laboratoire de l'Ecole Polytechnique, A. Léauté réalise d'abord un appareil préliminaire à une seule lentille, puis en 1946 il réalise le premier microscope de ce type avec capteur à multiplicateur d'électrons et reproducteur d'image à tube cathodique qu'il construit dans son laboratoire avec les conseils de R. Barthélémy. Il obtient ainsi les premQres micrographies.

Il est arrêté faute de crédits, mais son élQve F. Davoine, qui dans son laboratoire a poursuivi des études sur l'émission électronique secondaire des métaux et sur l'influence des contraintes et de l'écrouissage, sur celle-ci, poursuit ses recherches à la Faculté des Sciences de Lyon et grâce à l'appui du CNRS et à l'expérience acquise auprQs de A. Léauté, construit un nouveau microscope basé sur le même principe. Dans ce domaine, A. Léauté a joué le rôle de précurseur, car maintenant les microsondes électroniques Castaing–ONERA sont munies d'un microscope à balayage et à émission électronique secondaire et les deux grands principes de A. Léauté sont repris dans d'autres appareils: d'une part le balayage dans un microscope anglais à optique classique et à comptage électronique, et d'autre part l'émission secondaire dans un microscope électronique à photographie directe aux U.S.A."

No other publication in which this microscope is mentioned has been traced, apart from a passing reference to it in the discussion at the end of a paper by Charles Brachet (1946). The work of François Davoine at Lyon is, on the other hand, better known and referred to in Grivet's *Electron Optics*, for example, in the bibliography assembled by Oliver Wells (1972) and in the list of references included in Oatley *et al.* (1985). In his doctoral thesis, Davoine (1957) tells us "J'ai fait mes premiQres armes en microscopie par balayage, au laboratoire de l'Ecole Polytechnique où MM les Professeurs Léauté et Vignal m'ont accueilli avec beaucoup de bienveillance".

The chapter devoted to the scanning electron microscope built at Lyon opens with a brief account of earlier work on the SEM: Knoll's proposal of 1935, von Ardenne's and Zworykin's instruments and then a brief description of Léauté's microscope: "En France, au Laboratoire de Physique de l'Ecole Polytechnique, un microscope à balayage a été réalisé en 1944–46 sous la direction du Professeur Léauté. Dans cet instrument, les électrons secondaires étaient collectés par une simple plaque; les fluctuations dans le réseau d'entrée limitaient considérablement le pouvoir séparateur, bien que les vitesses de balayage aient été extrêmement réduites et l'enregistrement effectué sur l'écran rémanent d'un tube cathodique." No reference to any publication concerning Léauté's microscope is given.

A year later, the microscopes built at the CSF Laboratory and near to Toulouse were presented to a wider audience at an electron microscope conference held in Oxford in September 1946. Henri Bruck, a colleague of Grivet's, described the electrostatic instrument and Dupouy himself talked about the magnetic instrument. At the same meeting, the Dutch and UK efforts were presented by J.B. Le Poole (1917–1993), A.C. van Dorsten and W.J. Oosterkamp and by M.E.

Haine respectively, thereby encouraging cross-fertilization. Before leaving the immediate post-war years, we draw attention to three publications that had repercussions for many decades. The first is a book entitled *L'Intégrale de Fourier et ses Applications à l'Optique* (1946), published at his own expense by Pierre-Michel Duffieux (1891–1976). This extraordinary book, by a no less extraordinary man laid the foundations of what we now call Fourier Optics and of optical transfer theory. Duffieux (1970) has told us how the idea came to him: "Un matin où j'étais trQs libre, je reçus un ingénieur de l'Institut d'Optique qui représentait la maison Mader–Ott.

Il me montra des appareils de mathématiques et les catalogues de la maison Mader–Ott. J'ai pris tout de suite un trQs beau planimQtre d'Amsler.

Il me proposa un dispositif nouveau qui en faisait un analyseur harmonique, plus lent que le Corradi, mais beaucoup moins cher et en réalité plus rapide quand on ne demandait que les premiers harmoniques. Tandis qu'il montait les deux appareils, je feuilletais la notice qu'il m'avait ouverte et j'eus tout de suite une révélation hallucinante... J'ai écrit en 1963 dans *L'Education Nationale:* "Il y a eu deux parts de ma vie; j'ai d'abord cherché ma voie, puis un jour, comme cela, brusquement, je l'ai trouvée.

J'ai eu quelques secondes pour choisir, j'ai choisi, et depuis cet instant-là, je travaille toujours dans la même direction. C'est à la lecture de ce catalogue que je faisais allusion". For many years, his work was known to only a tiny number of opticians but in 1959, when Born and Wolf's *Principles of Optics* appeared, justice was at last done to Duffieux: "We shall now describe some general methods based on the techniques of Fourier transforms, relating to imaging of extended objects.

These methods were developed chiefly by Duffieux...", wrote Emil Wolf, and Duffieux's little-known book is cited. Many years later, the importance of the book was recognised in France and a second edition was published (1970); an English translation appeared in 1983.

The Société Française des Microscopies marked the fiftieth anniversary of the first publication with a Duffieux Symposium at its annual meeting, which was held in Rennes in 1996. The second publication is by Otto Scherzer, who had established the formulae discussed above in 1936. Here, he re-examined the proof and showed how, by relaxing one or other of the necessary conditions, aberration correction could be achieved. In particular, he suggested correctors based on departure from rotational symmetry, space charge, the use of high frequency excitation and discontinuity in the potential or field on the optic axis. Each of these suggestions has been investigated and we shall see that one at least of Scherzer's proposals has been successful, though it took just half a century to make it work! Lastly, we recall that it was in 1948 that Dennis Gabor (1900–1979) published his first paper on holography, which he intended as a method of circumventing the adverse effect of spherical aberration in electron microscopy.

He had already been thinking about this problem for some time but this completely new idea, of recording an image blurred by spherical aberration and restoring a sharp image in a second step, came to him as a bolt out of the blue: he was sitting with his wife on a garden seat beside tennis courts in Rugby, where he worked, when holography sprang into being in his mind. Theory and instrumentation progressed rapidly in the next two decades. In the 1950s and early 1960s, experimental evidence accumulated showing that the principle of Scherzer's corrector based on cylindrical lenses (subsequently, quadrupole lenses) and octopoles was sound.

However, this was not sufficient for the production of a working corrector that could be incorporated into an electron microscope. The first family of correctors employed four quadrupoles and three octopoles and the task of aligning and adjusting such a system proved too difficult in pre-computer days, for these correctors are inherently unstable: the objective lens of the electron microscope to be corrected is already a highly perfected optical component and the object of the exercise is to reduce the aberration coefficient of a lens on which considerable design skill has been expended.

But the principle of the corrector is to create asymmetry by means of quadrupoles, which have much larger and many more aberrations than the round lens and then correct the new (large) aberrations and the (much smaller) inherent aberration by means of octopoles. It is therefore disappointing but not surprising that none of the very ambitious attempts to build correctors during the 1970s and 1980s was successful. We shall return to aberration correction below but let us first return to the 1960s, a decade in which several major microscope projects came to fruition. In 1960, the first very high voltage electron microscope, built under the direction of Gaston Dupouy in the Laboratoire d'Optique Electronique in Toulouse, furnished its first images and it continued to operate with no serious difficulties for many decades. In about 1989, a new director of the laboratory chose to sell it for scrap. In a historical article, Dupouy (1985) wrote: "we planned to build a microscope working at one million volts, and more if it was possible.

This was a bold enterprise and forecasts on all sides were absolutely pessimistic. Many difficulties were advanced; lens design, stabilization, lack of contrast. Naturally we had, for our part, thought a lot about all these problems before we tried this bold experiment. Many people believed our plans would lead to failure; they turned out very well indeed." The main justification for striving for such high voltages came from the life sciences. Dupouy's ambition was to image living specimens, enclosed in an environmental chamber in which the conditions necessary for survival were maintained.

Only very high-energy electrons could penetrate the windows of the chamber, the atmosphere inside it and the specimen itself. Although the specimens examined did not in practice survive – the x-ray dose received was undoubtedly lethal – the ability of high-energy electrons to penetrate relatively thick layers was quickly appreciated in other areas of research and the success of the Toulouse instrument led several companies in the UK (AEI), the USA (RCA), France (GESPA) and Japan (JEOL, Hitachi) to put such instruments on the market.

In Cambridge, a series of research students under the direction of Charles Oatley had been perfecting prototype scanning electron microscopes and, after considerable hesitation, the Cambridge Instrument Company launched the first commercial scanning electron microscope, the Stereoscan, in 1965. Here, a fine electron probe explores a thick specimen, the incident electrons generating one or more signals characteristic of the topography of the surface or the chemical nature of the sample: secondary electrons, back-scattered electrons, x-rays, fluorescence.

The Stereoscan was the culmination of nearly twenty years of research and development in the Cambridge University Engineering Department and was at first regarded as a very risky venture, for it was far from sure that this new type of microscope would catch on. Very soon, however, demand outstripped the initial production capacity.

That same year, 1965, an English accelerator physicist working in the USA, whose name was unknown to the electron microscope community, caused a furore at a small Institute of Physics meeting on 'Non-conventional Electron Microscopy' held in Cambridge at which many of the SEM pundits were present. Here, Albert Crewe described his new scanning transmission electron microscope, an instrument in which, as in the SEM, a fine probe explores the specimen but now, the specimen is as thin as in a TEM and it is the electrons of the incident beam that emerge from the far side of the sample that are used to form the image. At the time, this was thought to be quite impracticable, as the time needed to record an image with the thermionic guns in regular use would be prohibitive.

Crewe's major innovations were the utilisation of a field-emission source, much brighter than the thermionic gun, and above all, attainment of the very high vacuum necessary for stable field emission. Three years later, the microscope was working well and three companies launched commercial models: Siemens, who put a fully engineered STEM (the ST104) on the market shortly before the company withdrew from the electron microscope market; AEI, who did not persist in their project; and VG Microscopes, who equipped many laboratories with STEMs before discontinuing production. The 1960s also witnessed other major developments in electron optics. The first holographic reconstructions were made by Akira Tonomura in Japan and H. Wahl in the laboratory of G. Möllenstedt, where the electron biprism had been introduced. Later developments in Tübingen and more recently in Dresden are described by Lichte.

Karl-Joseph Hanszen translated Fourier optics into the language of electron optics and introduced the notions of phase and amplitude transfer functions, which give a clear understanding of the meaning of resolution and are at the heart of much of today's digital image processing. In Cambridge, the first three-dimensional reconstructions were being made by de Rosier and Klug (1968) while in the early 1970s, Ralph Gerchberg and Owen Saxton obtained the first practical solution of the 'phase problem', the problem that arises from the fact that electron microscope samples are *phase* objects (no electrons are halted within the specimen) but the fluorescent screen or photographic medium display only *amplitude* information.

We now take a large step forward to the final stage in the history of aberration correction but first, we need to mention a new type of non-rotationally symmetric corrector that was discovered in the late 1970s by Beck and Crewe in Chicago. In order to use non-rotationally symmetric elements to correct the spherical aberration of a round lens, it is clearly essential that the corrector should suffer from the same aberration as the round lens but with opposite sign.

I had noticed many years earlier that sextupoles possess this property but since these have second-order effects as well as the third-order spherical aberration, we had dismissed them as potential correctors. However, by suitably combining two sextupoles with different orientations, a device that has the desired properties can be designed and the resulting corrector is much easier to align and excite than the quadrupole–octopole version. This type of corrector was studied and perfected and in 1995, Joachim Zach and Max Haider showed that the size of the probe in a low-voltage SEM could be reduced with the aid of such a device.

This was essentially a proof-of-principle experiment, since the lens being corrected was comparatively poor one, but it showed that it would be worth trying to use such a device to correct the spherical aberration of a high-resolution electron microscope objective. This was attempted and in 1998, the first results form a corrected TEM were published. For some years now, the corrected microscope at the Forschungszentrum in Jülich has been producing stunning micrographs. Delivery of commercial microscopes equipped with such correctors begins in 2004 (FEI–Philips, JEOL, ...).

Meanwhile, Ondrej Krivanek, working in the Cavendish Laboratory in Cambridge, had returned to the quadrupole–octopole configuration for correction not of a TEM objective, which is required to form an extended image, but of the probe-forming lens of a STEM. For this, a very flexible system was built, with all the excitations and alignment under computer control and, for the first time, correction of a high-quality lens was achieved. Krivanek announced his achievement at a meeting of the Electron Microscopy and Analysis Group (EMAG) of the Institute of Physics at their meeting in Cambridge to coincide with the 'discovery' of the electron in 1897. Many STEMs have now been fitted with an improved version of this corrector and the British SuperSTEM project will make two such instruments available to the scientific community.

We have seen that the widespread availability of fast computers with ample memory was an essential element in the successful development of these correctors. Computing power has also made it possible to calculate the properties of most of the electron optical elements with great accuracy in a very short time on desktop computers. Evidence for this can be found in the proceedings of the Charged-particle Optics Congresses and especially in the work of Eric Munro and Bohumila Lencová. The introduction of the sextupole corrector generated new work for the theoreticians of electron optics and new results of the highest interest have indeed emerged as a result. The importance of system symmetries to cancel intrinsic aberrations has become clear, and this is exploited in the omegashaped energy analyser, in which the device not only has symmetry about the mid-plane but also exhibits anti-symmetry about intermediate planes.

For a systematic account of this. Before leaving correctors, we mention that the defeat of spherical aberration has acted as a catalyst for work on overcoming the deleterious effect of chromatic aberration, notably by means of monochromators, and even more ambitiously, on correcting all the geometrical aberrations of rotationally symmetric systems. We have mentioned image processing in passing in this survey, which is confined to a small part of the history of the subject, and to conclude, we should like to draw attention to a mathematical tool that has been added to the image processing paraphernalia and to an unsolved but fascinating problem. A glance at any image processing textbook reveals immediately that such works are more like cookery books than scientific treatises and that the vocabulary of the subject is quite different in the widely separated areas of application (microscopy, astronomy, medicine, geology, forensic science,...).

In an attempt to harmonize all this work and to put it on a sound mathematical footing, an image algebra has been created, in terms of which the various image processing algorithms can be written compactly. The elements of the algebra are images (not individual pixel values) and the word 'image' is interpreted very broadly. An image is a *set* of values (pixel values in the simplest case) and addresses. It may be one-dimensional (an energy-loss spectrum, for example) or two-dimensional (an everyday image) or of higher dimensions. Several values may be associated with a given address (SEM signals, colour images).

In one extremely important case, each 'pixel value' may itself be an image; this is useful in the mathematics to handle such operations as convolutions and other linear transforms and is so ubiquitous that such image-valued images are given a special name: *templates*. They are, however, much more than mere mathematical constructs, for the scanning transmission electron microscope generates such images. Since the object and image planes are not conjugates in such instruments, a far-field diffraction pattern is formed in the detector plane of the STEM from each point of the specimen. Moreover, the use of templates provides a remarkable and unexpected unification of the families of linear convolutional filters and the nonlinear operations of mathematical morphology.

These have a very similar appearance in image algebra, the sum and product operators that appear in convolutions being replaced by max (sup) and sum, respectively. The structuring elements of mathematical morphology are represented by templates, just like the filter functions of the convolutional procedures. This algebra was introduced, as we have seen, to simplify and harmonize the many image processing algorithms. But why should we stop there? It would be very satisfying if we could express the whole chain – image formation plus image processing – in terms of this algebra.

The idea is all the more attractive in that a microscope contains stops that truncate the wavefunctions characterizing the light or electron beam and these are therefore functions with finite support; they can hence be represented by a (finite) set of sample values. It ought to be possible to describe the propagation of the wavefunction through the instrument by means of a *difference* equation (instead of a *differential* equation) and this difference equation should show us that the succession of planes of the microscope (source, specimen, diffraction pattern, image) are related by a sequence of finite Fourier transforms. Work on this is in progress.

NANOSCALE RESOLUTION IN LIGHT MICROSCOPY

In the early 1870s, the German physicist Ernst Karl Abbé formulated a rigorous criterion for being able to resolve two objects in a light microscope.

According to his equation, the best resolution achievable with visible light is about 200 nanometers. This theoretical resolution limit of conventional optical imaging methodology was the primary factor motivating the development of recent higher-resolution scanning probe techniques. The interaction of light with an object results in the generation of what is called 'near-field' and 'far-field' light components.

The far-field light propagates through space in an unconfined manner and is the visible light utilized in conventional light microscopy. The near-field (or evanescent) light consists of a nonpropagating field that exists near the surface of an object at distances less than a single wavelength of light. So called near-field microscopy beats light's diffraction limit by moving the source very close to the subject to be imaged. When the first theoretical work on a new technique called "scanning near-field optical microscopy" (SNOM or NSOM) appeared in the 1980's, Abbé's classical diffraction limit was overcome, and resolution even down to single molecule level became feasible. However, light microscopy is still the only way to observe the interior of whole, or even living, cells.

The use of fluorescent dyes makes it possible to selectively obtain images of individual cell components, for example, proteins. Today, the wavelength dogma has been overcome with the development of the stimulated emission depletion (STED) microscope. Now, the German team that developed STED is reporting layer-by-layer light microscopic nanoscale images of cells and without having to prepare thin sections with a technique called *optical 3D far-field microscopy.*

They use a chemical marker for fluorescence nanoscopy that relies on single-molecule photoswitching. "Well thought-out modifications of the fluorescence process have opened a back-door for circumventing Abbé's paradigm" explains Dr. Stefan Hell, head of the nanobiophotonics department at the Max Planck Institute for Biophysical Chemistry in Göttingen, Germany. "In a scanning microscope, improving the resolution is equivalent to decreasing the spatial extent of the fluorescent spot that is scanned through the sample. Hence, we are scrutinizing physical phenomena that bear the potential of cutting down the size of the fluorescence spot. The focal spot is termed 'the point-spread-function'. But non-scanning microscopes have Point-Spread-Functions, too. In fact virtually all of the physical ingredients that we use in a scanning microscope can be employed in a non-scanning CCD camera based system as well, with some modifications and advantages and disadvantages." Hell explains that the states of the fluorescent marker was not just used for generating the signal, but also for breaking the diffraction barrier.

"In fact, all the methods that have successfully outperformed diffraction have so far relied on selected pairs of molecular states – specifically, a 'bright' one to generate the signal and a 'dark' one to ensure that the measured signal stems from a subdiffraction-sized region." "An alternative way of using molecular photoswitching to break the diffraction barrier is to stochastically switch on, read out the fluorescence, and switch off isolated marker molecules such that simultaneously emitting ('on') markers are further apart than the minimal distance resolved by the microscope" says Hell. "In this case, the spatial confinement of the fluorescence is down to the size of a single molecule by definition." Hell's group has now been demonstrating the power of this concept. They report the use of molecules that are not only transferred but can be 'switched' from fluorescent to non-fluorescent and back.

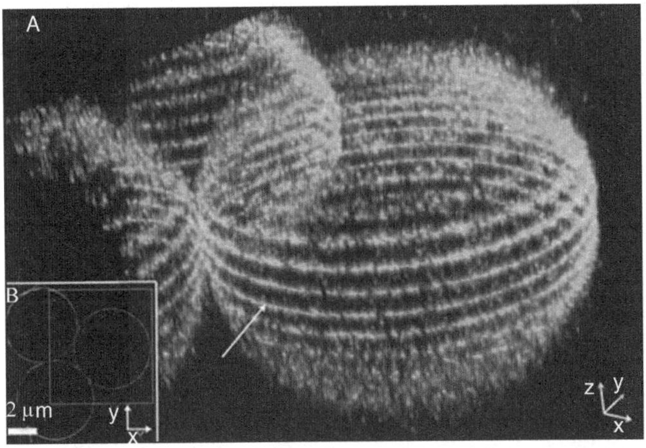

Fig. (A, B) Imaging of 5-mm Silica Beads Surface Stained with 5-NHSS. A) 3D Reconstruction from 17 Slices in the z direction; B) Equatorial Slice (arrow in (A)). For Clarity, a Smaller Area was Selected for the 3D Reconstruction (dotted-line box in (B)). (Reprinted with Permission from Wiley-VCH Verlag) In Contrast to STED, only Separate, Isolated Marker Molecules are Randomly Switched on at the Same Time. Their Fluorescence is Registered, and then they get Switched off again Automatically. In this way, the Simultaneously Fluorescing (Switched on) Markers are Farther Apart from each other than the Minimum Distance that the Microscope can Resolve. This is only Possible using Switchable Molecules that Emit many Photons, one after the other, when Switched on.

If these photons are captured with a camera, the centres of the individual fluorescing dots can be distinguished. After the exposure, the molecule becomes dark again (switches off), allowing further, neighbouring molecules to be photographed. This process is repeated many times, until many dots become a picture. The full distribution can be reconstructed – at a resolution not limited by the wavelength of light.

The researchers have now found a class of substances that fulfill all the requirements of this technique: rhodamine amides. At the core of these molecules lies a system of five rings. In this form, the compound is colourless and does not fluoresce. Irradiation with light induces an isomerization in which one of the rings is opened. This form of the molecule is red and can be excited several times. Most importantly: rhodamine amides can be switched on by either a UV photon or two photons in the red part of the spectrum. This two-photon excitation can be focused onto a thin plane, which allows biological samples to be photographed layer by layer. The individual images can then be reconstructed into a single multilayer image. The resolution reached in the focal plane is far beyond the diffraction barrier (10–30 nm). Hell has also published a comprehensive review of the state of far-field optical nanoscopy.

In it he concludes: "Along with improved fluorescence schemes, nonfluorescent schemes should enable us to take another step in the not-too-distant future: nanoscale 3D imaging at high speed. In any case, the works reviewed herein already broke the barrier of perception of what a lens-based light microscope is able to accomplish. With human perseverance focused on this matter, 3D imaging of live cells with electron microscopy resolution should be possible."

NEAR-FIELD MICROSCOPY OF SHG

Near-field SHG microscope can be exploited in both apertured and apertureless realizations. We will consider these two techniques in turn.

Aperture-Based SHG SNOM

There are three possible experimental configurations for near-field secondharmonic measurements with an apertured probe: (1) the secondharmonic (SH) collection mode when the excitation is achieved by direct illumination of a sample and SH light is collected with a fibre tip, (2) the SH excitation mode when a surface is illuminated with fundamental light passing through a fibre tip and SH light is collected in the far field, and (3) the double-pass mode when SH excitation is achieved through a fibre tip and second-harmonic light is collected in the near-field with the same tip. In addition to subwavelength resolution, SHG SNOM provides additional parameters for local optical characterization such as the polarization of fundamental and SH light. Thus, the described techniques allow one to excite SHG locally at a specifically chosen surface area (excitation mode) or to detect local SHG before the SH contributions from individual surface features are averaged in the far-field zone (collection mode).

Collection mode SHG SNOM is most often used because SH generation requires high-intensity excitation light, which is achieved by a focused laser beam. In most experiments, femtosecond Ti:sapphire lasers are used, which provide high excitation power at a low pulse energy, thus effectively generating SHG but introducing low heating and sample damage. The SH signal collected by a fibre tip is usually filtered by an interference filter and detected by a gated photon counter. Shear-force distance regulation is conventionally used to keep a tip in a near-field proximity and at a constant distance from the surface, thus providing simultaneous topographic imaging.

In most SHG SNOM experiments, uncoated fibre tips are used. The biggest disadvantage of metal-coated fibre tips in nonlinear optical measurements is a strong perturbation introduced by the tip in the local electromagnetic field distribution and SH generation from the metal coating. Uncoated fibre tips introduce a much smaller perturbation since the dielectric constant of a fibre is much less than that of a metal. However, even for fibre (glass) tips, some degree of the electromagnetic field enhancement is associated with a tip apex.

This is crucial in nearfield second-harmonic measurements for which the signal depends on the fourth power of the local excitation field. Due to the field enhancement, the surface area just below the tip dominates the detected SH signal. As a result, theoretically, the resolution of the SHG SNOM can be estimated to be as small as nm for 50 nm radius of a fibre tip apex. In fact, more severe limitations on the resolution of a SHG SNOM are imposed by the small number of SH photons collected by the tip and the related shot noise and thermal drifts due to long signal acquisition times.

The high repetition rate of pulsed Ti:sapphire lasers allows the image acquisition time to be kept in a reasonable range even at a low number of detected SH photons per pulse. The characteristic SH photon count rate in SNOM measurements is, on average, of the order of one SH photon count per 100–300 laser pulses. However, a long signal averaging time, which would help to register more photons, is not desirable due to possible drifts in the experimental apparatus. For that reason, the resolution of the SHG SNOM is in many cases limited by a low detected SH signal rather than the geometry of a probe tip and is determined by a trade-off in the choice of SH signal averaging time.

In the SHG measurements with uncoated fibre tips, the experimentally achievable spatial resolution is usually about 80–100 nm; however, it can be much higher in some specific cases. The resolution and perturbation introduced by a fibre tip depends on the sharpness and size of the taper via the tip-induced field enhancement. It is possible to reduce the influence of the tip by choosing an appropriate tip apex shape. With less sharp tips, the influence of the tip-induced effects will be smaller and SHG will be directly related to the surface.

Although the theoretical resolution in this case will be smaller than that for sharper tips due to a smaller electromagnetic field enhancement, the practical resolution will remain about the same (~100 nm) due to a better overall collection efficiency of such tips. Thus, the choice of the tip depends on the task at hand.

For example, for studies of SHG from metal surfaces, perturbation introduced by a tip should be as small as possible. On the other hand, for local crystal analysis with SHG, the field enhancement at a probe tip that depends on the excitation light polarization can be advantageous. In the case of polarization measurements with uncoated fibre tips, SNOM resolution down to 40 nm was reported, which is related to the strong field enhancement in the vicinity of high refractive index materials.

Apertureless SHG SNOM

Another type of second-harmonic SNOM makes total use of the effect of the field enhancement at a metallic tip to achieve a local enhancement of nonlinear optical processes in the vicinity of the tip, similar to apertureless scanning near-field optical microscopy (ASNOM) schemes based on two-photon excited fluorescence and Raman scattering. It provides much stronger light confinement and better spatial resolution than an aperture-based SHG SNOM. This technique, however, poses the problem of image interpretation.

In SHG ASNOM, the excitation light illuminates the tip-surface junction and the scattered SH light is detected in the far field. The signal related to nonlinear optical processes is determined by the superlinear dependencies on the intensity of the driving field (the fourth power of the local fundamental field in the case of SHG). This makes such techniques more sensitive to the field distribution over a surface than ASNOM based on linear optical processes. Not only the magnitude of the signal but also the spatial resolution of ASNOM is related to the field enhancement since it determines the effective surface area most strongly contributing to the measured signal.

As discussed by Kawata, Xu, and Denk, in contrast to the scattering or conventional fluorescence, the light associated with nonlinear processes can be confined (falling faster than with a distance from a tip apex) even without localized surface plasmon excitation. An approach based on SHG is advantageous compared to other ASNOM schemes because there are almost no restrictions on tip and sample materials since all materials exhibit surface SHG. Only one tip or sample needs to posses strong SHG to provide second-harmonic ASNOM imaging.

In different experimental realizations, the technique of apertureless second-harmonic SNOM allows one to probe the local second-harmonic field generated at the surface if the nonlinear susceptibility of the sample is higher than that of the probe tip (a nonlinear background-free analogue of ASNOM based on local scattering). In the opposite case, one can achieve a strongly confined second-harmonic light source at the tip apex with properties dependent on the topographic and dielectric properties of the surface under investigation.

RESOLUTION LIMIT IN LIGHT MICROSCOPY

What is Resolution

The term 'resolution' has to be discriminated from 'sensitivity', 'sampling' and 'precision'. Subsequently we will try to clarify the meaning of these terms.

One can have a very sensitive light microscopic system which makes it possible to see single virus particles or even single fluorescent molecules. This only means we would have single molecule sensitivity, but the size of such a molecule in the image could, for example, still correspond to 0.5μm in the sample coordinate system, which would be a relatively poor optical resolution.

Another common confusion is the difference between resolution and sampling, which relates to magnification. It is easy to magnify an image of the sample by optical means to any desired degree, however, the process of magnification does not increase the optical resolution; at best it preserves it. When an image of the sample is detected by our eyes or for example a CCD camera, it gets 'sampled' into a discrete set of measured intensity values, one for each detector element (such as a photosensitive cell in the brain).

These sampling points correspond to nominal positions in the coordinate system of the object under investigation, without a limit to their density. The magnification has to be adjusted such that the finest level of detail present in the image is still measured (sampled) by at least two such detector elements, but any denser sampling will not yield new information about the sample. Magnification significantly beyond this limit is sometimes called 'empty magnification'. When there is a noise attributable to the readout process of the detector ('read noise'), such empty magnification should be avoided as it deteriorates the image quality.

In many cases there can be very specific questions in biology which are to be answered by microscopic imaging. uch questions could be: 'What is the spatial distance between two specific genetic loci in the nucleus or between two molecules on the cell surface?' To answer these questions, there is no need for a resolution in the order of this distance, but the error of localization (the reciprocal of the localization precision) needs to be below this expected distance.

That means that the distance between the estimated and the true centre position of the object has to be below the distance between the two objects. Localization precision can be far higher than the optical resolution (*e.g.* the localization error of single molecules can be smaller than 10nm on a system of 200nm optical resolution), and we can use this even for relative position determination between multiple loci, as long as we are able to discriminate between the target genes (or other small structures) in our image. Such discrimination can be achieved by using multiple colours, differences in fluorescence lifetime, photo-bleaching, the individuality of statistical blinking events and photoactivation.

The higher the resolution the better the localization precision, but how much better than the optical resolution depends strongly (with a square-root dependence) on the number of photons collected from each target. An impressive application of the localization idea was to observe the molecular steps of Myosin V. Another interesting application was to follow the kinetics of the receptor-mediated entry of NTF2 and transport of NTF2 and transporting 1, with and without transport substrate, into the nucleus. The method of speckle microscopy uses precise localization for the tracking of molecular clusters to reveal their temporal behaviour. Nanosizing can be used to address the question of the size of biological particles of known shape.

Even when objects are not point-like structures, the position of features like straight edges (*e.g.* tubulin fibres) or planes (*e.g.* plasma membrane) can often be determined very precisely by microscopy methods. All of these methods are based on localization precision, which should not be confused with the term resolution as discussed subsequently. When we talk about high resolution of an optical instrument in this article, we mean the ability to see a structure at a high level of detail. There are various ways to define the term resolution more precisely which are discussed in the two separate boxed sections. Although, the definitions of resolution mentioned in the boxed section 'The PSF' may be useful in some cases, we would like to use a different limit called the Abbe limit within this article, as this limit has a very direct relationship to which light rays are captured by the objective lens of the microscope.

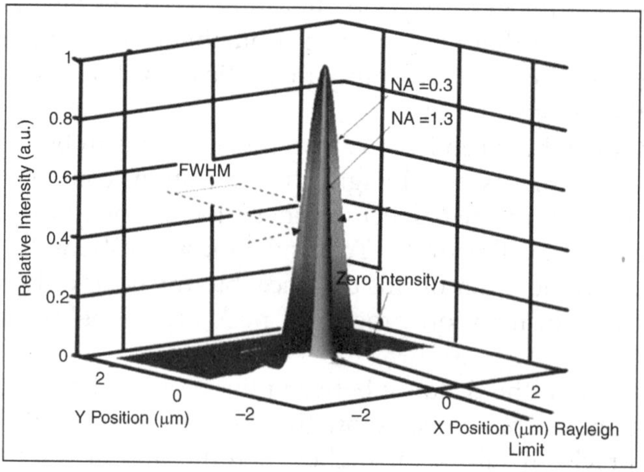

Fig. The Point Spread Function (PSF). Two PSFs are Shown, one for High Numerical Aperture (NA = 1.3 in grey) and one for NA = 0.3 (Slightly Transparent Shades). As can be Seen the Low NA PSF is Wide and has well-defined Positions of Zero Intensity, Leading to the Definition of the Rayleigh Limit.

It is interesting to note that the ability to precisely localize can be turned into a genuine 'resolution' when many closely spaced particles can be discriminated. From the precise localization of these particles a picture can be constructed by painting dots at each localized position. A method termed 'Pointillism'. A very successful way of discriminating closely spaced molecules is using a repetitive cycle of faint photo-activation followed by bleaching of the activated molecules.

In every cycle only very few molecules get activated, so the chance of PSF overlap is small, and the localization thus very precise. In fixed samples it was possible to reconstruct impressive images of cryo-prepared thin section from a COS-7 cell expressing CD63 (lysosomal transmembrane protein) tagged with Kaede and dEosFP-tagged cytochrome-C oxidase import sequence in mitochondria of COS-7 cells down to a resolution of about 20nm. Currently a problem is the rather long imaging time well above an hour for a single slice image, but further development should obviate this.

Detection Abbe Limit

A number of modern microscopy techniques such as confocal, 4Pi and structured illumination microscopy achieve their high resolution by a spatially non-uniform illumination. Non-uniform illumination implies that adjacent positions in the sample can receive different illumination intensity. Subsequently, we will describe in detail how the resolution enhancement is achieved for these methods.

In confocal microscopy the sample is illuminated with a focused beam, which is then raster scanned over it. The emitted fluorescent light is imaged onto a pinhole, whose diameter can be adjusted. Due to the combined effects of illumination and detection, this procedure does indeed enhance the resolution beyond that of a standard widefield fluorescence microscope. If the pinhole is imagined to be very small, the light distribution used for illumination is very similar (except for the usually small difference in wavelength caused by the Stokes shift) to the map of how sensitive the detection system (with a small pinhole) can detect from each position in the sample (called the detection PSF).

For light particles (photons) to make it to the detector a particular sample point needs to be illuminated (*e.g.* 1/10 probability) *and* light emitted from this point needs to be detected (*e.g.* 1/10 detection probability). This chance of illumination and detection (in the above example 1/10 * 1/10 = 1/100) in combination with a linear response of the sample leads to the resulting PSF of the overall system to be the product of the illumination and detection PSFs. Discussed in terms of periodic distributions inside the sample, this confocal system is now in principle able to pick up a 2-fold finer periodicity.

However, in practice this is almost never achieved as the fine details (high spatial frequency) are suppressed so much that they get lost in the noise. To make matters worse, a closed pinhole would not detect any light, which leads to pinhole diameters in the range of the diameter of an Airy disc (where the first theoretical minima occur in the image of a point) to be used in biological experiments.

In this case, even features slightly finer than the standard uniform widefield illumination Abbe limit cannot be resolved as they get lost in the noise. This reduces the in-plane performance of a confocal microscope back to that of a uniform illumination system, however the increased sectioning ability of confocal microscopes is important and lead to their success.

By altering the intensity and phase of the light which gets sent into or detected from the sample under different angles, a technique often referred to as 'apodization' or pupil plane filtering, the high spatial frequencies can be emphasized, getting the practical performance a bit closer to the theoretical limit of twice the widefield resolution with uniform illumination.

Nevertheless, to our knowledge, no such system has ever been successful enough to make it into the market. The I^5M and the 4Pi type C microscope take both the illumination aperture and the detection aperture to their extreme. As the name 4Pi suggests (a full sphere has a solid angle of 4π), the idea is to get as close as possible to illumination (and detection) from all sides of the sample.

For practical reasons the best approximation is achieved by combining two objective lenses opposed to another along the Z (optical axis) direction to act as a single lens.

To do so, the two optical paths to the sample (through the top and though the bottom objectives) need to be identical in length, both for the illumination and detection wavelengths (4Pi type C microscope). It can be seen that the maximal possible numerical aperture is obtained along the optical axis (Z) to which the two objectives are perpendicular (becoming identical to the refractive index), whereas little changes along the X and Y directions.

Thus for illumination the resolution limit is $\lambda_{ex}/(2n)$ and for detection it is $\lambda_{em}/(2n)$, with λ being the respective wavelengths of excitation and emission. This means the finest periodical fluorescent structure which can be detected in a 4Pi (type C with visible excitation) microscope is about $\lambda/(4n)$, which amount to approximately 100nm along the optical axis for glycerol embedding and a wavelength around 500nm.

However, considering the in-plane directions (X and Y) this realization of the 4Pi concept has not increased the numerical aperture (and thus the resolution) as compared to a standard confocal microscope. Noticeably the lateral resolution, albeit in theory improved by a factor of ~2, suffers also from the same problem as standard confocal microscopes, with practically no XY-resolution improvement in comparison to a uniform illumination microscope, detectable for reasonable pinhole sizes.

Nevertheless the enormous improvement along the axial direction is worthwhile. Features in only slightly different planes would normally blur together in one image plane, but are now separated nicely in a 4Pi microscope. This leads also to a much cleaner appearance of individual slices of a 4Pi data set. Similar arguments apply to the I^5M method, where incoherent illumination is used.

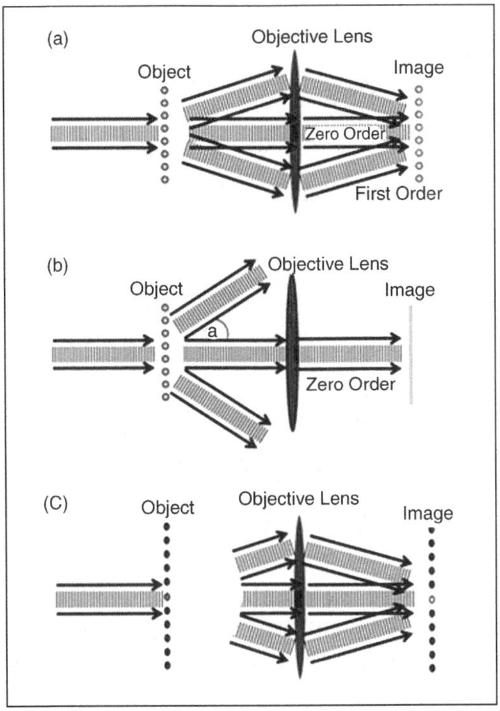

Figure: The Abbe Limit (a) Diffraction of a Parallel Light Beam from a Periodic Grating Structure (only the zero and first diffraction order are shown) (b) Diffracted Light from a Fine Periodical Structure already Fails to Fall within the Angular Acceptance Range of the Objective Lens. The Image can thus only be Formed from the Undiffracted Zero order Yielding a Flat Image of Equal Intensity. (c) For Fluorescence the Emitters have to be Considered Independently and Intensity will be added. Nevertheless, the Finest Resolution (Constituent of the PSF) is still Defined by the Biggest Angle of Emitted Light Rays that were Captured and Refocused.

The methods of structured illumination exploit the fact that the excitation and emission Abbe limit can be added. The highest possible spatial frequency is intentionally introduced during excitation via a fine sinusoidal illumination pattern in the sample volume.

This is achieved by projecting a fine grating into the sample under coherent illumination conditions, which allows the contrast of the excitation structure to be maximal. A high-resolution image can be reconstructed from a series of images under different positions (and orientations) of the excitation pattern.

In comparison to confocal microscopy, structured illumination captures high spatial frequencies of the sample much more efficiently, and thus allows for a true factor of two resolution improvement, albeit at the expense of requiring extensive computational reconstruction.

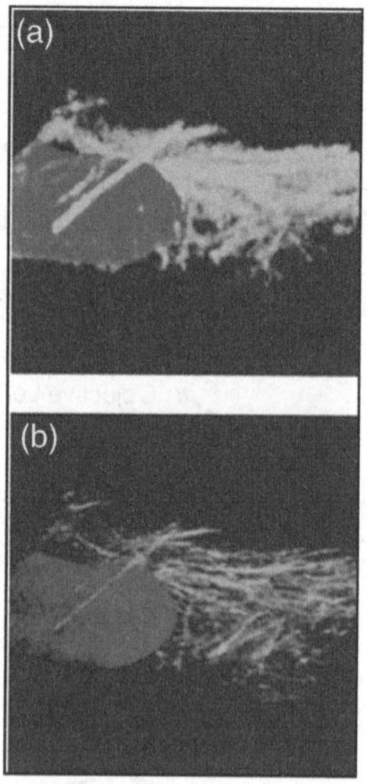

Figure : Confocal and 4Pi Image. Comparison of Volume Rendered Images of the Actin Cytoskeleton and the Nucleus of Muscle Support Tissue Endothelial Cells. The Nucleus was Stained with DAPI and the Actin Cytoskeleton with Alexa488-phalloidin. (a) 2-Photon Confocal View of the Sample (b) 2-photon 4Pi (type A) View (after 3-point Deconvolution) of the Sample. The Improved Resolution and Better Discrimination of Biological Structures is Visible. This Data Set Corresponds to a Width of 23.4 μm, a Height of 2.9 μm and a Depth of 30.0 μm. Sample Prepared by Elisabeth Ehler.

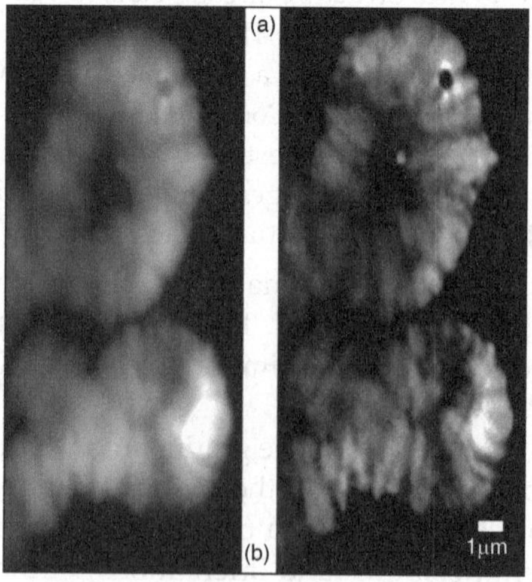

Figure : Linear Structured Illumination of *Drosophila* Polytene Chromosomes Stained with Sytox Green (a) Widefield Image Obtained by Summing all the Individual Images under Structured Illumination. (b) Reconstruction of the Sample from the Structured Illumination Data.

Methods Circumventing the Abbe Limit

The most advanced high-resolution microscopy methods use some form of nonlinearity to avoid the restrictions of the Abbe limit. In the following section, the importance of the non-linearity is stressed and exemplified with two-photon microscopy, nonlinear structured illumination and stimulated emission depletion microscopy.

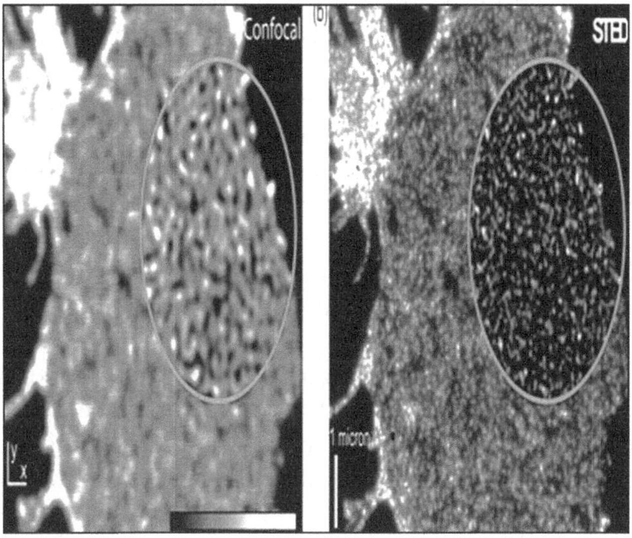

Figure : Stimulated Emission Depletion Image.

As mentioned in the boxed section 'The Abbe Limit', the linear response of the sample led us to define the combined illumination and detection Abbe limit as a fundamental limit in far field microscopy. This limit, however, does not apply if this linear relationship can be violated. Abbe's theory has to be modified if the emitted light can only be described by a nonlinear dependence of emission on the local excitation intensity. As it turns out, the highest exponent present in a polynomial expansion of the nonlinear dependence between excitation and emission defines the resolution limit. Every exponent in such an expansion of the nonlinear dependence of emission on local excitation can be thought of a number of additional multiplications, each adding another excitation Abbe limit to the dependence.

The simplest non-linear dependence is encountered in two-photon microscopy. Due to the probabilistic nature of the two-photon absorption process (both photons need to interact at the same time) the chance for a fluorescent molecule to be in the excited state is proportional to the square of the incident intensity at a specific position in the sample.

Thus also the emitted light intensity is proportional to this square of the excitation intensity. As a result two-photon microscopy indeed circumvents the Abbe limit by a factor of two in the excitation pathway. However, this improvement is bought at the expense of a roughly 2-fold longer excitation wavelength leading to essentially no additional gain, when compared with a standard single photon excitation in a confocal microscope. However, the significant advantage of the two-photon microscopy is the reduced absorption and scattering, allowing to image deeper into tissue and the inherent confocality due to the squaring of the excitation intensity, obviating the need for a pinhole on the emission path. Another advantage of the square dependence is the reduction of sidelobe height, which is particularly useful in 4Pi microscopy.

However, other processes can achieve this nonlinear dependency with the help of visible light, and are also not limited to a maximal exponent of two in the polynomial expansion. The simplest case is saturation of the excited state of fluorophores by intense illumination. Practically this can be realised as a microscopy method by combining the nonlinearity with the concept of structured illumination, or by detecting high harmonics in the time domain under temporally sinusoidally modulated illumination intensity in a confocal system.

Also the spatial non-uniformity of a depletion, bleaching or light-induced inactivation process (dynamic saturation optical microscopy,) can be used, which is again a nonlinear saturation effect. All of these methods are still in a developmental stage and some of them (nonlinear structured illumination, dynamic saturation optical microscopy) require a substantial amount of image processing, since the high-resolution information contained in the measured data has to be computationally extracted.

The most successful method so far is stimulated emission depletion (STED) microscopy. The method itself is slightly more complex, but has the advantage of obtaining a directly scanned image without the need of image processing. The idea is to deactivate fluorophores surrounding the centre of the scanning spot by stimulated emission by the help of a second beam (the doughnut-shaped STED beam), with a wavelength being within the emission spectrum of the fluorophore. If this was only done with a relatively dim STED beam this method would add another linear term, leading only to a moderate resolution improvement.

However, driving this STED beam into saturation establishes the required nonlinearity of high order. In the centre of the spot to detect this STED beam is expected to have zero intensity, leaving the fluorophores unchanged in their excitation state. If a little bit of STED beam is present at a location only slightly away from its centre (zero intensity) position, its long enough interaction with the fluorophore pushes the latter into the ground-state. This will force emission of light at the wavelength of the STED beam, which can be specifically blocked by appropriate filters allowing most of the spontaneous fluorescence emission to pass. Also a temporal discrimination is possible. It has recently been demonstrated that STED can be applied to imaging of endogenous markers such as green fluorescent protein (GFP). By allowing the triplet states to relax between successive excitations, the fluorescence yield in STED microscopy could be dramatically improved.

A practical rule of thumb can be established as alteration to Abbe's equation, with the aim to estimate a resolution criterion considering the limited signal to noise ratio available. However, such a rule of thumb should not be taken to be a new Abbe limit, as the Abbe limit sets a fixed limit to linear resolution methods, whereas the suggestion in is an approximation to the width of the directly obtained PSF, which may or may not be a good indicator of the smallest periodical structure that can be discriminated for a given noise level in the system. In principle all of the saturation-based nonlinear methods have infinite resolution, but the high-resolution information becomes obscured by the noise.

A Comment on Near Field Methods and Pendry's Lense

Obviously the distribution of emitting fluorophores in the sample can have any structure, and is not constrained by the Abbe limit. The sample is just as it is with no limit to its periodical constituents, but the light intensity is known to fall off by an inverse square law (in the scalar approximation) with the distance to each emitting molecule. However, such a light distribution within the sample would also not be limited by the Abbe limit. The discrepancy to the light detection discussed so far arises from the fact that in the aforementioned theory, we only considered light waves that are able to propagate (so called far field waves).

However, Maxwell's equations in a homogeneous medium allow also for a different kind of oscillating electro-magnetic fields called evanescent (or near field) waves. These solutions of the Maxwell equations in a homogeneous medium decay exponentially in a direction of space. Even worse; the finer the (diffracting) sample structure causing these evanescent waves is, the quicker their decay over distance. In the optical regime these evanescent waves become usually undetectable after about one micrometer.

Scanning near-field optical microscopy (SNOM) methods place a probe (a tip, a scatterer of particles that can fluoresce or resonantly be excited) close enough to the sample structure to measure these evanescent fields. From the aforementioned it becomes clear that near-field methods are constrained to imaging surfaces.

An additional problem is that at these small distances to the sample the interaction of the tip with the electromagnetic field can change the local field and the interpretation of the images may become difficult. Like with the described nonlinear methods, the Abbe limit does not apply to the near field. Practically SNOM achieves a resolution (expressed as half width of the direct PSF) of down to about 30nm in a watery environment. An interesting perspective is the measurement of tip-enhanced Raman scattering.

Another particular example of a near-field effect is evanescent wave microscopy. Here an exponentially decaying intensity distribution can be generated close to a surface by illumination above the critical angle of total internal reflection (TIRF) of a planar interface between an optically dense medium (*e.g.* glass, $n = 1.518$) and a less dense medium (*e.g.* water, $n = 1.33$). In contrast to SNOM, this evanescent wave just extends along Z and the resolution improvement gained is also just oriented along Z. However, the main advantage of TIRF is a relatively sharp discrimination between (usually membrane and vesicular) sample structures close to the optical interface (depending on the illumination angle closer than about 200nm) and structures further away from this interface. For cells which are adherently grown to this interface, this method is interesting as vesicles in the interior of the living cell can be suppressed, and only structures tethered to or fused with the plasma membrane are present in the image.

Recently the idea of the 'perfect lens' has arisen, which is also based on using non-propagating waves. Pendry realised that a medium of negative refractive index (*e.g.* a thin slab of silver) would lead to exponentially growing waves within this medium. This effect can, in theory, make up for the exponentially decreasing evanescent waves outside the medium such that they theoretically have similar levels at the focus position like in the sample space. This effect has recently been demonstrated experimentally for light at small distances, but it still remains to be shown how much the non-ideal media used for such lenses can re-amplify the 'lost' evanescent waves and to which maximal distance such an ideal lens could work.

8

Infrared Diode Laser Irradiation

INTRODUCTION

The morphology of polymer films doped with an infrared dye could be modified significantly with single pulses of a few microseconds from a diode laser with a wavelength of 839 nm. The infrared light was focused into a Gaussian spot having a power density in the centre of up to 200 kW/cm^2. Atomic force microscopy was used to characterize the influence of the infrared-laser power, the pulse width, and the concentration of the embedded infrared dye on the restructuring of the topography of the doped polymer films. Furthermore, the incorporation into the polymer matrix of a thermolabile sulfonium salt combined with acridine as an acid-sensitive fluorescent probe allowed the visualization of the thermally affected area of the irradiated film using laser scanning confocal fluorescence microscopy and scanning near-field optical microscopy.

The comparison between irradiated films with poly(methyl methacrylate) and poly(styrene-*co*-butyl methacrylate) revealed different morphologies in the restructured areas, indicating the important influence of the polymer matrix on the laser action. Experimental evidence was found for the ejection of small droplet like fragments from the irradiated film area. Furthermore, the impact of the infrared irradiation on the film morphology in this study shows close similarities with the restructuring of doped polymer films upon ultraviolet irradiation, suggesting similar mechanisms for the restructuring at these different irradiation wavelengths.

Many examples in the literature have shown important modification of the morphology of polymer films upon irradiation with several types of laser sources. The restructuring of various polymer films can be divided into three sets of experiments, depending on the absorbing entity of the film. In the first set, the functional groups belonging to the molecular structure of the polymer material absorb the incident light; in the second, a dopant molecule is added as a sensitizer for absorbing the excitation wavelength of the laser; and finally in the third, both the dopant and the polymer absorb at the irradiation wavelength of the excitation laser.

However, the impact of the irradiation on the polymer morphology does not depend solely on the absorption characteristics of the film at the irradiation wavelength. At least as important are parameters of the excitation source, such as the laser power (watt) and the pulse width (second), which determine the total energy (joule) absorbed by the film.

Another major factor, however, is the spatial distribution of the incident light. Only few reports in the literature, contain detailed information on the size and shape of the incident light spot on the film, although these are very important parameters that determine the incident power density (watt/meter^2) on the film.

The combination of the power density, the absorption of the film, and the film thickness determine the number of absorbed photons per unit of film volume and per unit of time. As the irradiation of the film occurs in a pulsed fashion, the time dependence of the number of absorbed photons per unit of film volume will be determined by the pulse width, the pulse shape, and the photophysical properties of the excited species. The impact of the pulsed irradiation on the film properties will be determined by the nature of the chemical reactions and physical processes, such as expansion and contraction6 and the ejection of gaseous and solid film fragments, that can be induced by the absorbed photons in the polymer film.

The understanding of these irradiation- induced processes is of great importance for many technological applications that rely on the space-resolved modification of the bulk properties as well as the surface properties of a polymer film, such as roughness, hydrophilicity, wettability, or adhesion. Laser irradiation from high-power UV lasers can drastically etch and modify a polymer film.

These modifications have been the subject of many reports in the literature, and the etching mechanism has been defined as laser ablation. Although ablation is mostly observed in combination with short (nanosecond) pulsed high-power laser irradiation, the light from a moderate-power pulsed argon-ion laser with micro- to millisecond pulse widths may be used to irradiated doped polymer films.

The results shows that polymers can be modified using these milder irradiation conditions, and the authors described the irradiation-induced observations in the polymer as "photokinetic etching", to distinguish the effect from that of the ablation process from highpower short pulsed irradiation. Only a few examples in the literature report on the impact of infrared-laser light on the properties of polymer films doped with infrared-absorbing compounds. This report mainly concentrates on the imagewise characterization of infrared-laser-irradiated films using atomic force microscopy (AFM), near-field scanning optical microscopy (NSOM), and laser scanning confocal microscopy (LSCM) to gain insight into the processes induced by the infrared laser.

A comparison will be made between the impact of the infrared irradiation and that of the ultraviolet irradiation on doped polymer films for similar irradiation conditions. The infrared-laser power density in this report ranges up to a maximum of 200 kW/cm^2, which is of the same order of magnitude as the conditions used during the ultraviolet irradiation for photokinetic etching.

EXPERIMENTAL PROCEDURES

(i) Preparation and Characterization of Polymer Films *Doped with an Infrared-Absorbing Dye*. The chemical composition of the films is given in Table below, and the molecular structure of the infrared-absorbing dye is given in Chart below.

Table. Chemical Composition of the Infrared-Sensitive Polymer Films

Compound	Wt. %
IR dye[a]	0.5-3.0
MDPST[b]	5
Acridine	2
poly(STY-*co*-BMA)[c]or or poly(MMA)[d]	rest

[a]Infrared-absorbing dye (molecular structure shown in Chart 1).
[b]Methyl diphenylsulfonium trifluoromethanesulfonate (acid generator).
[c]Copolymer of80%styreneand20%butyl methacrylate.
[d]Poly(methyl methacrylate).

The compounds were dissolved in 2-butanone and spin-coated on glass supports at 1500 rpm. The total concentration of dissolved material in the solutions for spin-coating was between 5 and 10 wt %. The film thickness was between 1 and 2 ím and was measured withAFMat the edge of a mechanically applied scratch on the film. The choice of acridine as the acid-sensitive fluorescent probe is based on results reported by Pohlers et al.25 on the spectroscopic characterization of an UV-light-sensitive acid generator in polymer films. Films of 20 ím thickness were cast on a poly(ethylene terephthalate) (PET) support from solutions containing 20 wt % solid material dissolved in 2-butanone.

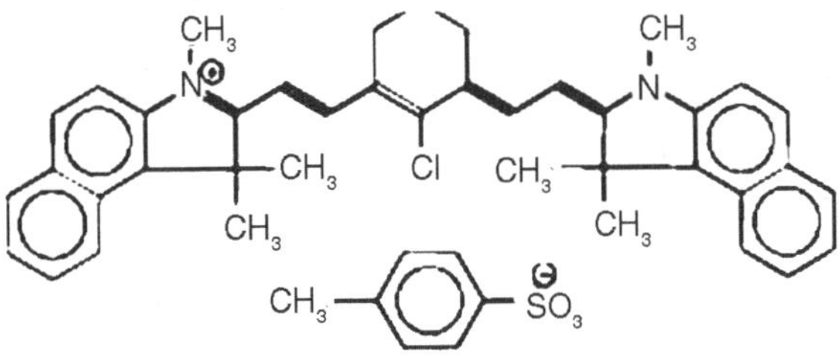

Fig. Chemical Structure of the Infrared-Absorbing Dye in the Polymer Film

(ii) *Infrared-Laser Irradiation of Polymer Films.* The films were irradiated with spatially separated single-shot pulses from an infrared diode laser with a wavelength of 839 nm. The infrared light from the diode laser was focused through a microscope objective lens on the film to create a Gaussian spot profile with a diameter at half the maximum down to 7.3 ím. The maximum power density in the centre of the spot reaches nearly 300 kW/cm2 and was calculated using Equation.

$$P_{(x)} = \frac{2P_0}{\pi(R_0)^2} \exp\left| -2(x/R_0)^2 \right|$$

(for $x = 0$ m). Single-shot light pulses were applied with a Gaussian pulse profile and a pulse width of 3 ís or longer, measured at half the maximum. The complete setup for infrared irradiation has been described elsewhere.

The power density $P(x)$ (watt/meter^2) of the incident light on the film at a distance x (meter) from the centre of the focused Gaussian light spot with radius $R0$ (i.e., 6.3 – 10^6 m) at $1/e^2$ of the maximum, was calculated using eq 1, with $P0$ representing the total incident power (watt). The total energy density $E(x)$ (joule/meter^2) at a distance x meter from the centre of the focused light spot, with a power density $P(x)$ (watt/meter^2), after a single-shot irradiation with pulse width $tfwhm$ (second) was calculated with Equation.

$$E_{(x)} = P_{(x)} t_{twhm}$$

(iii) *Atomic Force Microscopy.*AFM images were obtained with a TopoMetrix TMX 2010 controller system with Discoverer AFM unit, using standard silicon nitride AFM tips (TopoMetrix model 1520) in contact mode. The height of the sample and the signal of the feedback sensor were simultaneously recorded during the scanning of the sample. All of the topography data were leveled with the first-order plane-correction operation using the TopoMetrix software.

(iv) *Laser Scanning Confocal Fluorescence Microscopy.* The LSCM setup consists of a Bio-Rad MRC600 unit connected to a Nikon Diaphot 300 inverted microscope. The excitation light had a wavelength of 420 nm and was generated by a TiSa (Tsunami, SP model 3960) pumped by an Ar^+ laser. The fluorescence from the sample was collected above 465 nm. A detailed description of the setup can be found elsewhere.

(v) *Near-Field Scanning Optical Microscopy.* The NSOM setup consists of a modified commercially available TopoMetrix Aurora instrument, fully described elsewhere.

(vi) *Absorption Spectroscopy.* The absorption spectra of the films were recorded on a Perkin-Elmer Lambda 6 spectrophotometer.

(vii) *Fluorescence Spectroscopy.* The emission spectra and excitation spectra of the irradiation-induced fluorescence of the films were recorded on a SPEX fluorolog. A front-face configuration was used for the direction of the emitted light relative to the excitation of the film.

(viii) *Collection of Ejected Film Material.* The material that was ejected from the irradiated film was collected on the surface of a glass slide facing the irradiated film surface. The receiver glass was mounted 20 ím above the film surface, using a thin spacer between the film and the receiver glass.

RESULTS AND DISCUSSION

(i) *Atomic Force Microscopy. The Influence of the IR Dye Concentration.* The topography and feedback sensor images in Figure 1 represent the impact of single-shot

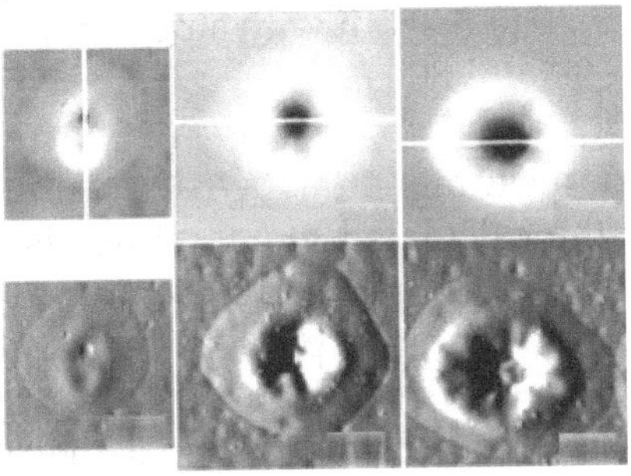

Fig. Atomic force microscopy (AFM) images of the topography and the feedback sensor signal of irradiated poly(STY-*co*-BMA) films with 0.5 wt %

(A), 1.5 wt % (B), and 3 wt % (C) infrared dye, after a single-shot laser pulse with 100 mW power and 10 ís pulse width (t_{fwhm}), using a focused laser beam with a 12.6 ím diameter ($d_{1/e2}$). The absorbance of the films at the incident wavelength of 839 nm amounts to 0.13 (A), 0.33 (B), and 0.55 (C).

The size of the images is indicated with a scale bar of 5 ím. laser irradiation of 839 nm with 100 mW power during 10 ís pulse width on three films with the poly(styrene-*co*- butyl methacrylate) [poly(STY-*co*-BMA)] matrix absorbing different amounts of the incident photons.

The film thickness was derived from AFM analysis of the edge height of an applied scratch on the polymer film. The film composition is given in Table 1. The topography images reveal a relatively large deformation of the film morphology as a result of the irradiation. The deformation is characterized by several interesting features:

1. The lateral shape of the modified film area reflects the shape of the incident focused light spot and
2. A crater in the centre is surrounded by an expanded rim. The shape of the profiles through these single dots qualitatively resembles the shape of topography profiles of UV-laser-induced

cuts in poly(methyl methacrylate)[poly(MMA)] films doped with 2% 2-(2′-hydroxy- 3′, 5′-diisopentylphenyl) benzotriazole, as reported by Srinivasan.

The main features of the irradiated films correspond rather well, although the experimental conditions from which they were obtained are different. The size of the IR-laser-induced features in the doped poly-(STY-*co*-BMA) matrix is about a factor of 10 smaller than that of the UV-laser-induced features in the doped poly-(MMA) films, probably mainly due to the different size of the focused laser spots in both experiments. The argonion laser exposures were performed while moving a focused continuous-wave (CW) beam with a controlled speed relative to the film surface.

The pulse width for these irradiation conditions was defined as the transit time of the laser spot over the nominal diameter of 60 ím. Channels were cut in the doped poly(MMA) film upon argon-ion laser exposure of as much as 335 consecutive pulses of 300-330 nm with a power density of 78 kW/cm^2 and a pulse width between 30 and 80 ís.

Thermal decomposition of the poly(MMA) into material of lower molecular weight was assumed to play a role in the etching process, aswasconcluded from the calculated temperature rise of the irradiated film and the experimentally determined enhancement of the solubility of the irradiated material. Discharge of vapour from the irradiated area during irradiation was also observed and attributed to the escape of the monomer. However, the authors did not explain in detail the physical processes that gave rise to the irregular topographical features in the irradiated area.

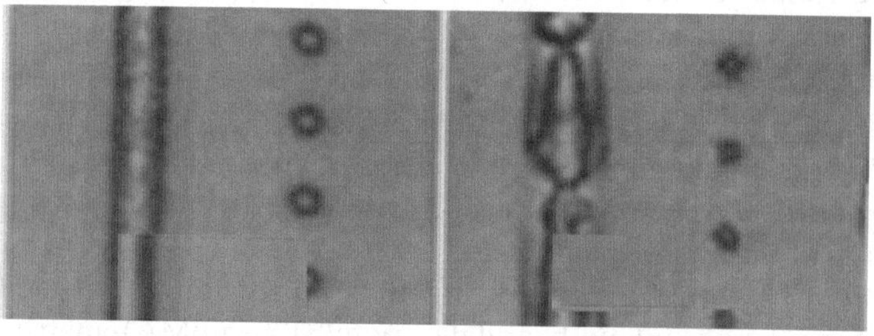

Fig. Optical transmission micrographs of poly(MMA) andpoly(STY-*co*-BMA)films of 20 ímthick films on PET support with 1.5 wt % infrared dye, after irradiation with 40 mW for continuous-wave (CW) and pulsed irradiation during 182 ís.

The translation speed of the laser spot [7.3 ím spot diameter (fwhm)] during theCWirradiation was 4 cm/s, which also leads to a pulse width (the transit time over the spot diameter) of 182 ís.

The optical transmission micrographs in figure show the impact of CW and pulsed irradiation on poly(MMA) and poly(STY-*co*-BMA) films, after infrared irradiation with similar conditions (40 mW during 182 ís). The comparison between the channels in figure and those for irradiated poly(MMA) published by Srinivasan et al. shows that CW irradiation of appropriately doped polymer films with UV and IR light can induce similar physical modifications of the film.

There is very striking similarity in shape between the irregularities in the channels in poly(STY-*co*-BMA) after 1 pass of the infrared laser with 40 mW during 182 ís and the channels in the poly(MMA) films reported by Srinivasan18 after 335 passes of the laser with transit times of 50 and 67 ís and a power density of 78 kW/cm2. The calculated incident energy density on the films was 13.8 J/cm2 for the irradiation with 839 nm and 1384 J/cm2 for the 300-330 nm irradiation with the argon-ion laser.

This absorbance of the doped films at the irradiation wavelength in both experiments was around 4 for poly(STY-*co*-BMA) at 839 nm (calculated from 0.2 ím-1 and 20 ím film thickness) and 100% for poly(MMA) at 300-330 nm. A more quantitative comparison between the results from both experiments is, however, impossible because of the lack of information about the penetration depth in the poly-(MMA) sample of the UV irradiation. The following experimental results and discussion will contribute to the understanding of the physical processes during these modifications.

A closer look in figure at the feedback sensor images, which emphasize the small surface structures, reveals small droplet like material around the irradiated area on the films with 1.5 and 3.0 wt % infrared dye. These material droplets are not present on the unirradiated film, and their location on the expanded rim close to the central crater suggests that they were expelled from this irradiated area during the creation of the crater. Furthermore, the droplet like shape of these surface features suggests that the polymer filmwasheated above its glass-transition temperature into a molten state upon laser irradiation.

To provide experimental evidence for the IR-laserinduced ejection of film material, a previously reported experimental approach was used. In our procedure, the polymer films were irradiated through a transparent receiver substrate,mountedabove the sample surface prior to laser irradiation, and separated from it by a 20 íc thin spacer. This experimental approach was performed on films with poly(STY-*co*-BMA) and with poly(MMA) as the polymer matrix, to test the influence of the nature of the polymer matrix on the irradiation-induced ejection process.

The laser power of the irradiation was adjusted according to the absorbance of the films, to allow an equal amount of absorbed energy in both polymer films. AFM was used to characterize and compare the impact of the IR light on the topography of the irradiated poly(STY*co*- BMA) and poly(MMA) films, as well as on the surface topography of the corresponding receiver substrates above these polymer films. The results of the topography analyses of the irradiated polymer films and the corresponding receiver substrates (compared with the topography of a blank substrate) are presented in figures, respectively. The receiver surfaces above both polymer films clearly contain features with a larger size than those of the artifacts on the surface of the blank receiver glass, confirming the ejection of material from both types of polymer films upon laser irradiation.

The comparison between the receiver surface for poly(MMA) and poly(STY-*co*-BMA) in figure clearly shows that the amount of collected material depends on the nature of the polymer matrix. Furthermore, the amount of collected material at the receiver surface is clearly correlated with the size of the crater in the irradiated polymer films. The receiver glass above poly(MMA) collected more ejected material compared to the receiver glass above poly(STY*co*- BMA), and the size of the craters in the poly(MMA) film is also larger than those in the poly(STY-*co*-BMA) film.

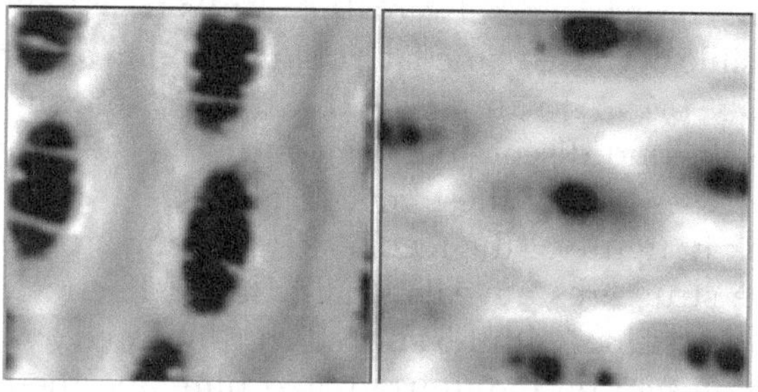

Fig. AFM topography images with 21 ím scan range of poly(MMA) (total height range, 890 nm) and poly(STY-*co*-BMA) (height range, 810 nm) after dotwise irradiation at a distance of 10 ím.

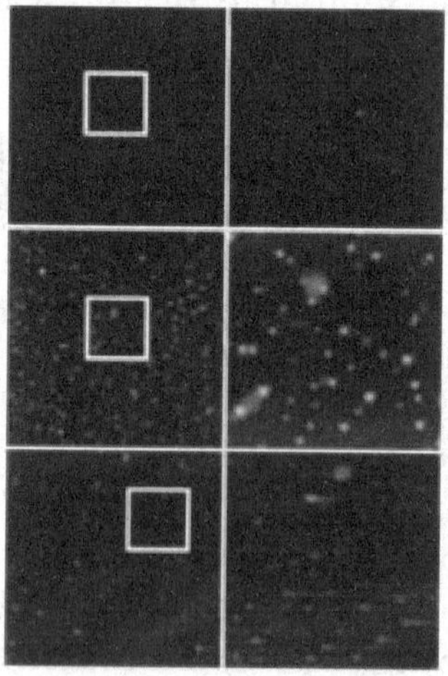

Fig. Atomic force topography analyses (an overview with a scan range of 10 ím and a zoom of 3 ím) of the surface topography of the receiver glasses facing the polymer film of poly(MMA) and poly(STY-*co*-BMA), after laser irradiation.

The irradiation conditions for poly(MMA) were 100 mW during 10 ís (fwhm), and for poly(STY-*co*-BMA), 88 mW during 10 ís (fwhm), for an OD839 nm of 0.25, respectively. The distance between the centre of the separate dots on the polymer films was 10 ím.

The full height range (from black to white) of all the images is identical from 0 to 90 nm. *The influence of the IR Power and Pulse Width.* The AFMtopography images in figure show the film surfaces after the irradiation-induced processes have ended and give no direct information on the time dependence of these processes. The time-dependent properties of laser-induced modifications of polymer films have been analyzed with nanosecond interferometry and time-resolved photography.

However, these techniques cannot be used to analyse the time-dependent properties during the IR-laser modifications of our samples because the size of the modified film area is too small to be visualized accurately. To obtain an indirect view of the nature of the laserinduced physical processes, we irradiated the polymer films with a set of increasing IR powers and increasing pulse widths and analyzed the impact of these irradiation conditions on the topography of the films.

These experiments allowed us to characterize the influence of the energy density of the incident light on the properties of the surface modification for various combinations of IR power and pulse width. The atomic force topography analyses of the irradiated films of poly(STY-*co*-BMA) and poly(MMA) are presented in figures, respectively. The irradiated area on the poly(STY-*co*-BMA) films in figure is not circularly symmetric because of a small out-of-focus deviation of the incident laser beam.

The surface of the film remains rather smooth after irradiation with the smallest total energy density (40 mW during 3 ís) for poly(STY-*co*-BMA) and for poly(MMA). The outer rim is expanded above the level of the unirradiated area, and the area within this rim is below this level. The height of the rim on poly(STY-*co*-BMA) varies between 30 nm (40 mW during 3 ís) and 100 nm (80 mW during 8 ís). The height of the rim on the poly(MMA) film tends toward

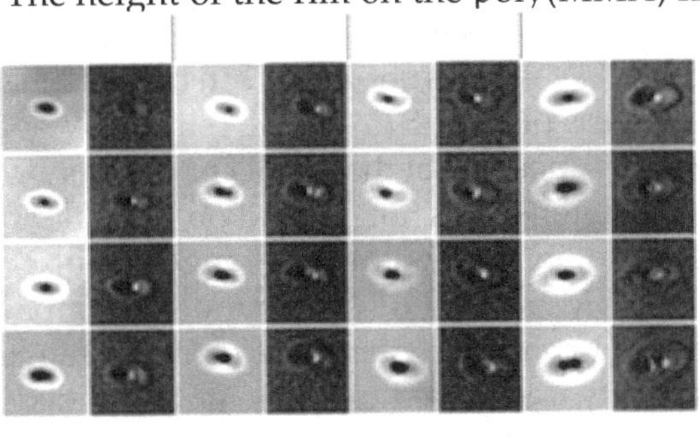

Fig. Topography and feedback sensor images of poly(STY-*co*-BMA) films after irradiation with various combinations of infrared-laser power and pulse width.

The height range and the sensor signal of the images have been normalized. The height range and the sensor signal of the images have been normalized a plateau value of 150 nm with increasing energy density of the irradiation.

For a laser power above 40 mW or a pulse width longer than 3 ís, an additional crater can be seen for both poly(STY-*co*-BMA) and poly(MMA), and the diameter of this crater increases with increasing infraredlaser power or pulse time. The depth of the crater in both polymers shows a relatively poor correlation with the energy density, is always smaller than the overall film thickness, and amounts to between 150 (40 mW during 3 ís) and 730 nm.

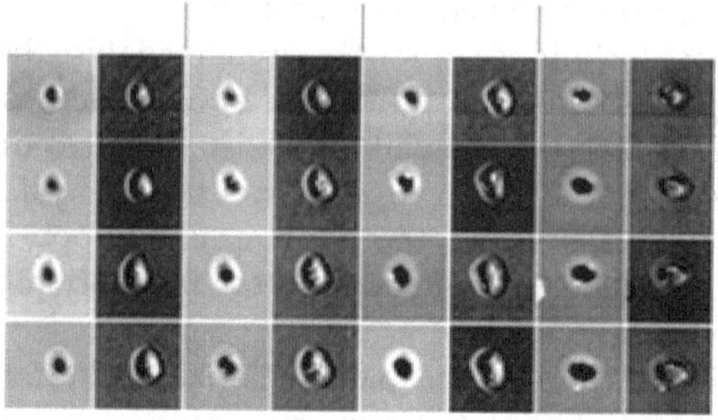

Fig. Topography and feedback sensor images of poly(MMA) films after irradiation with various combinations of infraredlaser power and pulse width.

The shape of the crater is rather different for both types of polymer, but the edges and the surface of the craters are very irregular for both polymers. This irregularity also correlates with the observation made by Srinivasan et al. concerning the shape of the channels cut with an argon-ion laser in doped poly(MMA) films.

The analysis above has shown that the height of the expanded rim and the depth of the crater correspond rather well for both polymers. However, a remarkable difference between both polymers can be observed concerning the influence of the energy density on the size of the expanded film area. A qualitative comparison of the results shows that, for a given energy density, the size of the expanded area on poly(MMA) does not depend on the combination of laser power and pulse width, whereas on poly(STY-*co*-BMA) it clearly does.

In other words, this means that the impact on the poly(MMA) topography of a givenamountof photons does not depend on the pulse time during which they were absorbed by the film (within the range of pulse widths studied).

For poly(STY-*co*-BMA) on the other hand, the impact of the incident photons on the film morphology is more effective at higher photon flux, so less total energy density is required to obtain an expanded area with a given diameter. Because the expanded area must be the result of one or more thermally driven processes, its size must reflect the temperature profile during the local heating of the film.

In turn, this temperature profile reflects the competition between the rate of heat transfer from the photoexcited infrared dye to its surrounding matrix and the rate of heat dissipation in the polymer matrix (and/or to the film substrate).

This means that, for a given irradiation condition, the size of the expanded area reflects the relative rate of heat generation compared to the rate of heat loss. The heat transfer from the infrared dye to the surrounding poly- (MMA) is competing better with the dissipation of heat within the polymer matrix or to the film substrate or both than in case of poly(STY-*co*-BMA).

The series of topography images in figures gave little direct information on the nature of the thermally induced processes that gave rise to these physical modifications of the surface topography. A small indication can be observed in the centre of the irradiated poly(MMA) for 40mW during 3 *ís* as a circular topography, resembling the remains of a bubble after it bursts and releases its content. According to this model, the rough and irregular topography features within the crater and at the edge of the crater could also be the remains of similar burst bubbles. Furthermore, the collected material on the receiver substrates in figure could originate from film material around these bubbles which was ejected by the pressure inside the bubble.

The topography images in figures show the final cooled state of only one of the heated areas on the polymer samples, representing the features of the majority of the irradiated areas. Occasionally, bulging of the surface of the irradiated area could be observed, supporting the mechanism of bubble formation beneath the film surface. A few examples of these features for two different films of poly(MMA) are shown in the topography images a-d of figure and analyzed using topography profiles along the lines (1-4) in figure.

The topography profiles show that the height of these features can be surprisingly large compared to the overall thickness of the irradiated film, which means that they cannot be due to a mere thermal expansion of the polymer film. The height of the feature in image a of figure almost amounts to 2 *ím* relative to the unirradiated surface, whereas the film thickness is even less than a third of this expansion. The bulged areas in the images of figure are undoubtedly the result of laser

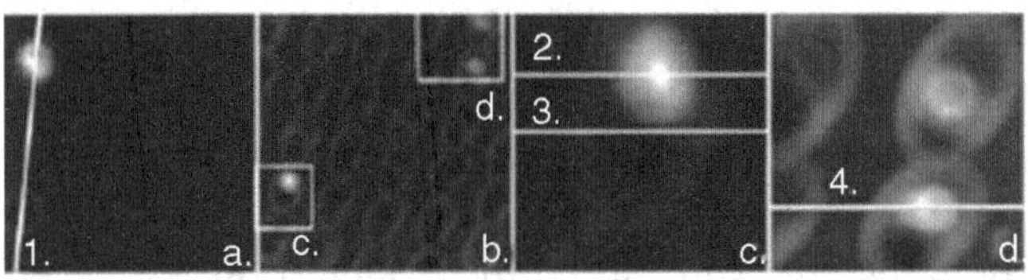

(A.) AFM images a-d of the topography of a poly(MMA) film with 0.5 wt % (a, 30 *ím* scan range) and 1.0 wt % infrared dye (b, 70; c, 20; and d, 30 *ím* scan range), after laser irradiation with 100 mW during 10 *ís*. The optical density at 839 nm for 0.5 wt % was 0.047 (film thickness, 0.57 *ím*) and 0.235

for 1.0 wt % (film thickness, 1.4 ím). The images c and d represent details from image b. irradiation of the film and merely artifacts in the film topography, because these structures only appear near the centre of the irradiated area and not on the unirradiated surface of the film. These bulged structures are presumably formed in the polymer film while heated to a molten state.

If the pressure in the bubble is not released by bursting at the surface, before the irradiated area has cooled below the glass-transition temperature of the polymer, this expanded structure can be rigidified in the film and appear in the topography.

LASER WORKS

A laser usually comprises an optical resonator (laser resonator, laser cavity) in which light can circulate (e.g. between two mirrors), and within this resonator a gain medium (e.g. a laser crystal), which serves to amplify the light. Without the gain medium, the circulating light would become weaker and weaker in each resonator round trip, because it experiences some losses, e.g. upon reflection at mirrors.

However, the gain medium can amplify the circulating light, thus compensating the losses if the gain is high enough. The gain medium requires some external supply of energy – it needs to be "pumped", e.g. by injecting light (*optical pumping*) or an electric current (*electrical pumping* '! *semiconductor lasers*). The principle of laser amplification is stimulated emission.

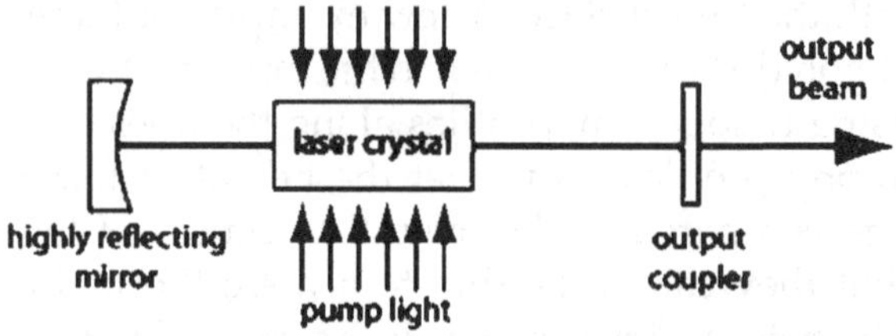

Fig Setup of a Simple Optically Pumped Laser.

The laser resonator is made of a highly reflecting curved mirror and a partially transmissive flat mirror, the output coupler, which extracts some of the circulating laser light as the useful output. The gain medium is a laser crystal, which is side-pumped, e.g. with light from a flash lamp. A laser can not operate if the gain is smaller than the resonator losses; the device is then below the so-called laser threshold and only emits some luminescence light. Significant power output is achieved only for pump powers above the laser threshold, where the gain can exceed the resonator losses.

If the gain is larger than the losses, the power of the light in the laser resonator quickly rises, starting e.g. with low levels of light from fluorescence. As high laser powers saturate the gain, the laser power will in the steady state reach a level so that the saturated gain just equals the resonator losses (' *gain clamping*). Before reaching this steady state, a laser usually undergoes some relaxation oscillations. The threshold pump power is the pump power where the small-signal gain is just sufficient for lasing.

Some fraction of the light power circulating in the resonator is usually transmitted by a partially transparent mirror, the so-called output coupler mirror. The resulting beam constitutes the useful output of the laser. The transmission of the output coupler mirror can be optimized for maximum output power. Some lasers are operated in a continuous fashion, whereas others generate pulses, which can be particularly intense. There are various methods for pulse generation with lasers, allowing the generation of pulses with durations of microseconds, nanoseconds, picoseconds, or even down a few femtoseconds (→ *ultrashort pulses* from *mode-locked lasers*).

The optical bandwidth (or linewidth) of a continuously operating laser may be very small when only a single resonator mode can oscillate ('! *single-frequency operation*). In other cases, particularly for mode-locked lasers, the bandwidth can be very large – in extreme cases, it can span about a full octave. The center frequency of the laser radiation is typically near the frequency of maximum gain, but if the resonator losses are made frequency-dependent, the laser wavelength can be tuned within the range where sufficient gain is available. Some broadband gain media such as Ti:sapphire and Cr:ZnSe allow wavelength tuning over hundreds of nanometers. Due to various influences, the output of lasers always contains some noise in properties such as the output power or optical phase.

Types of Lasers

Common types of lasers are:

- *Semiconductor lasers* (mostly laser diodes), electrically (or sometimes optically) pumped, efficiently generating very high output powers (but typically with poor beam quality), or low powers with good spatial properties (e.g. for application in CD and DVD players), or pulses (e.g. for telecom applications) with very high pulse repetition rates. Special types include quantum cascade lasers (for mid-infrared light) and surface-emitting semiconductor lasers (VCSELs and VECSELs), the latter also being suitable for pulse generation with high powers.
- *Solid-state lasers* based on ion-doped crystals or glasses (*doped insulator lasers*), pumped with discharge lamps or laser diodes, generating high output powers, or lower powers with very high beam quality, spectral purity and/or stability (e.g. for measurement purposes), or ultrashort pulses with picosecond or femtosecond durations. Common gain media are Nd:YAG, Nd:YVO_4, Nd:YLF, Nd:glass, Yb:YAG, Yb:glass, Ti:sapphire, Cr:YAG and Cr:LiSAF. A special type of ion-doped glass lasers are:

- *Fiber lasers,* based on optical glass fibers which are doped with some laser-active ions in the fiber core. Fiber lasers can achieve extremely high output powers (up to kilowatts) with high beam quality, allow for widely wavelength-tunable operation, narrow linewidth operation, etc.
- *Gas lasers* (e.g. helium–neon lasers, CO_2 lasers, and argon ion lasers) and excimer lasers, based on gases which are typically excited with electrical discharges. Frequently used gases include CO_2, argon, krypton, and gas mixtures such as helium–neon. Common excimers are ArF, KrF, XeF, and F_2.

Less common are chemical and nuclear pumped lasers, free electron lasers, and X-ray lasers.

Laser Sources in a Wider Sense

There are some light sources which are not strictly lasers, but are nevertheless often called *laser sources*:

- In some cases, the term is used for amplifying devices emitting light without an input (excluding seeded amplifiers). An example is X-ray lasers, which are usually *superluminescent sources,* based on spontaneous emission followed by single-pass amplification. There is then no laser resonator. A similar situation occurs for optical parametric generators, where the amplification, however, is not based on stimulated emission. Light from such devices can have laser-like properties, such as strongly directional emission and a limited optical bandwidth.
- In other cases, the term *laser sources* is justified by the fact that the source contains a laser, among other components.
  This is the case for combinations of lasers and amplifiers ('! *master oscillator power amplifier*), and also for sources based on nonlinear frequency conversion of laser radiation, e.g. with frequency doublers or optical parametric oscillators.

SAFETY ASPECTS

The work with lasers can raise significant safety issues. Some of those are directly related to the laser light, in particular to the high optical intensities achievable, but there are also other hazards related to laser y for details.

Laser Light

Laser light (laser radiation) is simply light generated with a laser device. Such light has some very special properties, which very much distinguish it from light with other origins:

- Laser light is usually delivered in the form of a laser beam, i.e. it propagates dominantly in a well-defined direction with moderate beam divergence. Such a laser beam has a high (sometimes extremely high) degree of spatial coherence. This means that the electric fields at different locations across a beam profile oscillate with a rigid phase relationship. Exactly this coherence is the reason why a laser beam can propagate over long distances without spreading very much in the transverse directions, and why it can be focused to very small spots (high *focusability* of laser beams).

- In many but not all cases, laser light also has a high degree of *temporal* coherence, which is equivalent to a long coherence length. This means that a rigid phase relationship is also maintained over relatively long time intervals, corresponding to large propagation distances (often many kilometers) or to huge numbers of oscillation cycles.
- The large temporal coherence, quantified with a large coherence time or coherence length, is associated with a narrow spectral bandwidth (or linewidth). For a visible laser beam, this means that it has a certain pure color, e.g. red, green or blue, but not white or magenta. Some lasers allow a degree of wavelength tuning (e.g. dye lasers). The large coherence length introduces a tendency for the phenomenon of laser speckle, i.e. a characteristic granular pattern which can be observed e.g. when the laser beam hits a metallic surface.
- In most cases, laser light is linearly polarized. This means that the electric field oscillates in a particular spatial direction (→ *polarization of laser emission*).

Depending on the case, laser light can have other remarkable properties:

- Laser light may be visible, but most lasers actually emit in other spectral regions, in particular in the near-infrared region, which human eyes cannot perceive.
- Laser light is not always continuous, but may be delivered in the form of short or ultrashort pulses. As a consequence, the peak power can be extremely high – for some amplified laser systems well above 1 TW (10^{12} W) – if the pulse duration is far below the temporal distance of the pulses, so that a large pulse energy is possible.
- The noise properties of lasers can also be very interesting (→*laser noise*). For example, the oscillation frequency of a laser can be stabilized to stay within an extremely narrow range.

For comparison, light from an incandescent lamp is partly visible and mostly infrared, has a very high spectral bandwidth, cannot be strongly focused (because of low spatial coherence), and cannot be generated in the form of short pulses.

Laser light can cause significant hazards, particularly for the eye, but also for the skin, apart from fire hazards and others. Lasers pulses are often particularly dangerous, because they can have enormously high optical intensities, and invisible laser beams add specific risks..

MANUFACTURING

Lasers are widely used in manufacturing, e.g. for cutting, drilling, welding, cladding, soldering, hardening, ablating, surface treatment, marking, engraving, micromachining, pulsed laser deposition, lithography, alignment, etc. In most cases, relatively high optical intensities are applied to a small spot, leading to intense heating, possibly evaporation and plasma generation. Essential aspects are the high spatial coherence of laser light, allowing for strong focusing, and often also the potential for generating intense pulses.

Laser processing methods have many advantages, compared with mechanical approaches. They allow the fabrication of very fine structures with high quality, avoiding mechanical stress such as caused by mechanical drills and blades. A laser beam with high beam quality can be used to drill very fine and deep holes, e.g. for injection nozzles. A high processing speed is often achieved, e.g. in the fabrication of filter sieves. Further, the lifetime limitation of mechanical tools is removed. It can also be advantageous to process materials without touching them.

Medical Applications

There is a wide range of medical applications. Often these relate to the outer parts of the human body, which are easily reached with light; examples are eye surgery and vision correction (LASIK), dentistry, dermatology (e.g. photodynamic therapy of cancer), and various kinds of cosmetic treatment such as tattoo removal and hair removal.

Lasers are also used for surgery (e.g. of the prostate), exploiting the possibility to cut tissues while causing minimal bleeding. Very different types of lasers are required for medical applications, depending on the optical wavelength, output power, pulse format, etc. In many cases, the laser wavelength is chosen such that certain substances (e.g. pigments in tattoos or caries in teeth) absorb light more strongly than surrounding tissue, so that they can be more precisely targeted. Medical lasers are not always used for therapy. Some of them rather assist the diagnosis, e.g. via methods of laser microscopy or spectroscopy.

Metrology

Lasers are widely used in optical metrology, e.g. for extremely precise position measurements with interferometers, for long-distance range finding and navigation. Laser scanners are based on collimated laser beams, which can read e.g. bar codes or other graphics over some distance. It is also possible to scan three-dimensional objects, e.g. in the context of crime scene investigation (CSI). Optical sampling is a technique applied for the characterization of fast electronic microcircuits, microwave photonics, terahertz science, etc. Lasers also allow for extremely precise time measurements and are therefore essential component of optical clocks which are beginning to outperform the currently used cesium atomic clocks. Fiber-optic sensors, often probed with laser light, allow for the distributed measurement of temperature, stress, and other quantities e.g. in oil pipelines and wings of airplanes.

Data Storage

Optical data storage e.g. in compact disks (CDs), DVDs, Blu-ray Discs and magneto-optical disks, nearly always relies on a laser source, which has a high spatial coherence and can thus be used to address very tiny spots in the recording medium, allowing a very high density data storage. Another case is holography, where the temporal coherence can also be important.

Communications

Optical fiber communication, extensively used particularly for long-distance optical data transmission, mostly relies on laser light in optical glass fibers. Free-space optical communications, e.g. for inter-satellite communications, is based on higher-power lasers, generating collimated laser beams which propagate over large distances with small beam divergence.

Displays

Laser projection displays containing RGB sources can be used for cinemas, home videos, flight simulators, etc., and are often superior to other displays concerning possible screen dimensions, resolution and color saturation. However, further reductions in manufacturing costs will be essential for deep market penetration.

Spectroscopy

Laser spectroscopy is useful, e.g. in atmospheric physics and pollution monitoring (e.g. trace gas sensing with differential absorption LIDAR technology). It also plays a role in medicine (e.g. cancer detection), biology, and various types of fundamental research, partly related to metrology.

Microscopy

Laser microscopes and setups for coherence tomography provide images of, e.g., biological samples with very high resolution, often in three dimensions. It is also possible to realize functional imaging.

Various Scientific Applications

Laser cooling makes it possible to bring clouds of atoms or ions to extremely low temperatures. This has applications in fundamental research and also for industrial purposes. Particularly in biological and medical research, optical tweezers can be used for trapping and manipulating small particles, such as bacteria or parts of living cells. Laser guide stars are used in astronomical observatories in combination with adaptive optics for atmospheric correction. They allow substantially increased image resolution even in cases where a sufficiently close-by natural guide star is not available.

Energy Technology

In the future, high-power laser systems might play a role in electricity generation. Laser-induced nuclear fusion is investigated as a alternative to other types of fusion reactors. High-power lasers can also be used for isotope separation.

Military Applications

There are a variety of military laser applications. In relatively few cases, lasers are used as weapons; the "laser sword" has become popular in movies, but not in practice. Some high-power lasers are currently developed for potential use as *directed energy weapons* on the battle field, or for destroying missiles, projectiles and mines. In other cases, lasers function as target designators or laser sights (essentially laser pointers emitting visible or invisible laser beams), or as irritating or blinding (normally not directly destroying) countermeasures e.g. against heat-seeking anti-aircraft missiles. It is also possible to blind soldiers temporarily or permanently with laser beams, although the latter is forbidden by rules of war. There are also many laser applications which are not specific for military use, e.g. in areas such as range finding, LIDAR, and optical communications.

Laser Design

The term *design* can have two different meanings. In some cases, it is meant to be a detailed description of a device, including e.g. used parts, how they are put together, and important operation parameters. In other cases, the term denotes the process leading to such a description.

Defining the Design Goals

Before a design is made, the *design goals* must be carefully evaluated. These should include not only the central performance parameters such as output power and wavelength; many more details can be relevant:

In Short: Details of the design of a laser product can have a strong impact on its performance, reliability, flexibility, and manufacturing cost. It is essential to be aware of all the design goals, to know the design properties required for reaching them, and to use a proper laser design (which is worked out at the beginning) as a vital part of an efficient development process.

- Optimum performance, e.g. in terms of output power, power efficiency, beam quality, brightness, intensity and/or phase noise, long-term stability (e.g. of the output power or the optical frequency), timing jitter, etc.

- Compact and convenient setup, ease of operation (e.g. simple turn-on procedure, simple wavelength tuning, no need for realignment)
- Maximum flexibility (e.g. for changing operation parameters)
- Reliability, low maintenance requirements, simple and cost-effective error analysis, maintenance and repair
- Minimum sensitivity to vibrations, temperature changes, electromagnetic interference, aging of components
- Low production cost, i.e., a small number of parts, simple alignment and testing, avoiding the use of parts which are expensive, sensitive, or difficult to obtain

It is certainly advisable to work out carefully the list of these requirements for the particular case before investing any significant resources in laser development, because it can easily be much more expensive and time-consuming to introduce additional properties into an already existing device.

IMPORTANT ASPECTS OF LASER DESIGNS

The properties of the designed laser device are largely determined by the design details, not only by the parts used. Some aspects are particularly important:

- General design parameters, such as resonator length (influencing compactness, tuning issues, frequency stability, etc.), pump intensity
- Selection of gain medium (e.g. a laser crystal) and pump source, suitable choice of geometry (e.g. rod or thin disk, side pumping or end pumping), doping concentration, crystal length, etc.
- Pump setup (e.g. for diode-pumped lasers), influencing output power and beam quality, long-term stability, and the ease of exchanging pump diodes
- Optimum type of laser resonator (e.g. as linear or ring laser, with monolithic or with discrete elements) and optimized resonator design, influencing aspects such as the number of parts, the output power and beam quality, alignment tolerances, sensitivity to thermal lensing, mechanical stability and drifts
- Selection and placement of laser mirrors and intracavity components for wavelength tuning, generation of short pulses via mode locking, dispersion compensation, frequency stabilization, etc.
- Mechanical housing, influencing mechanical stability, efficiency of cooling, temperature drifts, ease of maintenance, and safety issues
- Electronic equipment, e.g. for stabilizing the output power, controlling the laser wavelength, monitoring the status of pump diodes or temperatures, ensuring safe operation
- Proper documentation, including a part list (possibly with suppliers), mechanical designs, alignment and testing procedures, design ideas, possibly optional extensions and limitations for modifying operation parameters

This list, which is certainly not yet complete, shows that proper laser designs are not a trivial matter, but are essential for achieving full customer satisfaction, cost efficiency, and flexibility for future developments.

Needed for Designing Lasers

Designing a laser is a challenging task. The following are definitely required:

- A detailed understanding of the requirements, which may include some understanding of the application
- A detailed understanding of all the relevant physical effects, such as laser amplification, thermal lensing, resonator modes, laser noise, etc., and their interaction
- The essential data, e.g. of laser crystals
- The ability to reduce the complexity to a practical level without losing important details
- Flexible software for calculations and simulations
- Practical experience with lasers, enabling one to recognize typical problems, correctly interpret experimental observations, etc.

It is clear that software alone is by no means sufficient to work out good laser designs.

Role of a Design in a Development Project

It is common practice, but nevertheless generally not advisable, to consider a laser design as a result of a development process which is largely based on trial and error. The design then plays a minor role, just summarizing the results of a lengthy process. In such cases, the design is often not even properly documented, which creates a risk of losing a lot of potentially valuable information while saving only a minor amount of time at the moment.

In any non-trivial design project – and laser design projects are hardly ever trivial – it is very advisable to attribute a vital role to the laser design:

- The laser design *is made in the office,* not in the laboratory, and properly deals with all known issues which can be or become relevant. Although this process can be much faster and cheaper than a trial-and-error approach in the lab, it takes some considerable discipline and of course requires a comprehensive expertise.
- A proper design is not just a set of ideas, but a very specific description, including e.g. the list of required parts, a more or less detailed prescription on how to put them together wherever this is not trivial, and is ideally supplemented by a description of the underlying reasoning, a discussion of limitations, etc.
- The prototype is then fabricated according to the design, and not vice versa. This greatly speeds up the fabrication, thus making efficient use of costly laboratory resources.

- For any future development of similar kind, the carefully worked out design will be a very valuable input. If it does not exist, and particularly if in addition a vital person has left the company, future developments will be much less efficient.

Attempts to abbreviate this process carry the risk of obtaining reduced performance values and of large time delays due to unexpected technical problems. The later such problems are recognized, understood and solved, the larger can be the resulting damage.

Deriving Designs from Older Designs

In industrial development, it is common to derive some product design from an older design, rather than starting from scratch. Although this appears very economical, there are significant risks, particularly in cases where the first design has not be properly worked out and documented in a process as described above. A central challenge is that modifying some detail of a laser design may easily have unexpected side effects, introducing new problems which then require additional measures, which again can have side effects.

For such reasons, starting with some initial design, which works e.g. with some lower than desired output power, can be helpful, but it still requires a detailed understanding of that design and its limitations. A proper design document for the initial design can make it easy to produce a whole family of designs, which differ in e.g. output power or pulse repetition rate.

LASER RESONATORS

A laser requires a *laser resonator* (or *laser cavity*), in which the laser radiation can circulate and pass a gain medium which compensates the optical losses. Exceptions are only exotic cases where a medium with very high gain is used, so that amplified spontaneous emission extracts significant power in a single pass through the gain medium.

The laser radiation is automatically generated at one or several frequencies corresponding to resonances (resonator modes), possibly with small deviations caused by "gain pulling". No special measures are required for operating on the resonance; this is different for external resonators, e.g. resonant enhancement cavities.

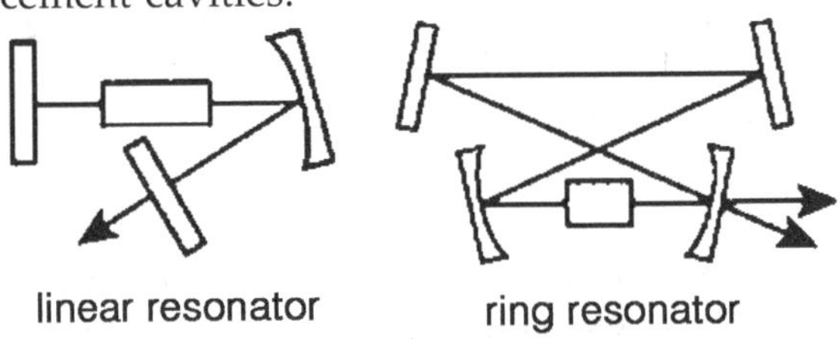

Fig. Two Simple Solid-State Laser resonators with a Laser Crystal as Gain Medium.

Output beams are generated where resonator mirrors are partially transmissive. The ring laser (right) can exhibit laser action in two directions, thus generate two output beams; unidirectional operation could be enforced, e.g. with an additional intracavity optical isolator.

LASER RESONATORS OF SOLID-STATE LASERS

Solid-state bulk lasers are usually built with several dielectric mirrors (laser mirrors), which may be plain or curved. A linear resonator and a ring resonator built in that way, and containing a laser crystal as the gain medium. In some cases, a dielectric mirror coating is placed on the gain medium itself on monolithic solid-state lasers. One of the mirrors, usually an end mirror, is the partially transmissive output coupler.

The design of the laser resonator (comprising optical elements, angles of incidence, and distances between the components) determines the beam radius of the fundamental mode at all locations along the beam, together with other important properties. For maximum beam quality (→ *diffraction-limited* output), the beam radius in the gain medium has to match approximately the radius of the pumped region.

For smaller beam radii, operation with multiple spatial modes is obtained, leading to a non-ideal beam quality; however, such multimode lasers have other advantages such as much wider stability zones and a lower sensitivity to misalignment. In many cases, the laser resonator design should have additional features. For example, it can be optimized

- For compactness
- For achieving certain values of the beam radius in other optical components (e.g. on a saturable absorber of a passively mode-locked laser)
- For avoiding too small beam radii in optical components (leading to optical damage particularly in Q-switched lasers)
- For minimizing adverse effects of thermal lensing and aberrations in the gain medium
- For minimizing the alignment sensitivity
- For accommodating multiple laser heads
- For a certain resonator length, determining the pulse repetition rate in a mode-locked laser or the pulse duration of a Q-switched laser.

Particularly for high-power lasers with good beam quality, thermal lensing in the gain medium is very important. The resonator design should be made so that changes of the thermal lens do not affect too much the mode sizes. Also, it should have a low sensitivity to thermal aberrations and misalignment . The importance of these factors should not be underestimated; there are cases where two resonators even with equal mode sizes in the gain medium lead to very different laser performance and are radically different in terms of alignment.

Although it is normally not that difficult to evaluate the properties of a given laser resonator, it can be challenging to find a resonator design which satisfies multiple criteria such as those listed above. Numerical optimization, using special resonator design software, can be the only way to find good solutions, particularly for some mode-locked lasers. Also, a solid understanding of resonator properties can help considerably when trying to find resonator configurations with special combinations of properties, such as large mode areas and short lengths. For advanced design issues, a great deal of experience is at least as important as a versatile design software.

Some high-power lasers (for example with slab designs) are operated with *unstable resonators*, allowing a reasonable (but typically not diffraction-limited) beam quality to be achieved despite the presence of strong thermal effects in the gain medium. Due to the high diffraction losses, such laser cavities require relatively high gain. 2There are various types of monolithic solid-state lasers which have the whole beam path within the laser crystal. Beam reflections are then typically realized either with dielectric coatings on crystal surfaces, or with total internal reflection.

Laser Crystals

Laser crystals are typically single crystals (monocrystalline material) which are used as gain media for solid-state lasers. In most cases, they are doped with either trivalent rare earth ions or transition metal ions. These ions enable the crystal to amplify light at the laser wavelength via stimulated emission, when energy is supplied to the crystal via absorption of pump light (→ *optical pumping*).

Compared with doped glasses, crystals usually have higher transition cross sections, a smaller absorption and emission bandwidth, a higher thermal conductivity, and possibly birefringence. In some cases, monocrystalline laser materials may be replaced with ceramic gain media, which have a fine polycrystalline structure.

Common Laser-active Dopants

The most frequently used laser-active rare earth ions and host media together with some typical emission wavelengths:

Table. Common rare Earth Ions in Laser–Active Crystals.

Ion	Common Host Crystals	Important Emission Wavelengths
neodymium (Nd^{3+})	$Y_3Al_5O_{12}$ (YAG), $YAlO_3$ (YALO), YVO_4(yttrium vanadate), $YLiF_4$ (YLF), tungstates ($KGd(WO_4)_2$, $KY(WO_4)_2$)	1064, 1047, 1053, 1342, 946 nm

ytterbium (Yb^{3+})	YAG, tungstates (e.g. KGW, KYW, KLuW), YVO_4, borates (BOYS, GdCOB), apatites (SYS), sesquioxides (Y_2O_3, Sc_2O_3) nm	1030, 1020–1070
erbium (Er^{3+})	YAG, YLF	2.9, 1.6 μm
thulium (Tm^{3+})	YAG	1.9–2.1 μm
holmium (Ho^{3+})	YAG	2.1, 2.94 μm
cerium (Ce^{3+})	YLF, LiCAF, LiLuF, LiSAF, and similar fluorides	0.28–0.33 μm

The following tables lists common transition-metal doped crystals:

Table. Common Transition Metal ions in Laser–Active Crystals.

Ion	*Common host crystals*	*Important emission wavelengths*
titanium (Ti^{3+})	sapphire	650–1100 nm
chromium (II) (Cr^{2+})	zinc chalcogenides such as ZnS, ZnSe, and $ZnS_xSe_{1"x}$	2–3.4 μm
chromium (III) (Cr^{3+})	Al_2O_3 (ruby), $LiSrAlF_6$ (LiSAF), LiC aAlF_6 (LiCAF), $LiSrGaF_6$ (LiSGAF)	0.8–0.9 μm
chromium (IV)	YAG, $MgSiO_4$ (forsterite)	1.35–1.65 (Cr^{4+}) μm (YAG), 1.1–1.37 μm forsterite)

These tables contain only the most common host crystals; many others exist, which are less frequently used.

Important Properties of Host Crystals

The host crystal is much more than just a means to fix the laser-active ions at certain positions in space. A number of properties of the host material are important:

- The medium should have a high transparency (low absorption and scattering) in the wavelength regions of pump and laser radiation, and good optical homogeneity. To some extent, this depends on the quality of the material, determined by details of the fabrication process.
- The host medium influences strongly the wavelength, bandwidth and transition cross sections of pump and laser transitions and also the upper-state lifetime. For example, Nd:YVO_4 has much higher cross sections, a larger gain bandwidth, and a smaller upper-state lifetime than Nd:YAG. Other neodymium hosts provide other transition wavelengths, e.g. 1047 or 1053 nm from Nd:YLF.
- Nonradiative transitions (e.g. multi-phonon transitions) are also strongly influenced by the host, in particular by its maximum

phonon energy. Some of these transitions are very detrimental, leading to quenching of the upper-state population (thus lowering the quantum efficiency). Others are essential for laser operation, e.g. for removing ions from the lower laser level.
Energy transfer processes are also dependent on the host material.

- The maximum possible doping concentration can depend strongly on the host material and its fabrication method.
- Different crystalline materials are very different concerning their hardness and other properties, which determine with which methods and how easily they can be cut and polished with good quality.
- Some materials are chemically not stable, e.g. hygroscopic.
- Particularly for high-power lasers (but often enough also for medium and low powers), a high thermal conductivity low thermo-optic coefficients (for weak thermal lensing) and a high resistance to mechanical stress are desirable.
- Optical isotropy can be beneficial, but in other cases birefringence (reducing thermal depolarization) and possibly polarization-dependent gain is preferable.
- A high damage threshold in terms of pulse fluence or peak intensity can be important for high-energy amplifiers.

It is apparent that different applications lead to very different requirements on laser gain media. For this reason, a broad range of different crystals are used, and making the right choice is essential for constructing lasers with optimum performance.

Common Crystalline Laser Host Media

There is a wide range of crystalline media, which can be grouped according to important atomic constituents and crystalline structures. Some important groups of crystals are:

- Garnets such as $Y_3Al_5O_{12}$ (YAG), $Gd_3Ga_5O_{12}$ (GGG), and $Gd_3Sc_2Al_3O_{12}$ (GSGG): hard and chemically inert materials, optical isotropic, with high thermal conductivity
- sapphire (Al_2O_3) (e.g. for titanium–sapphire lasers) and aluminates such as $YAlO_3$ (YALO, YAP) for neodymium doping: high hardness and thermal conductivity, anisotropic
- Sesquioxides such as Y_2O_3, Sc_2O_3: isotropic, high hardness and thermal conductivity
- Vanadates such as YVO_4 and $GdVO_4$: very high laser cross sections of Nd^{3+}, anisotropic
- fluorides, e.g. $YLiF_4$ (YLF): good UV transparency, birefringence, large energy storage capability of Nd:YLF; also LiCAF, LiLuF, LiSAF as chromium-doped broadband gain media

- Silicates, e.g. $MgSiO_4$ (forsterite): broad gain bandwidth
- monoclinic double tungstates such as $KGd(WO_4)_2$ (KGW) and $KY(WO_4)_2$ (KYW): combination of relatively high Yb^{3+} laser cross sections, large gain bandwidth, and high thermal conductivity
- Disordered tetragonal double tungstates such as $NaGd(WO_4)_2$ (NGW) and $NaY(WO_4)_2$ (NYW): particularly large gain bandwidth of ytterbium
- Chalcogenides such as ZnS or ZnSe for mid-infrared lasers

Geometries of Laser Crystals

Different geometric forms can be used in lasers:

- A common form is that of a cuboid. The crystal can be, e.g., a thin coplanar plate, with transverse dimensions (perpendicular to the laser beam) and a thickness of a few millimeters. It may be oriented for near perpendicular incidence of the laser beam, or at Brewster's angle. It can be fixed in some solid mount which also acts as a heat sink. Larger crystals are usually used for side pumping e.g. with high-power diode bars.
- In some cases, extreme angles between the end faces are required, e.g. if one end face has to be Brewster-angled while the other one is made for perpendicular incidence.
- Slab lasers are based on relatively flat slabs, which may or may not be of cuboid form.
- Many side-pumped lasers use relatively long cylindrical laser rods, e.g. made of Nd:YAG. Particularly for lamp-pumped lasers, the rod length can be several centimeters, whereas the rod diameter is much smaller (a few millimeters).
- Thin-disk lasers require a disk, often with circular cross section, having a thickness of only e.g. 100–200 ìm and a relatively high doping concentration.
- Special geometries are required for monolithic solid-state lasers, such as nonplanar ring oscillators.
- For various reasons, composite crystals are becoming popular. These have a spatially varying chemical composition and can be made with special shapes.

Bulk Properties

For a given dopant and host medium, the doping concentration is the most important parameter. Other issues of interest are the uniformity of doping (influencing the tendency for quenching), the level of impurities (e.g. unwanted other rare earth ions), and the optical homogeneities. Several of these factors influence the absorption and scattering losses of the material, and/or the strength of thermal lensing.

Of course, it is very desirable that a given crystal quality is produced consistently, although different laser designs can have a different sensitivity to material parameters.

Optimization of Geometry and Parameters

Which geometry, dopant and doping concentration of the gain medium are most advantageous depend on several factors. The available pump source (type of laser diode or lamp) and the envisaged pumping arrangement are important factors, but the material itself also has some influence.

For example, titanium–sapphire lasers have to be pumped with high intensities, for which the form of a transversely cooled rod, operated with relatively small pump and laser beam diameter, is more appropriate than e.g. a thin disk. As another example, Q-switched lasers reach a higher population density in the upper laser level and are therefore more sensitive to quenching effects and energy transfer processes; therefore, a lower doping density is often appropriate for these devices.

For high-power lasers, lower doping densities are often used in order to limit the density of heat generation, although thin-disk lasers work best with highly doped crystals. Many laser products do not reach the full performance potential because such details have not been properly worked out.

Optical Surfaces

Those surfaces which are passed by the laser beam are normally either oriented at Brewster's angle or have an anti-reflection coating. Even AR-coated crystals are often slightly tilted against the beam so as to prevent back-reflections staying in the laser resonator. This is important for, e.g., mode-locked lasers and tunable single-frequency lasers.

A high surface quality is of course important. Specifications of surface flatness are often better than $\lambda/10$. This helps to avoid both scattering losses and wavefront distortions which can degrade the laser's beam quality. In addition, *scratch and dig specifications* (*cosmetic surface quality*) limit the density of small-scale surface defects; they may read e.g. "80–50" for medium quality mass production, or "10–5" for particularly demanding laser applications. Proper surface treatment also influences the damage threshold, which is important e.g. for high-energy pulse amplifiers. Finally, a high degree of end face parallelism can be important for avoiding changes of beam direction in a crystal.

Diode-pumped Lasers

Virtually all optically pumped lasers fall into one of two categories:

- *Lamp-pumped lasers,* having some kind of discharge lamp as the pump source
- *Diode-pumped lasers,* pumped with some kind of laser diodes

The latter category, for which the term all-solid-state lasers is also used.

Types of Diode-pumped Lasers

Most diode-pumped lasers are solid-state lasers (DPSSL = DPSS lasers = *diode-pumped solid-state lasers*). These are either bulk lasers, using some kind of laser crystal or bulk piece of glass, or fiber lasers (although the term DPSSL is less common for fiber lasers). Both categories span a range of output powers from a few milliwatts to multiple kilowatts (→ *high-power lasers*).

Less common are optically pumped semiconductor lasers (particularly VECSELs = vertical external cavity surface-emitting lasers), and there are also some relatively exotic types of diode-pumped gas lasers, e.g. alkali vapor lasers.

Types of Laser Diodes for Diode Pumping

There are different types of laser diodes which can be used for diode pumping:

- Low-power lasers (up to roughly 200 mW) can be pumped with small edge-emitting laser diodes. These exhibit a diffraction-limited beam quality and make it fairly easy to achieve the same for the solid-state laser.
- Broad area laser diodes typically generate several watts and are suitable for pumping solid-state lasers with output powers up to a few watts. Their beam quality is substantially asymmetric, but normally still sufficient for achieving a diffraction-limited laser output without using complicated optics.
- High-power diode bars emit tens of watts (or even >100 W), allowing for higher output powers, particularly when several bars are combined. Their beam quality is strongly asymmetric and fairly poor, so that their brightness (radiance) is much lower than that of lower-power diodes. Various types of beam shapers are used to symmetrize the beam quality. This makes it easier either to pump a bulk laser or to couple the light into a fiber.
- For the highest powers, diode stacks are often used. These have a still worse beam quality and lower brightness, but can provide multiple kilowatts.

In most cases, the pump diodes are operated continuously. This applies to all continuous-wave and mode-locked lasers, and also to many Q-switched lasers. However, quasi-continuous-wave operation with higher peak power for limited time intervals (e.g. 100 ìs) is sometimes used for Q-switched lasers with a high pulse energy and low pulse repetition rate.

Depending on the type of laser diode, different kinds of pump optics are used. It is also possible to use fiber-coupled diode lasers, which make it possible to separate the actual laser head from another package containing the pump diodes, so that the laser head can become very compact.

Advantages of Diode Pumping

The main advantages of diode pumping are:

- A high electrical-to-optical efficiency of the pump source (of the order of 50%) leads to a high overall power efficiency (→ *wall-plug efficiency*) of the laser. As a consequence, small power supplies are needed, and both the electricity consumption and the cooling demands are drastically reduced, compared with those for lamp-pumped lasers.
- The narrow optical bandwidth of diode lasers makes it possible to pump directly certain transitions of laser-active ions without losing power in other spectral regions. It thus also contributes to a high efficiency.
- Although the beam quality of high-power diode lasers is not perfect, it often allows for end pumping of lasers with very good overlap of laser mode and pump region, leading to high beam quality and power efficiency. In the domain of slab lasers, it allows edge pumping instead of face pumping, which brings important advantages.
- Diode-pumped low-power lasers can be pumped with diffraction-limited laser diodes. This allows the construction of very low-power lasers with reasonable power efficiency, i.e., with a very small electrical pump power, as is important for battery-powered devices.
- The lifetime of laser diodes is long compared with that of discharge lamps: typically many thousands of hours, often even well above 10 000 hours. However, various factors can lead to drastically reduced lifetimes. Also, the exchange of laser diodes is much more costly than that of discharge lamps.
- The compactness of the pump source, the power supply and the cooling arrangement makes the whole laser system much smaller and easier to use.
- Diode pumping makes it possible to use a very wide range of solid-state gain media for different wavelength regions, including e.g. upconversion lasers. For many solid-state gain media, the lower brightness of discharge lamps would not be sufficient.
- The low intensity noise of laser diodes leads to low noise of the diode-pumped laser.

Achievements

The benefits of diode pumping have lead to amazing achievements. Some examples are:

- There are miniature solid-state lasers with excellent efficiency, beam quality, spectral purity, and stability. Some of those may even be battery-powered.

- Diode-pumped high-power solid-state lasers can deliver kilowatt output powers with fairly high beam quality. This applies particularly to thin-disk lasers, but also to high-power fiber lasers and amplifiers.
- Diode pumping is also essential for a large variety of mode-locked lasers, generating e.g. up to 80 W of average output power in sub-picosecond pulses, or picosecond pulses for telecom applications with a pulse repetition rate up to 50 GHz.

Limitations

In the early years of diode pumping, the output powers achievable were very limited – smaller than those of lamp-pumped lasers. In the meantime, however, high-power diode bars and diode stacks have become very powerful, and the highest output powers are now usually achieved with diode pumping.

The main disadvantage of diode pumping (as compared with lamp pumping) is the significantly higher cost per watt of pump power. This is severe for high powers. For this reason, lamp pumping is still used in cases where high powers are needed, particularly when the power is used only for short times. For example, lamp-pumped Q-switched Nd:YAG lasers are still widely used for laser marking, and will not soon be replaced with diode-pumped lasers.

Laser diodes are electrically less robust than discharge lamps. They may e.g. be quickly destroyed by excessive drive currents, or by electrostatic discharges. In conjunction with properly designed electronics, however, this should not happen. Problems can also arise from optical feedback.

Applications

Diode-pumped solid-state lasers have a very wide range of applications. Indeed, they are used in all of the areas mentioned in the article on laser applications.

FLUORESCENCE MICROSCOPY

Surface restructuring is obtained through fluorescence imaging of the irradiated areas. The fluorescence in these areas originates from the acidsensitive acridine probe that was added to the films in combination with a thermolabile acid-generating sulfonium salt. This experimental design allows visualization of the laser-irradiated areas that were sufficiently heated for thermal decomposition of the acid generator and protonation of the acridine.

Light with a wavelength of 420 nm can selectively excite the fluorescent acridinium ion. The corresponding topography images of one area on the same films have been presented in figure. The shape of the fluorescence excitation spectra from these irradiated films is not shown but closely resembles the absorption spectrum of the acridine probe in the presence of added trifluoromethanesulfonic acid and proves the generation of acid by the laser irradiation. Two different scanning microscopy techniques were used to analyse the heated fluorescing areas of the laser-excited films: laser scanning confocal fluorescence microscopy and near-field scanning optical microscopy. The latter technique offers the additional possibility of spatially correlating the laser-induced modification of the topography with the laser-induced heating of the film area.

LASER SCANNING CONFOCAL FLUORESCENCE MICROSCOPY

The AFM analyses of irradiated films of poly(STY-*co*-BMA) in figure show the significant impact of the laser irradiation on the topography of the film, together with the influence of the infrared-dye concentrations.The lateral distribution of the thermally generated acid, detected by the acid sensitive fluorescent acridine probe, was analyzed with LSCM. The influence of the infrared dye on the fluorescence images of the irradiated film is presented in figure. Laser scanning confocal fluorescence images (38′ 25 *í*m) of irradiated films of poly(STY-*co*-BMA), for different concentrations of infrared dye (0.5, 1.5, and 3 wt %), after laser irradiation with 100 mW during 10 *í*s. The intensity of the images was normalized.

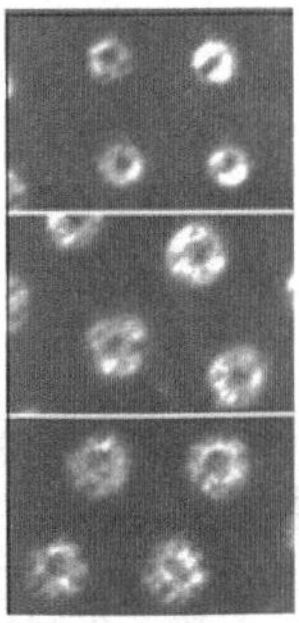

The irradiated area is composed of a fluorescent rim around a darker central area. The fluorescence in the rim is not homogeneous and contains regions of differing brightness. These darker areas fit into the model of irradiation-induced bubble formation, if we assume that these areas indicate the places where the heated film material was expelled by the formation of bubbles followed by resolidification of the film upon cooling.

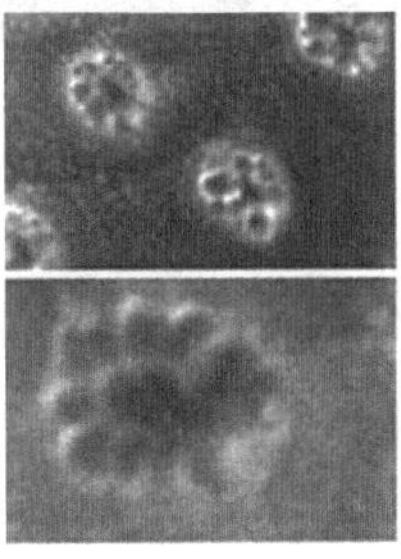

Laser scanning confocal fluorescence images (33 × 22 and 11 ´ 8 ím) of an irradiated poly(STY-*co*-BMA) film, with 1.5 wt% infrared dye, after laser irradiation with 100mW during 10 ís. The intensity of the images was normalized.

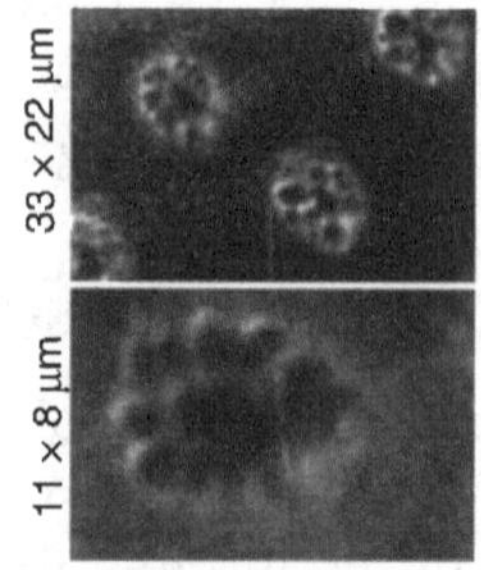

Laser scanning confocal fluorescence images (33 × 22 and 11 × 8 ím) of an irradiated poly(STY-*co*-BMA) film, with 1.5 wt%infrared dye, after laser irradiation with 100mW during 10 ís. The intensity of the images was normalized. This figure presents two additional fluorescence images with more detailed views of some bubble structures within the fluorescent rim in poly(STY-*co*-BMA). Different features are observed for poly(MMA), as shown in figure. The fluorescence in the irradiated areas on poly(MMA) is situated in a rim around a dark area. Even the most detailed analysis of 23 ´ 15 ím in figure does not indicate additional structure within this fluorescent rim.

These samples of poly(STY-*co*-BMA) and poly(MMA) were further analyzed with NSOM, because this technique allows the direct spatial correlation between the irradiation-induced topography changes and the fluorescence.

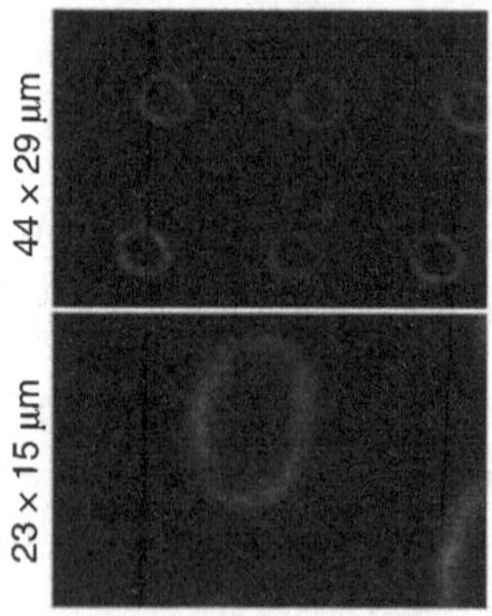

Laser scanning confocal fluorescence images (44 × 29 and 23 × 15 ím) of irradiated poly(MMA) films with 1.5 wt % infrared dye, after laser irradiation with 100mWduring 10 ís. The intensity of the images was normalized.

NEAR-FIELD SCANNING OPTICAL MICROSCOPY

The images in figure show an overview with a 35 ím scan range and a detailed view at a 15 ím scan range of one of the irradiated areas of the topography and the corresponding fluorescence for poly(STY-*co*-BMA) and poly(MMA). The main features in the fluorescence images in figure resemble those obtained with LSCM in figure.

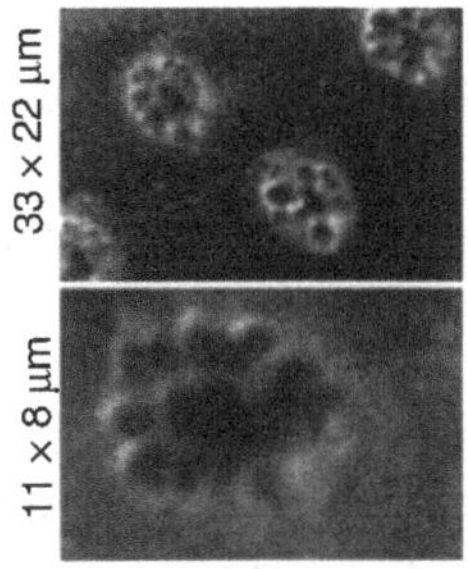

Laser scanning confocal fluorescence images (33 × 22 and 11 × 8 ím) of an irradiated poly(STY-*co*-BMA) film, with 1.5 wt%infrared dye, after laser irradiation with 100mW during 10 ís. The intensity of the images was normalized. In the case of poly(STY-*co*-MMA), the NSOM results reveal an interesting correlation between the structure of the irradiation-induced topography and the fluorescence intensity.

The crater in the centre of the irradiated area corresponds to the dark area in the fluorescence image, whereas the expanded rim around the crater is fluorescent with several darker areas having round edges. The centre of these darker regions corresponds with the lower places on the expanded rim, and the higher places of the rim are more brightly fluorescing. The different features of the topography of poly(STY-*co*-BMA) are most probably the result of several bubbles that were created in the polymer film upon laser irradiation.

Because of the Gaussian geometry of the incident infrared spot, heating is most pronounced in the centre of the irradiated area, and as a result, the bubbles are most likely to be formed and to burst in the centre, resulting in a crater. Away from the centre, bursting is less probable, and the shape of the bubbles remains more preserved. The laser-induced heightening of the surface as well as the formation of the crater, seem to be due mainly to bubble formation.

In case of the irradiation of the poly(MMA) film, the increase of the fluorescence intensity is almost 5 times less intense than that of poly(STY-*co*-BMA) and mainly originates from a small rim at the edge of the crater in the centre. Furthermore, no additional structure can be seen in the fluorescent rim of poly(MMA). Atomic force microscopy, near-field scanning optical microscopy, and laser scanning confocal fluorescence microscopy were applied in the characterization of the modified surface and bulk of doped polymer films after single-shot diode-laser irradiation.

The AFM technique revealed important topographical modifications of the irradiated polymer film and allowed a quantitative characterization. The results from AFM showed that the processes following the single-shot excitation of the infrared dye embedded in the polymer film give rise to a volcano-like structure consisting of an expanded rim around a central crater.

TheAFM results also showed the influence of several parameters of the film as well as of the irradiation conditions on the impact of the IR light on the film topography. A quantitative analysis of the irradiation-induced features revealed that their size is not simply proportional with the energy density of the incident light on the film.

For a constant energy density within the experimental range of irradiation parameters, the impact on the polymer film of the irradiation depends on the pulse time (and the IR power) for the films with poly(STY-*co*-BMA) as the polymer matrix. This observation suggests that the mechanism behind the restructuring is controlled, at least partly, by local temperature changes in the film, as a result of the nonradiative deactivation of the photoexcited infrared dye embedded in the film matrix.

Complementary information about the irradiated film was obtained with LSCM. The introduction of a thermolabile acid generator in combination with an acid-sensitive fluorescent probe in the film allowed the visualization and the quantification of the thermally affected area of the irradiated film. Direct spatial correlation between the irradiation-induced topography changes and the thermally induced fluorescence was obtained with NSOM.

A combination of the information obtained from AFM, LSCM, and NSOM suggested that the irradiation of the IR sensitive film also stimulates the release of volatile species from the heated polymer film. At least part of the expansion of the film at the rim could be due to the formation of bubbles below the film surface.

The crater in the centre of the expanded rim could be due to the displacement of heated molten film material during bubble formation, followed by the bursting of the film around and above the bubble.

Upon the explosion of the bubble, some material can be ejected, as indicated by small droplet like material around the irradiated area, which is not visible on the unirradiated film. An additional indication for the ejection of material was obtained from the film material collected on the surface of a receiver glass that was mounted above the film during the irradiation.

Example: INM300 DUV

Automated 300mm High-Resolution UV/DUV Review Microscope

Product Description: The KLA-Tencor INM300 DUV microscope includes the latest enhancements in optical imaging capabilities offering the flexibility to include 248nm DUV, 365nm UV, visual brightfield, darkfield and DIC illumination modes. With resolution capabilities down to 80 nm, advanced automation and a stage travel of 300 × 300 μm, this high-performance benchtop optical microscope can be used for critical materials, coating and process defect review for wafers, masks, optoelectronics and hard disk applications.

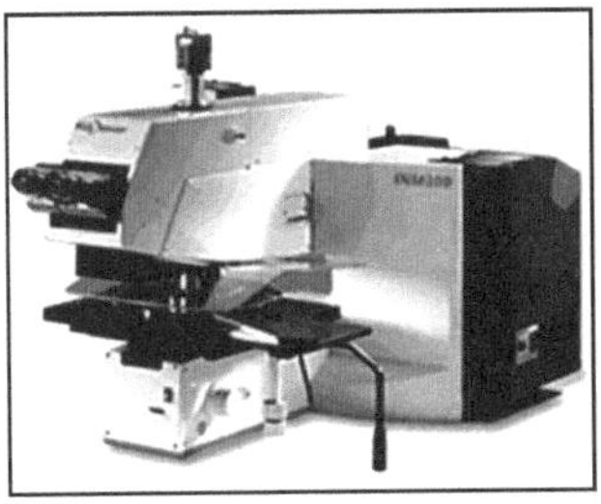

The opportunity to use various imaging modes to best suit the optical properties of the materials offers a superior advantage in being able to visualize, detect and analyse the greatest variation of process defects.

- Superb HCS visual, UV and DUV optics provide crisp and high contrast images with highest resolution.
- Real-time autofocus with upper end-switch for sample protection
- Imaging modes from brightfield, darkfield, DIC, 365nm UV and 248nm DUV with automated controls.
- 150x/0.90 UV (365nm) objective with 130nm resolution.
- Cement-free 150x/0.90 DUV (248nm) objective with 80nm resolution.
- Long Working Distance L50x DUV cement-free objective for thorough pellicle inspection of masks.
- Cement-free 200x/1.25 DUV water immersion objective with 60nm resolution.
- Upper limit focus stop and programmable nosepiece for safety and ease of use.
- Modular system configurations for all wafer sizes with manual or scanning stages.
- User-friendly VISCON software packages for semi-automated inspection and review.

9

Protocols for Light Microscopy Specimen Preparation

Simple method for preparing thin sections of biological material for light microscopy. These sections give better high resolution information and provide better specimen preservation than even the thinnest cryostat section. May even be better than confocal optical sections!

LIGHT

Light may be defined subjectively as the sense-impression formed by the eye. This is the most familiar connotation of the term, and suffices for the discussion of optical subjects which do not require an objective definition, and, in particular, for the treatment of physiological optics and vision. The objective definition, or the "nature of light," is the *ultima Thule* of optical research. "Emission theories," based on the supposition that light was a stream of corpuscles, were at first accepted.

These gave place during the opening decades of the 19th century to the "undulatory or wave theory," which may be regarded as culminating in the "elastic solid theory "- so named from the lines along which the mathematical investigation proceeded - and according to which light is a transverse vibratory motion propagated longitudinally though the aether.

The mathematical researches of James Clerk Maxwell have led to the rejection of this theory, and it is now held that light is identical with electromagnetic disturbances, such as are generated by oscillating electric currents or moving magnets. Beyond this point we cannot go at present. To quote Arthur Schuster, "So long as the character of the displacements which constitute the waves remains undefined we cannot pretend to have established a theory of light."

Optical and electrical phenomena are co-ordinated as a phase of the physics of the "aether," and that the investigation of these sciences culminates in the derivation of the properties of this conceptual medium, the existence of which was called into being as an instrument of research.' The methods of the elastic-solid theory can still be used with advantage in treating many optical phenomena, more especially so long as we remain ignorant of fundamental matters concerning the origin of electric and magnetic strains and stresses; in addition, the treatment is more intelligible, the researches on the electromagnetic theory leading in many cases to the derivation of differential equations which express quantitative relations between diverse phenomena, although no precise meaning can be attached to the symbols employed.

It is the fastest in the universe, and has to be the fastest, since nothing can even reach its speed. It can split into colours, and can be recombined from colours. In fact, a universe without it cannot even be imagined.Light, or "Adustum", as it is known in Latin,is a most apt title for our web site. Light is the ultimate form of energy. It is not just about the visible wavelengths, but even more about the numerous others, that sustain life on earth.

White light received from the Sun has all these wavelengths, and forms a much wider electromagnetic spectrum, that the visible spectrum. Well, that is what our site is all about - not just visible light, but light as a whole.Each of the six main links in the site come with a set of sub-links that deal with individual topics. Besides text, there is a huge collection of diagrams and images that have been prepared and collected to make the matter more lucid and interesting.

Now what exactly is light? A Scottish physicist named James Clerk Maxwell showed that electric and magnetic fields fluctuating together can form a propagating wave, which was named an electromagnetic wave. Light is this type of wave. Maxwell knew that a changing electric field produced a magnetic field. He hypothesized that a changing magnetic field would produce an electric field because most things in nature are balanced. Based on the work of Farady and Hertz, this was discovered this to be true.

A common way to generate an electromagnetic wave is to have an antenna connected to an alternating current which will cause a changing magnetic field which will cause a changing electric field which will cause another changing magnetic field and so on. These waves are also produced in stars in the form of UV rays, X rays and Gamma rays.

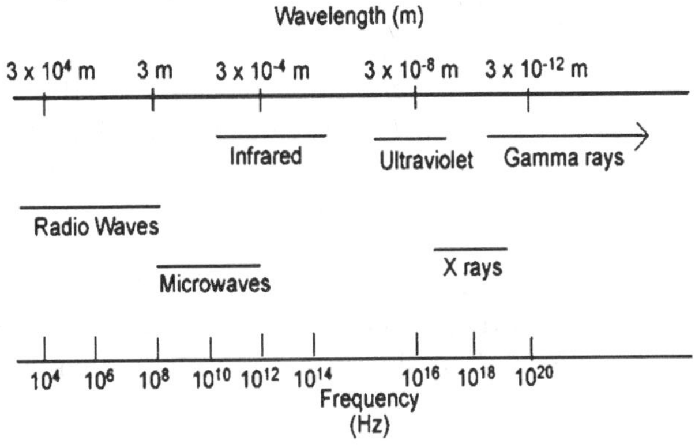

An electromagnetic wave, like other waves, has a frequency f, a speed v and a wavelength λ, which are related by the equation $v = f\lambda$. In vacuum or in air, to a good approximation, the speed $v = c$ (speed of light: 3×10^8 m/s) so the relationship would be $c = f\lambda$

Above shows the electromagnetic spectrum, which depicts electromagnetic waves over a huge range, for values less than 104 Hz to those greater than 1022 Hz. Because all of these waves travel at the same speed, this drawing can also be used to figure out the wavelengths. The grouping of the spectrum into radio waves, infrared, ultraviolet, x and gamma rays. Shown here the boundaries are sharp but in actual practice the boundaries are not so well defined and the regions overlap.

Beginning on the left are the radio waves. The lower frequency radio waves are generally produced by electric oscillator circuits while the higher frequency waves (microwaves) are generated using electron tubes called klystrons.

The next set, called infrared or loosely referred to as heat waves, generally originate with the vibration and rotation of molecules within a material. The next set,which we are most familiar with, is the visible light waves, represented by a white stripe. These are generally emitted by hot objects, such as the sun or the filament of an incandescent light bulb where the temperature is high enough to excite the electrons within an atom. Ultraviolet rays are generally produced from the discharge of an electric arc and x-rays are generated by the sudden deceleration of high-speed electrons. And last, gamma rays are radiation from nuclear decay.

Of all the frequencies, the most familiar set, that of light, is the smallest range indicated in the spectrum. The human eye can perceive only the frequencies between 4.0×10^{14} Hz and 7.9×10^{14} Hz as visible light. For the most part, visible light is discussed in terms of wavelengths with the unit being nanometers (nm) where 1 nm = 10^{-9} m rather than in frequencies. Our brain recognizes these wavelengths as different colors where the 750 nm is approximately the longest wavelength of red light and 380 nm is approximately the shortest wavelength of violet light. In between these two wavelengths are all of the other visible colors.

Wavelength also plays a very important part on other parts of the magnetic spectrum besides defining color for visible light. One important aspect about wavelength is that it determines how much a wave diffracts, or bends around an obstacle. Longer wavelengths bend around an obstacle more than shorter wavelengths. An example of why this is important is with radio waves. AM waves are significantly longer than FM waves and therefore they have a greater ability to bend around buildings than FM waves do.The inability of FM waves to diffract around obstacles is one of the major reasons why FM stations broadcast their signals from high areas in a "line-of-sight" manner.

HISTORY OF LIGHT

Many of the early scientists who studied sound also studied light. Pythagoras, for example, believed that light came from visible objects toward the eye. However, in addition to this basically correct thought, Plato and many other Greeks, also held the mistaken belief that vision issued out from the eye. Despite this, many of the ideas of the ancient Greeks were accurate. The philosopher Empedocles correctly believed that light traveled with finite speed. Aristotle, too, conjectured about light as well as sound. He rightly explained rainbows as a sort of reflection off of raindrops. The mathematician Euclid worked with mirrors and reflection, and many other thinkers observed refraction, though they did not know how to express it mathematically.

Ptolemy is the first recorded person to experiment with optics and collect data, but he believed in Plato's mistaken thought that vision extended out from the eye. Ptolemy's work was further developed by the Egyptian scientist Ibn al Haythen, who was known to Europeans as Alhazen. It was Alhazen who first drew ray diagrams, and who managed to discount Plato's theory, through a mixture of logic and experimentation. His work, done during the time of the Moorish Empire, was influential in later studies of light.

Many advances in the study of light were made in the 16th and 17th centuries by such renowned scientists as Galileo Galilei, Johannes Kepler, and Renes Descartes. Both Descartes and Dutch mathematician Willebrord Snell independently developed the law of refraction, basing much of their work on the earlier work of Alhazen. Snell discovered it earlier, however, and the law is now known as Snell's law.

During the late 17th century, a debate grew over whether light behaved as a particle or a wave. Sir Isaac Newton believed in a particle theory of light, in part because he did not observe diffraction in light, a property it should have had, had it been a wave. Though he did not know it, this was due a lack of coherency in the light when he was experimenting. Coherency is important, because without a steady, coherent beam, it is impossible to observe such things as diffraction of light. These days, coherency is generally achieved with a laser, a modern invention that was unavailable to Newton. There were those among his contemporaries, however, who believed in a wave theory of light. The most notable among these was the Dutch scientist Christiaan Huygens, who first wrote of light as a wave.

Though Newton had difficulty explaining certain phenomena of light, it seemed that the wave theory also had difficulty explaining certain optical phenomena. Because of Newton's prestige, the particle theory was accepted for almost a century.

It was not until the early 19th century that the wave theory of light became widely accepted. This acceptance came in large part due to the work of the English doctor Thomas Young. In 1801 he reported his famous double-slit experiment, which clearly showed light to diffract, and thus travel as a wave. By passing light through a single slit before passing it through a double slit, he managed to emulate a point source of light, and to achieve the coherency that had eluded Newton. In the 1850s Fizeau and Foucault showed through measurements that light traveled slower through denser media. In the same century, Augustin Fresnel and later James Clerk Maxwell, working within a wave theory of light seemed to explain phenomena that Newton had been unable to explain in terms of a particle theory of light, such as polarization, interference, and diffraction.

They were also able to address the question of what determines which part of the light will be reflected and which transmitted when light is reflected at a surface such as glass or water. This question was opened back up in the 20th century with the rise of quantum theory. It was the Irish mathematician Sir William Hamilton who developed a theory that joined optics and mechanics, thus elucidating the relationship between wave and particle viewpoints. His theory helped lay the foundation for the later development of quantum mechanics

James Clerk Maxwell did much more to aid the study of light. In essence, he was the father of our modern perception of light. Working from Michael Faraday's discoveries in electricity and magnetism, he managed to fully explain electromagnetic phenomena, developing a theory that described how electromagnetic waves traveled through space. His theory was later experimentally proven correct by Heinrich Rudolph Hertz, the man whose name gives us the unit hertz, one of which is equivalent to one cycle per second.

The velocity of the electromagnetic waves that Maxwell had described was found to be the same as that of light, essentially proving that light was an electromagnetic wave.Maxwell's work seemed to have finally settled the issue of whether light was a wave or a particle, but the whole debate was reopened in the 20th century. Scientists such as Albert Einstein, who described the Doppler effect for light, brought the particle theory back into the picture with quantum theory. This time they postulated that light did not just behave as a particle or a wave, but had properties of both.

PROTOCOLS FOR TEM SPECIMEN PREPARATION

Theoretical, technical, and practical explanations on how to examine biological specimens with microscopes. Includes negative staining, embedding, colloidal gold production, immunocytochemistry, crossectioning and more.

The Light Microscope: An explanation of how the light microscope works, how to use it, and how to get optimal results when using it.

The Transmission Electron Microscope (TEM): An explanation of how the TEM works.

EM visualization techniques: An introduction to the routine methods used when preparing specimens for transmission electron microscopy.

TEM SPECIMEN PREPARATION

Preparative Techniques for the TEM

For routine transmission electron microscopy (TEM), it is generally accepted that specimens should be thin, dry and contain molecules which diffract electrons. Biological specimens, which are large and consist of large amounts of water, also do not defract electrons and are therefore difficult to see in the TEM. Preparing biological specimens for the TEM, whilst retaining the structural morphology of the material, is a challenge. However, researchers have been looking at biological material for many years, and many protocols exist which allow us to look at biological material in many different ways. Below is a brief outline of some of the more common ways of looking at biological samples in the TEM.

Whole mounts

Small or very thin objects can be examined directly by mounting them onto a support film and introducing them directly into the electron beam. Contrast is provided by heavy metal precipitation in one of three ways.

1. *Positive staining:* The object is chemically stained with a solution of the metal salt and appears dark on a bright background.
2. *Negative staining:* The object remains unstained but is embedded in a dried film of the heavy metal salt. The specimen appears light on a dark background. This method of visualization has been used extensively in the study of virus particles but is also useful for cell fractions (e.g. coated vesicles). More details can be found *here.*
3. *Shadowing:* A thin layer of heavy metal atoms is deposited on the specimen by evaporation in a vacuum chamber. Shadowing from one direction only produces a pseudo-three-dimensional image. Rotary shadowing, where the specimen is uniformly coated with heavy metal, is used to visualize nucleic acids and proteins.

Ultrathin Sectioning

The most popular technique for examining biological materials is to embed the material under study in plastic and cut ultrathin sections that can be examined in a TEM. The material is stabilized by chemical fixation (usually with aldehydes such as formaldehyde or gluteraldehyde), contrasted with solutions of heavy metal salts (osmium tetroxide and uranyl acetate), dehydrated in ethanol or acetone, and embedded in plastic (epoxy resin).

Ultrathin sections (60 nm) cut with glass or diamond knives using an ultramicrotome are floated on water, transferred to specimen support grids and examined in the TEM. Often the sections are further contrasted with uranyl acetate and lead citrate prior to examination in the microscope. In some cases, macromolecules can be specifically labelled prior to embedding and sectioning. For example, the location of some enzymes can be visualized by incubating the tissue with a substrate whose reaction with the enzyme leads to the local deposition of electron opaque material. Alternately, antibodies can be coupled to such enzymes, and the electron opaque reaction product is used to localize the antigens recognized by the antibodies.

Some embedding resins (e.g. Lowicryl resins and LR White resin) have been designed to enable antibodies and electron opaque markers (such as colloidal gold particles) to be applied to the ultrathin sections. In this way, subcellular antigens recognized by the antibodies can be localized with the TEM.

Another sectioning technique that is increasing in popularity is *cryosectioning* (the sectioning of vitrified, frozen material). After chemical fixation, the tissue is immersed in cryo-protectant (usually sucrose) and then quickly frozen in liquid nitrogen. The cryo-protectant allows the biological material to be frozen without the formation of ice crystals, which would damage ultrastructure. This type of freezing, or vitrification, is possible in the absence of cryo-protectants but is technically demanding. Sections cut from the vitrified block can be thawed and incubated with antibodies specific to subcellular antigens. Electron opaque markers allow the antibodies to be seen in the TEM.

Colloidal gold coupled to protein A (a protein from bacterial cell walls which binds to the Fc portion of some immunoglobulins) has been used extensively in recent years to localize antibodies on resin and *frozen sections* of biological materials.

The ability to produce homogeneous populations of colloidal gold with different particle sizes has enabled researchers to use these probes to colocalize different structures on the same section.

Cryofixation

It is possible to freeze biological material fast enough to vitrify the water present inside the cells. Vitrification of water occurs when the freezing has occurred so fast that ice crystals have no time to form. Vitrified biological material can be sectioned at low temperatures. Thin films of vitrified water and sections of vitrified material can be examined in transmission electron microscopes that are equipped with specimen stages that can be kept cold.

Rapid Freezing Methods

There are seven main rapid freezing methods presently available. They are:

1. *Immersion freezing:* The specimen is plunged into the cryogen.
2. *Slam (or metal mirror) freezing:* The specimen is impacted onto a polished metal surface cooled with liquid nitrogen or helium.
3. *Cold block freezing:* Two cold, polished metal blocks attached to the jaws of a pair of pliers squeeze-freeze the specimen.

4. *Spray freezing:* A fine spray of sample in liquid suspension is shot into the cryogen (usually liquid propane).
5. *Jet freezing:* A jet of liquid cryogen is sprayed onto the specimen.
6. *High pressure freezing:* Freezing the specimen at high pressure to subcool the water.
7. *Excision freezing:* A cold needle is plunged into the specimen, simultaneously freezing and dissecting the sample.

Freeze-fracture Followed by Ffreeze etch and Replication

If, for some reason, the object to be studied cannot be examined in the TEM, then a thin replica can be made. This is usually made by evaporating a thin layer of a heavy metal (usually platinum) onto the specimen and then coating this with a thin layer of carbon.

The object and the replica are separated either by floating off the replica or by digesting away the object. There are four basic steps to follow

1. The specimen is frozen (often without regard to *vitrification*).
2. The specimen is fractured, while still frozen, under vacuum.
3. The fractured specimen can then be etched by leaving it frozen and under vacuum. Depending on the time of exposure, more or less water sublimes from the specimen (freeze drying).
4. A replica of the fractured surface is made which is then examined in the electron microscope.

A recent modification of this method employs rapid freezing achieved by slamming cells against a copper block cooled to -269°C with liquid helium. If these frozen cells are then exposed to extensive freeze drying (deep etching), very impressive images of the internal structures of cells are uncovered.

NEGATIVE STAINING

Introduction

Specimens that are to be examined in the transmission electron microscope (TEM) usually have to be thin, dry and contain contrast (usually from a heavy metal stain.

One of the easiest ways of preparing biological samples, of small size, for the TEM is by negative staining. This preparation method is useful for visualizing suspensions of small particles, which includes viruses, purified proteins, liposomes and small vesicle fractions.

Negative staining is a simple technique for routine examination of structure. It does not allow for high resolution examination of samples - more technically demanding methods, using sample vitrification (or rapid freezing) are used for this.

As with most techniques, there are levels of complexity which yeild increasingly better results. The major differences between the techniques is in the choice of sample support.

Staining Samples on Formvar-Carbon Coated Grids

The sample is suspended (or diluted) into water (if possible) and adsorbed onto a carbon-coated formvar film which is attached to a metal specimen grid. The carbon surface of the grid becomes contaminated when stored, and thus hydrophobic, so it is best to glow discharge the grid surface (making it hydrophilic) prior to use. This is usually done in a vacuum chamber of a vacuum evaporator.

Once the specimen has been adsorbed onto the film surface, the excess sample is blotted off and the grid is covered with a small drop of stain solution. This is left on the grid for a few minutes and then blotted off. The sample is dried and examined in the TEM. If carbon-formvar films are prepared on specimen grids with small mesh size, the films, with adsorbed sample, can usually be immunolabeled without them breaking.

This protocol is described in more detail in the section on *immunolabeling*. Briefly, the grids are floated onto drops of diluted antibody. washed by floating on drops of buffer, and then floated on drops of diluted visualization probe. We normally use *colloidal gold* coupled to staphylococcal protein A (protein A-gold) or other affinity markers such as antibodies.

Methods Using Carbon Films Alone

Carbon-Coated Grids

The grids are prepared by first preparing formvar-carbon coated grids and then removing the formvar support. This is done by placing the grids in an atmosphere of solvent vapour, which dissolves the formvar (which is a plastic). The grids are placed on a wire mesh in a glass perti dish, the solvent (chloroform or carbon tetrachloride) is placed in the dish below the wire mesh and the dish is closed by replacing the lid.

If the vapour alone does not remove the film (which should take a few hours), the process can be accelerated by dipping the coated grids into the solvent prior to closing the dish and placing them on the mesh to soak.

Allow the solvent to completely evaporate before removing your grids, which will now have only carbon on them. Remember that the carbon must be thick enough to be self supporting. Advantages include high stability in the electron beam and high resolution examination of the adsorbed specimen. Disadvantages include a hydrophobic surface which is not of uniform thickness.

Carbon Films

For this technique, carbon film is evaporated onto a freshly cleaved mica surface and the sample is applied to the surface of the carbon film attached to the mica. The carbon film and the sample is then picked up onto a specimen grid and examined in the TEM. The carbon films can be stored on the mica for long periods without them becoming contaminated, the films are thin but tough so can be easily examined in the TEM without any noticable specimen movement. However, they are fragile and must be manipulated with care. They are more stable if supported on specimen grids with small mesh size (e.g. 400 mesh). Hexagonal grids seem to offer a more stable support.

Making the Carbon Film

Take a piece of mica and cut off a square or rectangular piece that is approximately 2 × 3 cm. Cleave the mica with a razor blade or scalpel that has been cleaned with acetone. Once the mica has begun to separate along a cleavage plane, forceps may be used to pull the mica completely apart. It is better to do this than to scratch the surface of the mica with the razor blade. Mica, as purchased from most EM supply companies, is rather thick and a new piece may be split around 4 times. Attach the mica to a filter paper, with sticky tape, with the freshly cleaved plane facing upward. Place the filter paper in a vacuum evaporator for carbon coating and deposit a film of carbon onto the mica surface.

This mica should be cleaved immediately before coating as the freshly cleaved surface is clean and hydrophilic, but it becomes contaminated (and thus hydrophobic) over time time. After coating, the filter paper should be light gray in colour (compare the filter paper behind the mica with that which was exposed, and therefore carbon coated). This is a good indicator of the film thickness.

Negative Staining

Prepare the sample (the staining works best if the sample is in water) by making several serial dilutions of the preparation. The best negative stain preparations are those that have a single layer of individual, separated particles adsorbed onto the film. This is achieved by dilution of the sample.

With a pair of scissors, cut off a small square of mica, about 4 x 5 mm, about twice the size of an EM grid. Put the square on a piece of parafilm, carbon side up. With a pipetman, gently squirt 5 - 10 ml of the sample under the carbon by placing the pipette tip to the side of the mica square. The sample will do one of two things: 1) It will flow between the carbon and mica, in which case you will need only a small amount of sample. 2) It will flow between the mica and the parafilm, in which case just keep applying sample to the side of the square until it goes under the carbon.

Place a 400 mesh specimen grid onto the carbon. (Wash the grid in 0.5% acetic acid and then acetone prior to use) Break the carbon film to free the specimen grid, lift the grid and place it on a drop of stain solution for about 30 seconds (sample side down). Blot dry, and examine in the TEM when completely dry. Staining and washing times can be varied. A better support for the carbon film can be obtained by using specimen grids coated with a holey film of formvar.

This is a film containing lots of small holes which are covered by the carbon film. There are many ways to produce these films but one quick, simple way is to breath onto a film of formvar before it dries onto the glass slide. The moisture droplets will displace the film, leaving small holes. Althernatively, the wet film on the glass slide can be placed in a cloud of steam.

Adenovirus Particles

These adenovirus particles have been adsorbed onto a carbon film that was deposited onto a freshly cleaved mica surface. The film was picked up onto a clean, 200 mesh specimen grid coated with a holey formvar film. The preparation was stained for 1 minute with neutral 1% aqueous phosphotungstic acid and photographed in a transmission electron micrograph.

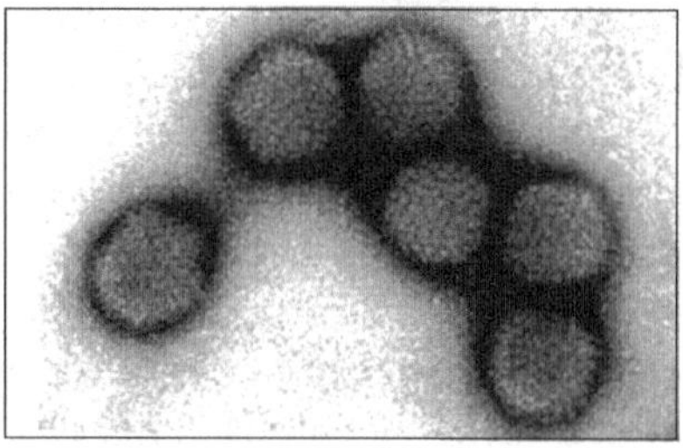

Negative Stain Solutions

Aqueous Uranyl Acetate: A 1% to 3% solution of uranyl acetate dissolved in water can be used to negatively stain many samples. The stain has a low pH so this solution is not recommended for particles that are unstable in acid conditions. *Neutral Phosphotungstic Acid:* A 1% to 3% solution of phosphotungstic acid is made up in water and the pH is adusted to 7 using sodium hydroxide. This is a useful stain for many samples but is especially good for viruses that dissociate at low pH. The stain produces less contrast than the uranyl acetate. *Ammonium Molybdate:* Make up a 1% solution of ammonium molybdate in water. This solution has also been used to negatively stain thawed, thin cryosections of fixed cells.

B. EMBEDDING IN EPOXY RESINS

Epon, Araldite and mixtures

Resin Embedding

Basic Processing Protocol for Tissue Culture Cells Grown in Plastic Dishes.

This method can easily be applied most biological samples without problem.

- Wash cells with PBS twice. Remove the PBS and replace it with 3% gluteraldehyde in 100 mM sodium cacodylate (pH 7.4). After 1 hr, remove the fixative and replace with buffer containing either 1% BSA or 1% FCS. *The cells can be washed with 100 mM sodium cacodylate buffer instead of PBS. This may alter the morphology of the mitochondria.*

- Carefully scrape the cells from the surface of the plastic, preferably using a wedge-shaped piece of teflon held in artery forceps, transfer to Eppendorf tubes and spin them down into a pellet. The presence of the protein in the buffer will stop the cells from sticking to the sides of the centrifuge tube.
  Cells may be scraped as soon as the aldehyde is added and the cells left to fix as a pellet. This may cause the cell pellet to stay together during processing.
- If the pellet of cells will hold together, remove them from the Eppendorf tube and transfer them to glass vials for further processing.
  If the cell pellet is too small or fragile for this, leave in the Eppendorf tube. Large pellets cannot be easily processed in the bottom of these tubes. If the pellets are too large but also fragile, the cells can be split into smaller pellets and processed more easily in their tubes.
- Wash pellets with 200 mM sodium cacodylate buffer (3 x 5').
  At this stage, the samples will be fixed so can be washed in any sutable buffer, and even water. Sodium cacodylate is suggested because it is often the buffer used during the first aldehyde fixation.
- Post fix in 2% osmium tetroxide in 100 mM cacodylate buffer (1 hr on ice).
  Take care with the osmium tetroxide, even the fumes are dangerous.
- Wash in water (3 × 5').
- Wash in 50 mM sodium maleate pH 5.2 (3 × 5').
- *En bloc* stain in 1% uranyl acetate in maleate buffer pH 5.2 (1 hr RT).

Using uranyl acetate at low pH will improve membrane contrast. Analternative contrasting method, which adds even more membrane contrast, is to soak the samples iovernight at 4°C in a saturated solution of uranyl acetate in70% methanol.

If using this method omit the next water wash and continue with the dehydration steps.

- Wash in water (3 × 5').
- Dehydrate: 70% ethanol (3 × 5'). 95% ethanol (3 × 5'). 100% ethanol (3 × 10').
- Replace the ethanol with propylene oxide (2 × 5'). Cells that have not been scraped from the culture dishes and pelleted can be removed from the plastic surface of the dish at this stage.

The propylene oxide will dissolve the plastic and the cell layer will float off. Remove the cells quickly because the propylene oxide continues to dissolve the plastic.

Transfer the cell layer, in propylene oxide, to Eppendorf tubes, pellet them down and continue. These pellets do not hold together very well so tke care that they do not fall apart in the resin.

- Leave in fresh propylene oxide for 20' then in propylene oxide/ Spurr's resin (1:1 mix) overnight. Mixing during the incubation will help infiltrate the Epon into the pellet.
  A slow turning wheel will achieve the correct amount of mixing.
- Remove the cap from the tubes and leave open for 2 hr. Remove Spurr's resin and replace with fresh resin, return the sealed container to the mixing wheel and leave for 4 hr in resin.
- Transfer the cell pellets to fresh Spurr's resin in molds, label the blocks with paper labels and leave at 60°C overnight. After one night in the oven, the pellets should be hard enough to trim and section.

If you are impatient for your results and find that this method is too long. Embedding using microwave ovens is a useful alternative.

IMMUNOCYTOCHEMISTRY

Introduction to Immunocytochemistry
A brief overview of common available methods.

Introduction to Immunocytochemistry

Immunocytochemistry is the name given to methods that use antibodies to detect the location of proteins within cells. The antibodies bind specifically to the protein being investigated. With electron microscopes this method is used with electron opaque markers which bind to the antibodies and visualize their location within the cell. One useful electron opaque marker is *colloidal gold*.

For electron microscopy it is important to have preparation methods that open the cells and allow access of antibodies to the proteins (or antigens) they bind to without destroying the normal cellular organization (or morphology).

Over the last ten years, specimen preparation techniques, especially those for immunocytochemistry, have reached high levels of sophistication and allow cells to be opened, labeled with *antibodies* and colloidal gold and yet retain good morphology.

At the moment, the very best way to gain access to the inside of cells is by cutting sections through them. In this way the morphology is retained but antibodies gain total accessibility to the cut face of the cell.

To cut sections thin enough to be examined in the electron microscope (approximately 40 - 100 nm), the cells must be made hard, either by embedding in A*resin* or by freezing them in a cryoprotectant. Both methods are used routinely in the Centre for Cell Imaging at Yale. For antigens which are only present in small amounts in the cell the most sensitive method is usually to label *sections of frozen material.*

Antibody labeling: Labeling protocols using affinity markers.

Embedding in resins for immunocytochemistry: Describes the use of the specialized and not so specialized resins.

Lowicryls, LR White, LR Gold etc.

Embedding in Resins for Immunocy to Chemistry

Biological specimens are embedded in resin so that they can be thin sectioned.

Many different resin formulas are currently available and all have been used for post-embedding labeling (i.e. where antibodies are applied to thin sections of the tissue) with variable results.

Specialized resin formulas have been developed specifically for use in immunocytochemistry. These include the Lowicryl resins, LR White and LR Gold, Unicryl and MonoStep.

They all have low viscosity and can be polymerized, at low temperatures, with ultra violet (UV) light.

The most common embedding protocol for these resins is called the progressive lowering of temperature (PLT) method where the sample is progressively cooled during the dehydration steps.

The specimens are then transferred to resin and the blocks are polymerized, at low temperature, by exposure to UV light. An alternative method for embedding specimens in these resins is by *freeze substitution*. This method consists of freezing the specimen and then substituting the frozen water in the specimen with a liquid solvent. Follow the following links for more details on the *theory* of freeze substitution and for a *simple protocol* of freeze substitution.

Freeze Substitution for Embedding in Resins

A simple, inexpensive way of embedding in resin using freeze substitution methods.

Freeze Substitution

This method has been used successfully in our laboratory since 1994. The protocol assumes all users will be familiar with routine TEM preparation methods. An important application of this method was made by *Dr. A. Whitney* who used it to examine MDCK cells grown on filters.

Aldehyde Fixation

Fix cells or tissues in buffered aldehyde solution. We use either 4% formaldehyde (made up from paraformaldehyde) or a mixture of 4% formaldehyde and 0.2% gluteraldeyhde, buffered in 100mM phosphate buffer. Fixation times vary but routinely we use 1 hr. Whole organs are best fixed by perfusiing the with fixative.

Cryoprotection

Incubate fixed material in 2.3M sucrose. This cryoprotection method is also used for material that is to be frozen for *cryosectioning*. Small pieces of the material are left for 1 hour to infiltrate.

Freezing

Mount the small, cryoprotected sample pieces onto metal pins and immerse in liquid nitrogen. The pins are useful for manipulating the frozen samples during the early stages of the protocol and ensure the sample is submerged during freezing.

Substitution

Quickly transfer the samples into glass vials of methanol on a bed of dry ice in a styrofoam box. Alternatively, the vials can be left in a REVCO freezer. Seal vials well and leave for 8 - 48 hours. Smaller samples will substitute faster than larger samples. Increased contrast can be achieved if 1% uranyl acetate or osmium tetroxide is added to the substitution mix. These heavy metals do not interfere with resin polymerization. Dry acetone can be used as a substitution medium instead of methanol. However, it is important for the acetone to be dry.

Washing

Wash in fresh changes of substitution medium prior to embedding in resin. Usually 2 - 3 changes over a few hours is sufficient.

Embedding in resin (using Lowicryl HM20)

Time	MeOH:resin
1 hr.	1:1
1 hr.	1:2
1 hr.	pure resin
overnight.	pure resin

Resin Polymerization

Remove sample pieces from the glass vials and place in precooled BEEM capsules or small Eppendorf tubes containing fresh HM20 Lowicryl resin. Keep the tubes on dry ice and seal them as soon as the sample pieces have been added. Place only one sample per tube. Arrange the tubes on a rack in a styrofoam box, covered on the inside with aluminum foil, and pack with dry ice.

Cover the tubes with a triangular shaped cardboard box that has been covered with aluminum foil. Place a UV light above the box and leave for up to 3 days to polymerize.

The resin polymerizes in indirect 360 nm UV light. Too rapid polymerization will result in shrinkage of the block and uneven polymerization. Once the blocks have hardened, polymerization can continue at room teperatures using direct sunlight or the UV light used to sterilize cell culture hoods

Sectioning

Lowicryl resins do not crosslink well at tissue-resin borders. It is important, when trimming the block, to make sure the sample is not dislodged. The best way to trim the block is by using either a glass knife or one of the trimming devices currently available (supplied by Diatome US or Harris Diamond Co.).

Once the block is trimmed, it is fairly easy to section using routine sectioning methods. Diamond knives are recommended. The HM20 resin will not change shape if water is splashed onto the block. Sections can be picked up from the water surface with metal specimen grids.

Cryosectioning

A detailed description of how to produce high quality cryosections of fixed, cryoprotected biological material. Using resin sections for immunofluorescence. A protocol for applying antibodies with fluorescent markers onto sections of resin-embedded material. The thin sections help increase resolution.

IMMUNOCYTOCHEMISTRY

Preparation of Tissues and Cells for Immunocytochemistry

Successful immunocytochemistry is only possible when antibodies (or other affinity markers) are able to bind to their target ligands. For extracellular antigens this is not a problem, but for molecules inside cells the affinity markers must gain access.

The best way to gain access into cells, for electron microscopy, is to cut sections through them. To prepare thin sections through biological material is not a simple process. Usually the water within the material is replaced with a medium that can be hardened, either by freezing or by polymerization.

The cryosection technique is now one of the two most important techniques for subcellular immunocytochemistry, the other being resin embedment.

Cryosectioning has been applied to the localization of many different antigens and for some antigens it appears to be the most sensitive detection method available. An extra advantage is that semi-thin sections cut from the same block as that used for thin sections can be labeled with fluorescent or gold markers for a light microscopic overview of the labeling pattern.

Preparation of colloidal gold probes: See how easily, and inexpensively, these expensive probes can be prepared. Excellent information if large volumes of gold probes are required.

Preparation of Colloidal Gold Conjugates.

Colloidal gold has been used for centuries in the preparation of stained glass for windows and fine glassware. In recent years, colloidal gold particles have become a useful tool for microscopists. Colloidal gold particles are especially useful for biological electron microscopy.

- Homogeneous preparations of particles varying in size from 3 nm to 20 nm can be easily prepared.
- Colloidal gold suspensions are inexpensive to prepare.
- Most proteins can be easily coupled to colloidal gold particles.
- Proteins coupled to gold particles do not appear to lose their biological activity.
- The colloidal gold particles can be easily seen in the electron microscope.
- Colloidal gold probes can be used for light microscopy. The larger gold particles can be directly observed by the light microscope. Smaller particles are detected by silver enhancement or epipolarized illumination.
- The same probes can be used for both LM and TEM immunocytochemistry.

Preparation Procedure for Producing Gold Sols

This page will explain a simple, published protocol for preparing colloidal gold particles, how to couple these particles to protein A and how to purify the probes after they have been made.

To make 100 ml of gold sol, two stock solutions have to be prepared.

- *SOLUTION A:* 80 ml distilled water and 1 ml 1% aqueous gold chloride.
- *SOLUTION B:* 4 ml 1% tri-sodium citrate $2H_2O$ + 16 ml H_2O + variable amount of 1% tannic acid.

When 1 ml or more tannic acid is needed, add an equal amount of 25 mM potassium carbonate for pH adjustment. Warm up solutions A and B to 60°C and mix them while stirring. When the red colour has formed heat up to 95°C and cool the solution on ice. The larger particles (where lower concentrations of tannic acid are used) take longer to form and the red colour can take up to 1 hr to develop.

Coupling the Protein A to the Gold Particles.

A gold sol will bind proteins more efficiently when the pH of the solution is close to the pI of the protein (for protein A the pI is pH 5.1). A tighter binding occurs at higher pH but this may have a denaturing effect on the protein making the probe less effective. Having too much protein coupled to the gold particles may also be disadvantageous: some of the weakly bound protein may detach from the particles.

This will make the probe less effective because the free protein will compete for binding sites with the gold-labelled protein. Check the pH of the gold sol with pH paper (the gold sol will block a pH electrode) and adjust the pH with 0.1N sodium hydroxide.

Protein A (Boehringer Mannheim) is dissolved in distilled water at 1mg/ml. A microtitration assay will show the correct amount of protein A to add to the gold solution (between 4-6μg/ml).

Add the protein A while stirring the gold sol. After 5 min add 10% bovine serum albumin (BSA) in PBS to a final concentration of 0.2% (2ml/100ml) to maximally stabilize the sol.

Purification of the Protein A-gold

The protein A-gold is centrifuged in a Ti 70 rotor in a Beckman ultracentrifuge at the appropriate speed for 30 min at 4°C.

At the correct speed the gold particles will settle to the bottom of the tube as a loose pellet. Remove the supernatant without disturbing the pellet and re-suspend the loose part of the pellet in PBS containing 0.2% BSA.

A second centrifugation step down a gradient will remove any gold particles of the wrong size. This is done on a 10-30% continuous glycerol gradient at 4°C.

Layer 1-2 ml of protein A gold onto the top of the gradient, spin in an SW40 rotor at the appropriate speed for 45 min. at 4°C. The dark red band in the middle of the gradient is collected. All aggregated protein A gold particles will have been removed.

Storage

Immunogold probes may lose activity within weeks, due to the dissociation of the proteins from the gold particles.

Dialyse the protein A gold against 50% glycerol in PBS and store at -20°C or freeze down small aliquots in liquid nitrogen and store at -70°C.

To determine the concentration at which to use the protein A gold, measure the optical density (OD) at 520 nm of a 1:100 diluted solution in PBS. Use a dilution with an O.D. of between 0.05 and 0.1 where there is no significant background.

If using a primary antibody then the optimal dilution of this first antibody must be known. Sections can be treated with protein A gold alone to determine the background labelling.

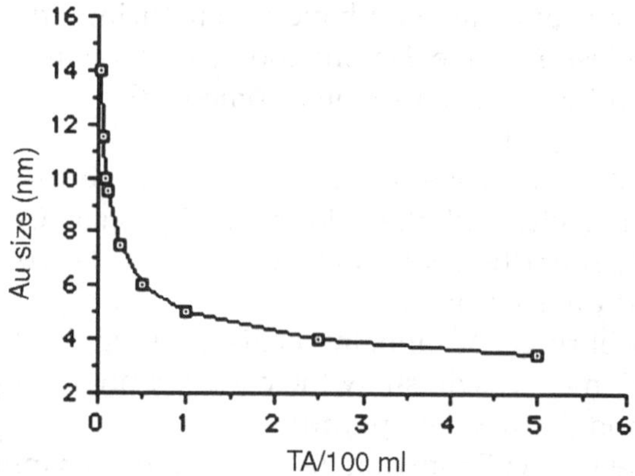

Table. The influence of the Tannic Acid Concentration, During Gold Sol Formation, on the Size of the Gold Particles.

Gold particle size	1% Tannic acid amount (for 100ml final sol)
3.5 nm	5.000 ml
4.0 nm	2.500 ml
5.0 nm	1.000 ml
6.0 nm	0.500 ml
7.5 nm	0.250 ml
9.5 nm	0.100 ml
10.0 nm	0.080 ml
11.5 nm	0.050 ml
14.0 nm	0.025 ml

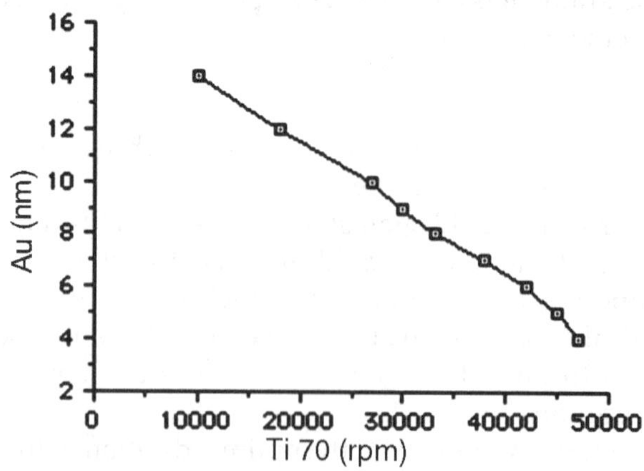

When 1 ml or more of tannic acid is used an equal amount of 25 mM potassium carbonate must be added to neutralize solution B.

Table. First Centrifugation, using a Ti 70 Rotor for 30 min at 4°C. This Step will Concentrate the Colloidal Gold Probe in the Bottom of the Tube

Gold particle size	Ti 70 rotor speed	g (r max)
4 nm	47000 rpm	225000
5 nm	45000 rpm	210000
6 nm	42000 rpm	185000
7 nm	38000 rpm	155000
8 nm	33000 rpm	105000
9 nm	30000 rpm	92000
10 nm	27000 rpm	75000
12 nm	18000 rpm	37000
14 nm	10000 rpm	7000

Table. Second Cntrifugation Using SW 40 Rotor for 45 min at 4°C. This will Separate the Different Sizes of Gold Particle Along the Gradient

Gold particle size	SW 40 rotor speed	g (r max)
4 nm	40000 rpm	284000
5 nm	37000 rpm	240000
6 nm	32000 rpm	170000
7 nm	30000 rpm	150000
8 nm	26000 rpm	115000
9 nm	22000 rpm	85000
10 nm	19000 rpm	65000
12 nm	11000 rpm	20000
14 nm	5000 rpm	5000

The 4 nm gold must be centrifuged for 1 hr for best results. The 5nm gold can also be centrifuged at 35000 rpm for 1 hr.

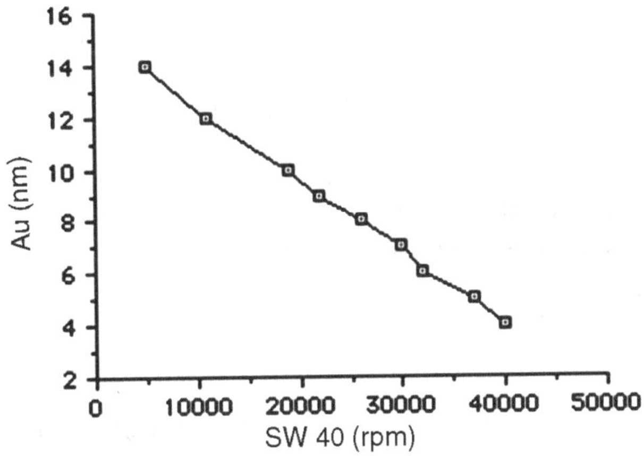

Coupling Bovine Serum Albumin (BSA) to Gold Particles

BSA-gold is a useful marker for studying the endocytic processes in mammalian cells. Typically, the living cells are incubated in a BSA-gold suspension with a final OD, at 520 nm, of 5, meaning that a large amount of gold probe is needed for these experiments. The above protocol for preparing protein A-gold can be followed for preparing BSA-gold but BSA is substituted for the protein A (we use 2.4 μg/ml). The centrifugation values are the same. After the gold probe has been concentrated and purified it is dialized against PBS, or culture medium, before being added to the living cells.

Silver enhancement of colloidal gold: How to precipitate silver onto small colloidal gold particles so that the signal can be detected by light microscopy.

TROUBLESHOOTING

Morphology of Cryosections

Cryosections do not have the same appearance, in electron micrographs, as conventionally embedded resin sections. Sometimes this different appearance is mistaken for poor morphology, sometimes the fine structure morphology is not good. Learn the difference and find out how to avoid poor morphology in these sections.

Problems with Immunolabeling

When labeling cells or sections with antibodies or other affinity markers, the results are not easily interpreted. Look here for some help.

IMMUNOCYTOCHEMISTRY: COMMON PROBLEMS AND SOME ANSWERS

Immunocytochemistry at the Ultrastructural Level (TEM)

Results obtained from immunocytochemical labeling can often differ from what was expected. Sometimes this may result in the formulation of new theories but often good results are ignored because of faulty interpretations.

- There are three major problems encountered when immunolabeling. These are:
  1. Little or no label detected.
  2. Too much label present.
  3. Contamination on the specimen.

No Label Detected

Possible causes

1. *Antigen destruction, masking or extraction*: Antigen may not have been fixed enough allowing it to be washed away during processing.

Demonstrate antibody labeling by light microscopy and use the same preparative conditions for electron microscopic labeling. Some antigens are sensitive to glutaraldehyde but not formaldehyde. Other antigens may be masked by other proteins leaving the antigens inaccessible to the primary antibody, protein A-gold or both, unless the cells are extracted with detergents.

2. *Primary antibody is not working:* Causes of this problem may be due to a wrong (i.e. too weak) dilution of the antibody. Check this by applying serial dilutions of antibody to the specimen. Use freshly made working dilutions for each experiment and do not store diluted antibody, rather, only make up the amounts needed for each experiment.
   Inadequate storage procedures producing repeated freeze-thawing, or growth of contaminants may also result in a deactivation of the antibody. Store antibodies frozen in small aliquots or in 50% glycerol containing 0.01% sodium azide to prevent contamination. If antibodies must be stored at 4¡C then include sodium azide or chloroform to prevent microbial contamination.
   Prepare positive controls using other methods to demonstrate specific labeling with the antibodies being used. Change the antibody if necessary.
3. *Low amount of antigen:* One of the advantages of using protein A-gold to detect antibody binding is that it can be quantitative. In some systems the amounts of antigen present will be very low and result in the presence of only small numbers of gold particles. Most often an estimation of antigen amounts can be obtained from biochemical experiments. If these amounts are low then the actual amount of labeling detected will also be low. Increasing the signal will be difficult in these cases unless antigen masking is occurring. Increasing antigen accessibility by detergent treatment may help.
4. *Magnification of the image is too small to detect colloidal gold particles:* It sounds obvious, but 5 nm is very small and requires that the screen magnification in the electron microscope be approximately 18,000x to 25,000x. It is very difficult to see small gold particles at low magnification. Small gold particles are also difficult to see at high magnification if the electron microscope is astigmatic.
   Primary antibody does not bind protein A-gold Many antibodies do not have binding sites for protein A. If this is so then include a bridging antibody (one that will recognize the primary antibody and bind protein A) between the primary antibody incubation and the Protein A-gold step.
5. *Image contrast is too strong:* Small gold particles cannot be easily seen on images of resin sections that have been contrasted with large

amounts of heavy metal stains. To improve the visibility of the smaller gold particles reduce the specimen contrast as needed. This can be a problem for sections through resin-embedded material.

6. *The diluted gold preparation has been centrifuged:* Unlike the primary antibodies, diluted gold suspensions should not be centrifuged prior to use. Centrifugation will remove the gold probe from the buffer and result in a dilution of the reagent. If over centrifuged, or if the gold particles are large, then most of the reactive probe will be removed from the buffer. This will result in lower levels of labeling than expected, the amount of label varying between experiments. In extreme cases no labeling will be detected.

Background Label (More Gold Particles than Expected)

Background labeling is that signal produced by the primary antibody or visualization probe which is judged to be non-specific labeling.

Possible Causes

A. *Antibody preparation:* Many background problems arise from the primary antibody. Suitable negative controls will identify this. Suggested control experiments include

1. Labeling in the absence of primary antibody only.
2. Label with antibodies from the same species as the primary antibody but which do not react with the cells or tissue under study.
3. Label cells or tissue that do not express the antigen
4. Adsorb out specific label by incubating the antibody with purified antigen prior to use. Bands cut from Western blots can be used for this.
5. Adsorb out possible non-specific label by incubating the antibody with material similar to the test sample but which does not contain the antigen. This should remove background label.

If the antibodies are producing background labeling then it may be sufficient to treat the sections with blocking agents dissolved in PBS to prevent non-specific binding. These include 1% gelatin (calf skin or fish skin), 2% bovine serum albumin, 5-10% fetal bovine serum or 1% ovalbumin.

Additionally, specimens that have been fixed with aldehydes can be treated with primary amines which will quench any free aldehyde groups which could cross link antibodies. These also can be dissolved in PBS an include glycine (0.15%) and ammonium chloride (50 mM). A 10 min incubation is sufficient.

If background persists then try diluting the antibody a little more. The best antibody dilution is that which produces as strong a specific signal as possible with a low a background signal. Often this means using the primary antibody as concentrated as possible and accepting some background labeling.

Background labeling can also be caused by using sera instead of purified immunoglobulin fractions. The primary antibodies that usually produce the best results are affinity purified IgG fractions.

If the background labeling cannot be removed then it is possible that the label is a specific signal. Check the results with other labeling procedures to confirm this.

B. *Non-specific binding to resins:* When using sections of resin embedded material, the primary antibody or the gold probe may stick to the resin. Determine which of the reagents is causing the non-specific binding using control experiments and eliminate by pre-incubating the specimen with a blocking agent. The ideal blocking agent and the incubation time may have to be determined empirically.

C. *Background caused by blocking agents*: Sometimes non-specific binding of protein A-gold can be caused by using protein A-binding molecules as blocking agents.

Rabbit or pig serum, if used as a blocking agent, will bind protein A-gold and produce a non-specific signal. Immunoglobulins will also occasionally bind specifically to blocking agents (e.g. do not block samples with FCS if using anti-BSA antibodies).

D. *Antigen migration:* If the antigen has not been properly immobilized by the fixation step it may become redistributed throughout the cell prior to its being washed away. Try a different fixation protocol.

E.*Gold probe dilution*: Background labeling can be caused by using too high a concentration of the gold probe that is being used for visulaization of the antibody binding. Diluting the probe will remove this background. Some preparations of probe will produce background if they are not diluted in PBS containing blocking agents. Addition of a suitable blocking agent will remove this background.

Background is Specific Signal

If all attempts to remove background labeling have failed then it is possible that the signal is specific. Re-examine all the scientific data.

Contamination of the Specimen

Contamination of the specimens by dirt can be easily avoided by using filtered solutions and being scrupulously clean during all the handling procedures. However, even if the most careful precautions are taken it is still possible to contaminate the specimens with fine precipitates which make evaluation of the labeling impossible. Possible causes.

Buffers Reacting with Specimen Grids

If copper specimen grids are being used then neither the antibodies nor the gold probes can be diluted in Tris-HCL buffer. The HCL will react with the copper. If this is happening then the antibody droplets will turn blue after the incubation. Change either the buffer in which the antibodies are stored or diluted or use nickel or gold grids on which to mount the specimens.

Some copper grids will react with PBS and produce small amounts of a fine precipitate over the specimen. This problem is characterized by a concentration of the precipitate close to the grid bars. To avoid this, either the source of the grids can be changed, or grids made from other metals can be substituted for the copper ones.

GENERAL PROTOCOLS

A recipe for Moviol, an inexpensive mounting medium that is easy to prepare.

A Semi-permanent Mounting Medium for Immunofluorescence Microscopy

Materials

- 6 gm glycerol
- 2.4 gm mowiol
- 6 ml distilled water
- 12 ml Tris buffer 0.2 M at pH 8.5

Method

Weigh out the glycerol in a 50ml plastic tube and add the mowiol. Stir thoroughly. Add water and leave for 2hr at room temperature. Add the Tris buffer and incubate at approximately 53¡C until the mowiol dissolves, stirring occasionally. Clarify by centrifugation at 4000-5000 RPM for 20 minutes and aliquot into small tubes.

Stored frozen it should last for up to12 months. The solution is stable at room temperature for 1 month. Small amounts of DABCO or other fluorescence stabilizer (anti-fade agent) added to each aliquot immediately before use will reduce signal fading under UV light.

How to Coat Metal Specimen Grids with Formvar

Strong, thin support films are essential for routine TEM. Here is a reproducable method for making thin films of formvar with details on how to mount them on metal specimen grids.

Formvar Coating Specimen Grids: All successful TEM experiment relies totally on the quality of the support film covering specimen grids. These films must be strong, clean and remain attached to the specimen grids even through many manipulations. As there is still some mystery in the methods used to prepare these grids, with some very unscientific methods being used, this protocol will clearly define a reproducible method.

1. Clean 100 mesh, hexagonal specimen grids. Place them in a 200 ml glass flask, rinse with 2% aqueous acetic acid, multiple washes with distilled water and finally with 3 changes in acetone. Dry over a warm plate.
2. Clean a glass slide by immersing it in ethanol, instantly removing it and drying with lens tissue. Use "precleaned 25 x 75mm micro slides selected" Catalogue #48312-002 from *VWR Scientific Inc.*
3. Dip slide into a solution of formvar dissolved in chloroform (2% w/v) and withdraw carefully. The speed of withdrawal will affect film thickness. Alternatively, semi-automated withdrawl devices can be purchased from *EM supply companies.*
4. Leave slide to dry vertically.
5. Score around slide on all four sides with a razor blade to define limits of film to be dislodged from slide.
6. Fill a deep glass dish with distilled water and place a light so that it reflects on the water surface.
7. Clean the water surface by passing a piece of lens tissue over the surface.
8. Hold the formvar coated slide at a 45¡ angle at the edge of the glass dish, with the water just touching the bottom of the slide. The formvar film should start to come away from the glass.
9. Slowly push the slide into the water, watching as the film as it lifts off the glass and floats on the water surface.
10. Remove the glass slide from the water and examine the floating film. The incident light should show interference colors in the film. These should be silver or gray. Yellow, purple or green colors indicate that the formvar solution should be diluted more by adding chloroform. Films that are too thin mean that the formvar solution is too dilute. To increase the thickness either remove the glass slide from the solution more slowly or increase the concentration of formvar.
11. Place cleaned specimen grids onto the floating film until it is filled. Place the grids so that the shiny surface of the grid is facing up and the dull side is on the film.
12. Dry the glass slide used for casting the film and attach a 1" by 2.5" white address label to it.
13. Hold the glass slide vertically over one edge of the floating film and push it into the water so that the film, with grids attached, stick to the address label.
14. Once the glass slide has been totally immersed remove it, with the grids attached and leave to dry.

15. Coat the grids with carbon using a vacuum evaporator.
16. Although not essential, the grids can be made hydrophilic prior to use using a glow discharge apparatus in a vacuum evaporator.

METHOD OF SPECIMEN PREPARATION

Improvements in both methods and instrumentation for EM are now allowing the structure of organelles and cellular subsystems to be characterized with unprecedented detail and reliability. It is widely acknowledged that EM of macromolecules is entering a new phase of productivity. It is less well known that comparable innovations are afoot in cellular EM. Studies of biological "ultra-structure" were of key importance in the early years of cell biology, but more recently their importance has waned significantly.

This change has been largely because much of what could be seen by classical techniques for EM has long since been described; it has also derived from the cost of EM in both time and money and the innovations that have recently revolutionized light microscopy, making it ever more powerful for the study of living cells.

However, recent improvements in both methods and instrumentation for EM are now allowing the structure of organelles and cellular subsystems to be characterized with unprecedented detail and reliability. Thanks to tomography, the three-dimensional (3-D)*1* structure of cells can now be visualized with 5–8-nm resolution. Thanks to high pressure freezing, cellular specimens of considerable size can now be well frozen, even without the addition of chemical cryoprotectants.

With this and other methods for rapid freezing, samples become solidified within milliseconds, whereupon they are embedded in glass-like ice. The cellular milieu is still aqueous, but rapid freezing has immobilized all the cell's constituents before significant rearrangement is possible. Under these conditions, biological structure is trapped in an essentially native state, and ice crystals, which would deform the physiological organization, have had little time to grow. More rapidly frozen specimens retain impressive preservation of intracellular detail.

Frozen-hydrated cells can be examined directly in the EM by using a low temperature specimen holder, or they can subsequently be fixed by "freeze-substitution," during which cellular water is replaced at -80°C to -90°C by an organic solvent that contains chemical fixatives to stabilize the biological structure before it is embedded in a matrix suitable for microtomy.

Freeze-substituted samples appear similar to material prepared for EM by conventional methods, but they have a greater likelihood of displaying structures in their native state. This review will describe recent progress in the 3-D imaging of both frozen-hydrated cells and those that have been preserved by freeze-substitution. We will then evaluate the impact of the resulting data on our understanding of cellular mechanisms.

EM Images of Frozen-hydrated Specimens, Purified from Cells, Can Reveal Functionally Important Conformational Changes in Large Assemblies of Macromolecules

Several fields of cell biology have profited from information obtained by cryo-EM of specimens isolated from whole cells, spread as very thin layers on supporting films, and then frozen by plunging into a "cryogen," such as liquid ethane. For example, our appreciation of microtubule dynamics was significantly advanced by the realization that polymer end morphology is different for assembling and disassembling microtubules.

Likewise, the stroke-cycles of motor enzymes, bound to either microfilaments or microtubules in the presence of different nucleotides, have been defined by cryo-EM. The resulting images have allowed both a correlation with atomic resolution information from x-ray diffraction and an exploration of the structural changes that accompany changes of the bound nucleotide. Much of this imaging has used isolated cellular subfractions that are small, stable, and symmetric enough to allow image averaging to facilitate high-resolution EM.

The beam sensitivity of a frozen-hydrated sample requires that the total exposure to the electron beam be kept low (e.g., <10 electrons/A^2 for a resolution of <1 nm). At this dose, the quality of a single image is severely limited by "shot noise," i.e., the poor statistics with which the electron-scattering properties of the specimen are sampled.

Symmetry in the structure allows a simple image averaging to increase the image signal-to-noise ratio, even when the dose per unit area is kept low. Image quality improves with the square root of the number of areas averaged, and several thousand can be used. Alternatively, a large number of images of identical, individual particles can be collected, the images oriented relative to one another, and their averages computed to accomplish the same task with particles that possess no symmetry. Image averaging based on symmetry or on galleries of distinct but identical particles (or in the case of viruses and some other objects, on the use of both approaches together) has made it possible to improve signal-to-noise ratios to the point that resolution has become as good or better than 1 nm.

With this much image detail, it has been possible to integrate information from x-ray crystallography with images from cryo-EM and build up models for changes in the structure of complex molecular assemblies that accompany their biological function. This important field of structural biology is rapidly expanding our knowledge of complex molecular mechanisms.

Images of Frozen-hydrated Cells Are Revealing New Aspects of Cell Structure but Pose Important Challenges to the Microscopist

The study of frozen-hydrated cells raises two issues that are not encountered during microscopy of isolated, subcellular assemblies:

(a) Cells are large and tremendously complex, so a huge amount of information is necessary to characterize them, and

(b) No two cells, or even two large organelles like mitochondria, are sufficiently similar that their images can be simply averaged to enhance image signal-to-noise.

Thus, the methods for detailed characterization of cellular structure differ from those that are appropriate for macromolecular assemblies. To tackle the problem of cell size, one must either select tiny cells or find a way to collect appropriately small pieces of the cell of choice. Such pieces of a frozen-hydrated cell must also be thin enough for view in the EM without too many of the beam's electrons suffering multiple scattering events, which seriously degrades image quality (less than 0.5 μm thick for 300 KeV electrons).

One approach for the preparation of such specimens has been direct thin sectioning of frozen cells or tissues. However, it has been very difficult to cut useful sections of samples that have simply been frozen. The embedding ice tends to crack during sectioning, and sample compression is usually significant, though improvements may be on the horizon.

When thinking about cryomicrotomy it is natural to recall the widespread success that has been achieved by cutting cryosections from samples that have first been chemically fixed and infiltrated with a cryoprotectant (usually sucrose) and then frozen. This method is commonly used to prepare samples for immunolabeling, but it fails to take advantage of the power of rapid freezing to preserve cell structure with the greatest possible fidelity. Thus, in spite of its value for antigen localization, the approach is not an ideal way to study cell morphology. Its reliance on both chemical fixation and infiltration with an osmotically active cryoprotectant reduces the likelihood that the resulting images will reflect the condition in a living cell. Current work in the lab of Jan Slot (University of Utrecht, Utrecht, Netherlands) is directed at improving this situation, but much remains to be done.

Since sectioning frozen-hydrated samples has been difficult, several scientists have sought cellular samples that can be viewed without it. For example, the thin margin of a cell spread on an electron-transparent substrate has been imaged directly by high voltage electron microscopy using samples maintained at -165°C.

The resulting images show how much detail is preserved in a rapidly frozen cell, but with a single projection of so much 3-D structure, the images are hard to interpret. Even thin processes that extend from the cell's margin are complex, though in such a region microfilaments can readily be identified.

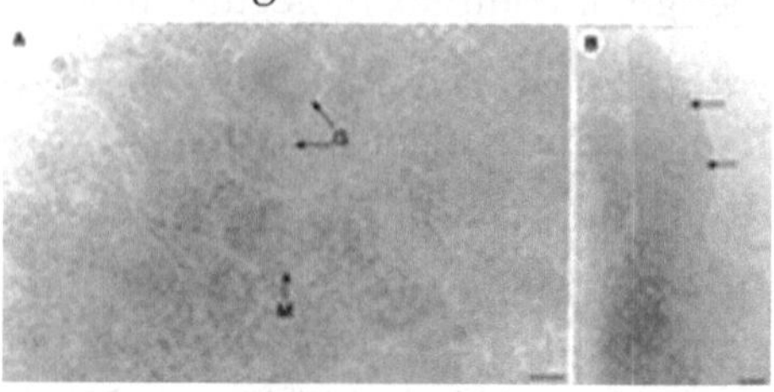

A platelet from human peripheral blood, frozen by plunging into liquid ethane and imaged at -165°C, using 1,000 KeV electrons. A shows platelet granules (G) and mitochondria (M) whose boundaries are clearly defined by membranes. Nonetheless, the density of organelles is sufficiently high that positional relationships among them are hard to discern. B shows a slender projection from the margin of such a platelet. Its component microfilaments are evident (arrows). Bars: (A) 200 nm; (B) 100 nm.

The problem of superposed detail in frozen hydrated samples has been tackled by using multiple tilted views and the mathematics of tomography to build 3-D images of small cells and isolated organelles. Figures show slices extracted from a tomographic reconstruction of the archaebacterium, *Pyrodictium abyssi.*

Here, automation of data collection has been used to keep the electron dose-per-view to a minimum, and the principle of dose-fractionation has allowed multiple tilted views to be assembled into a single 3-D volume whose signal-to-noise ratio is defined approximately by the total dose for the series, rather than the dose/image (subject to the constraint that each image must show enough detail to permit its accurate alignment with other tilted views). This work shows tremendous promise, given both the nearly native state of the cells at the time they are imaged and the 3-D detail that is visible.

Objects separated by only 6–8 nm are visible in the best of these cryotomograms. At this resolution some large protein complexes can be recognized by their structure alone, suggesting that cryotomography followed by appropriate image analysis may provide information about the localization of specific proteins. However, note that the promising results in the latter two figures have so far been achieved with frozen protein solutions, not with cytoplasm itself.

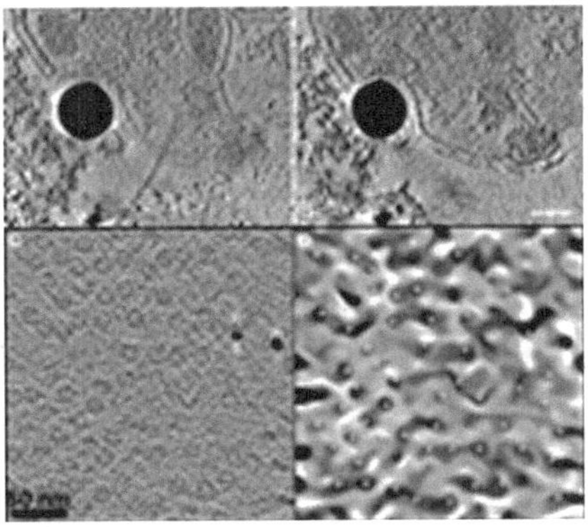

Electron tomographic reconstruction of frozen-hydrated specimens. A and B are two 12-nm slices extracted from a reconstruction of the archaebacterium, *Pyrodictium abyssi.* The distance between the slices is 90 nm. The boundary of the cell is evident, including both the plasma membrane and the "S-layer," which is composed of ordered protein subunits.

There are also crystalline structures in the cytoplasm (square arrays in A and B). Vesicles within the cell are evident. The dark circle in each image is a 250-nm latex sphere, which is apparently being internalized by endocytosis. Reprinted with permission from the author and *Trends in Cell Biology*. C displays a slice from a tomographic reconstruction of isolated, frozen-hydrated thermosomes, which are chaperonins with eightfold rotational symmetry. The proteins are arranged in a top view orientation. Image processing methods have been used to search for sites of correlations between the tomogram shown in C and a template generated from high resolution cryo-EM micrographs of the thermosome. Positions of strong correlation are marked by white crosses in D. This method for feature recognition holds promise as a way to identify the location of specific proteins in tomographic images of cytoplasm. Bar, (B) 200 nm.

Technological Developments Have Opened New Possibilities, but There Are Still Many Challenges to Overcome

The emergence of EM tomography has depended on many technological advances. Examples include:

(a) Microscopes that operate at higher than conventional voltages. These reduce the deleterious impact of multiply scattered electrons on the quality of images from specimens thick enough to be suitable for tomography.
(b) Field emission guns that are bright sources of comparatively coherent electrons. These increase the contrast and signal-to-noise in images of unstained specimens.
(c) Eucentric stages that allow smooth tilting between images; these ease and speed the process of data collection.
(d) Combinations of hardware and computer-facilitated controls that allow automated collection of the large numbers of tilted views necessary for tomographic reconstruction.

These devices also minimize the beam damage to the region being recorded. Finally, there are indications that an electron energy loss filter, developed by several companies to remove electrons that were inelastically scattered by the specimen before an image is formed, can enhance the quality of images from thick, frozen-hydrated specimens.

Although these technologies have led to significant progress in imaging frozen cells, there are still formidable problems to overcome before this approach will yield the most useful results. Both microscopes and cameras will have to be optimized, so every imaging electron is detected with high efficiency and negligible noise.

This is a particular problem with higher electron energies (e.g., 300 KeV), where electron scattering and x-ray generation commonly reduce a camera's detector quantum efficiency. Even with improvements, though, the inherent radiation sensitivity of frozen hydrated samples will make it hard to push the resolution of cellular cryotomography as low as 2 nm. Specimen holders that maintain liquid helium temperatures and permit tilting about two orthogonal axes may provide improved data, but such big improvements in resolution are unlikely to be straight-forward.

Nonetheless, this goal is worth significant effort, because at 2 nm one should be able to recognize the shape of many macromolecules, building on the pioneering work from the Baumeister laboratory (Max Plank Institute for Biochemistry, Martinsreid, Germany) to "dock" atomic resolution structures determined by crystallography or other methods into the lower resolution images available from EM. Such an advance might ultimately obviate labeling, e.g., by immuno-EM, because at least some large proteins could be recognized directly by their shapes. However, identifying small proteins will require improved imaging methods and probably novel ways of comparing cellular and molecular structures. In another vein, cryotomography is now applicable only to cells that are small or thin. Cryomicrotomy of frozen hydrated cells will have to be improved, so specimen thickness can be controlled by the experimenter, not by the nature of the sample.

Several laboratories are working on this and related issues, so progress can be expected. In addition, there are limitations to current methods for examining significant volumes from large frozen-hydrated cells. Even assuming that good cryosections can be cut, getting good SERIAL cryosections will present an additional hurdle. Moreover, the radiation sensitivity of frozen specimens, combined with current limitations on the size of the images that can practically be recorded, make it difficult to sample large areas from any one section. Thus, there will be serious difficulties in reconstructing volumes that correspond to a significant fraction of an entire frozen-hydrated, eukaryotic cell.

Images of Rapidly Frozen, Freeze-substituted Cells Reveal the Substructure of Extensive Cytoplasmic Volumes at 6-nm Resolution in 3-D

Given the complexities of cellular cryotomography, several labs have sought practical alternatives that would permit 3-D imaging of well-preserved samples that are structurally as close as possible to the living state, even if they are not frozen-hydrated. For such work, one would like samples that permit convenient microtomy for serial sections, exhibit good contrast, and are sufficiently robust in the electron beam that large sample areas can be imaged. Specimens prepared by rapid freezing followed by freeze-substitution fixation and embedding in plastic (RFFSE), fulfill these requirements and provide specimens suitable for detailed study of cells in 3-D.

3-D reconstruction of cellular samples has been accomplished for several years by tomography of chemically fixed, plastic-embedded samples stained with heavy metals. However, the well-known limitations of chemical fixation for preserving cellular ultrastructure have limited the information that can be derived from such work. These problems have recently been addressed by combining the power of RFFSE with tomography to obtain 3-D views of cellular structures that should closely resemble the living state. The accuracy of these specimen preparation methods has been tested in three ways:

(a) Significant fractions (1/20–1/5) of rapidly frozen yeast cells are viable upon rewarming, showing that rapid freezing itself preserves the native structure of cells very well.

(b) RFFSE samples have been compared with equivalent ones prepared by freeze-fracture EM, a method that visualizes replicas of fractured cellular surfaces formed by metal shadowing at very low temperatures.

These studies have found no significant difference between the structures seen by the two methods. (c) The images from samples prepared by RFFSE conform to expectations that are based on a broad range of knowledge about subcellular components: cytoskeletal fibers usually run straight; membrane profiles of the ER and of Golgi cisternae appear turgid and smooth; most vesicles are round; cytomatrix is even; even chromatin appears structured.

Thus, studies that have melded RFFSE with tomography are likely to be showing us cellular structures that are essentially native, revealing organelle morphology in its normal context and in a situation where the parameters of cell physiology, like cell cycle stage, can be manipulated by the experimenter. Indeed, it is possible to freeze cells or parts of cells at well-defined times after a stimulus, allowing 3-D snapshots of cellular processes, even fast ones, like muscle contraction.

This composite technology has been used to examine the architecture of mitotic kinetochores in mammalian cells, the centrosomes of budding yeast, and the Golgi apparatus and associated organelles in mammalian cells. The results from mitotic systems show that the methods preserve not only spindle microtubules, but also the fine structure of microtubule ends and the organelles that initiate them.

These tomograms contain an impressive amount of 3-D detail. They even allow investigators to examine a given structure sliced computationally in several different orientations to get the most informative views for answering each specific question. In addition, a highly asymmetric structure, like a spindle microtubule, can be sampled with a tomographic slice that exactly contains its axis, but that is only a fraction of the structure's thickness. Such a feat is currently impossible by microtomy.

A shows a 3-nm slice from a tomographic reconstruction of one end of a forming mitotic spindle. Six layers of the spindle pole body are indicated by numbers (1, outer plaque [OP]; 2, inner layer 1 [IL1]; 3, inner layer 2 [IL2]; 4, central plaque [CP]; 5, inner plaque [IP]; 6, the minus ends of the spindle microtubules [MT]). The six small panels to the right of A are slices cut from the tomogram perpendicular to the slice shown on the left, and the placement of the slices corresponds to the positions of the labeled lines in A on the left. Order is quite evident in the arrangement of subunits in IL2. However, the shape of the plaque is distorted by the collapse of the section under the influence of the electron beam.

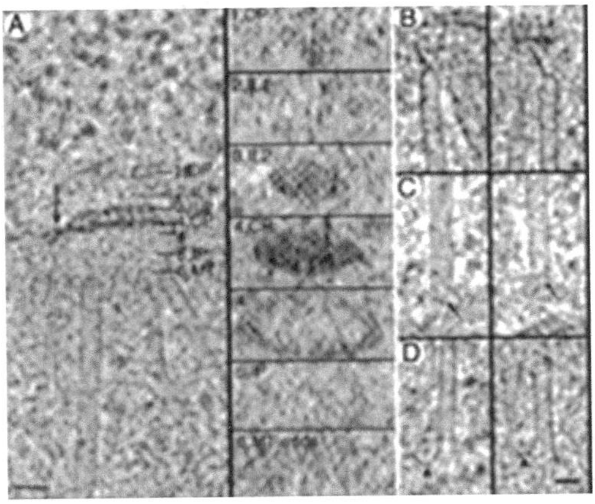

Fig. A Spindle Pole Body and its Associated Microtubules from the Budding Yeast, *Saccharomyces Cerevisiae.*

B and C show minus ends of microtubules at IP and OP respectively (arrows). Both spindle and cytoplasmic microtubules are capped, though the shapes of the caps are not the same. D shows the plus ends of two spindle microtubules, revealing their flared shape (arrows). Bars: (A) 50 nm; (D) 25 nm. Application of the same techniques to the study of membranes in the Golgi apparatus has revealed several features of cisternal organization that had escaped detection by conventional EM.

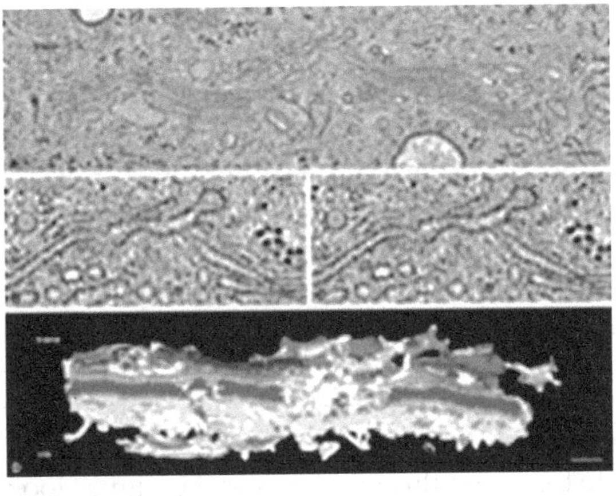

Fig. Tomographic Reconstruction of a Portion of the Golgi Ribbon from a Cultured Normal Rat Kidney Cell, Fixed by Plunge Freezing and Freeze Substitution.

For example, tomographic reconstruction of four serial, 1/4-μm sections has provided enough cellular volume to reveal continuity among all of the Golgi elements that bud clathrin; these elements lie at the transmost side of a Golgi stack.

Tubular projections, which are rarely seen in conventionally fixed and imaged material, form from the edges of several cisternae; these extend significant distances in either the cis- or transdirection. Such reconstructions have also revealed the pleomorphism of the membranous compartments that lie in noncompact regions of the Golgi region.

Moreover, they have provided quantitative information about the numbers of different types of vesicles in particular cytoplasmic domains and on the proximity of membrane-bounded organelles with each other and with elements of the cytoskeleton.

A shows one tomographic slice 4 nm thick from the 257 that comprise the entire reconstruction. The area shown is 3.8 x 1.9 μm, and the thickness of the whole tomogram (assembled from four serial 250-nm sections) is 1 μm. This volume is large enough that many features of Golgi morphology can be seen with greater clarity and detail than had previously been possible. B shows a small region from A, providing enough magnification to show that the two leaflets of a unit membrane can be resolved (6 nm).

C shows the same area with graphic objects laid on to mark the positions of membranes in that slice. Similar modeling on every slice, followed by triangulation and surface rendering, permitted the construction of the model shown in D. The trans side of the Golgi apparatus is at the top, the cis at the bottom, and a noncompact region lies at the middle of the reconstruction. Bars, 250 nm.

An additional advantage of EM tomography is that many cellular structures are visualized at once with a single imaging technology. This is a significant difference from fluorescence light microscopy, where one sees only that which has been made fluorescent.

Indeed, it is this feature of immunofluorescence that allows one to be satisfied with fixation protocols that preserve some cellular details very badly. EM tomography of well-fixed cells, on the other hand, reveals the relationships among diverse cellular structures at comparatively high resolution. The stereo images presented here show a 1.2-μm slab of cytoplasm from a cultured mammalian cell.

The cisternae of the Golgi region and the ER are displayed with all the nearby vesicles, microtubules, and ribosomes. This view reveals just how crowded cytoplasm really is.

However, note that even though the model in figure looks densely packed with organelles, fully 65% of the volume modeled lies outside of organelles, allowing plenty of space for cytosol.

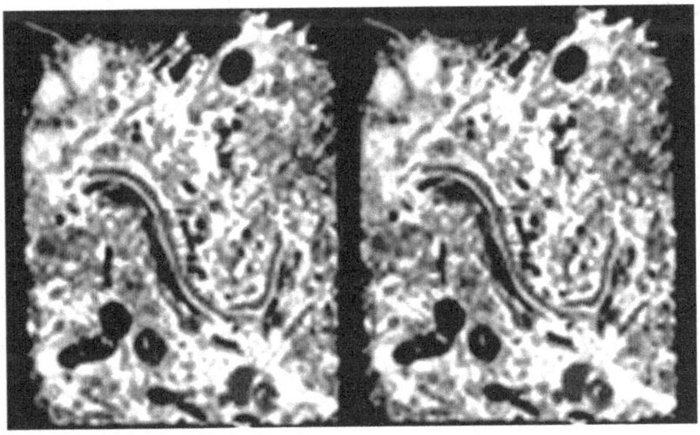

Fig. Stereo Views of a Model Generated by Tomographic Reconstruction of a Slab of Cytoplasm from a Cultured Cell, Strain HIT-T15.

Three serial 400-nm sections were reconstructed by dual axis tomography, and the IMOD software package was used to model all visible objects within a volume $3.1 \times 3.2 \times 1.2\ \mu m^3$.

The model displays the Golgi complex in the context of surrounding organelles: ER, yellow; membrane-bound ribosomes, blue; free ribosomes, orange; MTs, bright green; dense core vesicles, bright blue; clathrin-negative vesicles, white; clathrin-positive compartments and vesicles, bright red; clathrin-negative compartments and vesicles, purple; mitochondria, dark green. Bars, 500 nm.

Tomography of Cryofixed, Freeze-substituted, and Stained Specimens Is Currently Subject to Its Own Share of Limitations

For all its strengths, the composite method described above has limitations of its own that need to be considered. Plastic sections collapse under the action of an electron beam. Shrinkage in the section's plane is modest (5–10%), but along the direction of the beam it is 40–50%.

Reduced temperatures provide some protection against this effect, but in our hands there is some rapid initial collapse, even at low temperatures. Ice does not show this shrinkage, so if cryotomography can be made practical for large cells, this kind of sample deformation by the beam should be eliminated. Until that time, either we must correct for beam-induced distortions or simply live with them.

There are also limitations to the resolution that can be expected from stained, plastic-embedded samples. Currently, these are 6 nm, probably set by the granularity of the stain after exposure to so much radiation and by the accuracy with which stain represents the positions and structures of underlying macromolecules.

One fact that supports this resolution claim is that one can see two layers in a cytoplasmic membrane that is oriented appropriately relative to a tomographic slice. However, one cannot resolve subunits in a microtubule wall, which are separated by 4–5 nm. Improving on current resolution may require real breakthroughs in techniques for fixation, embedding, staining, and imaging at low dose. Another current limitation of tomography is the difficulty of identifying particular macromolecules within the samples. When a protein assembles to form a characteristic shape, like tubulin or clathrin, it is easy to identify and localize in a tomogram. However, less idiosyncratic proteins are simply stain-binding regions of the sample.

Postembedment antibody labeling of sections thick enough to merit tomography is unlikely to be useful, because immunoglobulins will not diffuse into the embedding plastic. Methods that remove the embedding resin tend to be harsh and therefore limit image resolution and reliability. Labeling before embedding usually requires cells to be "permeabilized" before antibody treatment, which results in some extraction of cytoplasm and the possibility of antigen movement.

Therefore, such methods are intrinsically less reliable than ones that use optimal fixation (e.g., RFFSE). Developing methods that get identifying tags onto proteins or nucleic acids in samples suitable for tomography is a standing challenge.

Cellular Tomography Should Have Profound Effects on Future Directions of Cell Biology

It is encouraging how much can be seen with the new kinds of cellular EM. The cryotomograms of small cells are sufficiently promising to motivate a real attack on ways to get comparably thin samples from any frozen cell or tissue. The tomograms of RFFSE samples already show enough detail and quality to prompt serious study in their own right, and it is realistic to think about using these methods to view major cellular subsystems, like an entire nucleus or a Golgi complex.

Results already obtained show that many structural details can be characterized, like the morphology of microtubule ends at their sites of cellular attachment. Future work should display changes in these structures as a function of interphase physiology or mitotic state. Volumes now under study comprise >10 μm^3, permitting the description of extensive tubular projections from individual Golgi cisternae and of the endoplasmic reticulum as it penetrates the Golgi region and lines up with regular Golgi elements. Future work should elucidate the connectivity in 3-D of all the components of the endocytic pathway and their structural relationships with each Golgi compartment. One can even envision the reconstruction of an entire eukaryotic cell.

In spite of its far-reaching value, EM tomography is no longer oppressively time consuming. Using a modern microscope equipped with a digital camera, a good eucentric tilting stage, and software to control data collection, a skilled operator can collect the 140 images required for a complete tilt series of 70°C in

<2 h. If one uses montages of digital data to enlarge the area per image, the time increases somewhat, and dual axis tomography essentially doubles the time investment.

Nonetheless, in our laboratory we are routinely collecting data for two and more tomograms per day. Comparatively modest computers, like a UNIX workstation or a fast PC, can align these images and calculate a tomogram in another few hours, so the preparation of a tomogram is now surprisingly fast. Using montages and serial thick sections, it should be feasible to image the entirety of a small eukaryote, like a budding yeast, in 3-D at 6-nm resolution (3×10^{10} bytes of image data) in <1 wk. However, our lab has concentrated on organelles and cellular subsystems in order to retain a functional focus for structural work.

Analyzing tomographic data is more time consuming than tomogram construction. Models like the ones shown in figures have taken months to prepare, so there is significant room for improvement in the ways that features of interest in 3-D images are abstracted from their backgrounds (a process called segmentation). The coordination of computer-assisted feature recognition with hand–eye efforts by a trained biologist offer many avenues for productive work, and we anticipate that this aspect of structural cell biology will soon move much faster. Of more concern is the issue of recognizing macromolecules of interest within cellular tomograms.

There is great promise in the approach under development by Baumeister's group, but the utility of this approach for smaller proteins in real cytoplasm remains to be demonstrated. Our lab and others are working to develop labeling methods that will be suitable for thick section EM reconstruction, but this work is still in its infancy. The concept of 3-D maps of specific antigens, localized at 6 nm in 3-D, is certainly attractive when thinking about data that could discriminate between competing models for complex cellular events, like mitosis, secretion, and endocytosis.

It seems likely that EM tomography will provide an important complement to the beautiful and informative images that are now emerging from work on live cells based on a range of fluorescent stains, including protein chimeras with green or other fluorescent proteins.

The spatial resolution of EM tomography is 40-fold better than most confocal microscopy, allowing the visualization of detail that is otherwise vague. When such methods are applied to cytoplasmic mechanisms, like endocytosis, they should reveal the connectivity among all relevant structural compartments. Applied to chromatin, they may identify aspects of higher order DNA structure that help to control gene expression.

When used to view nuclear pore complexes in situ, aspects of protein and RNA traffic that have previously escaped our notice may become obvious. Host–pathogen relationships may also appear in a new light, and some cellular subtleties of developmental processes may emerge from the morass of molecular detail in which they are currently buried. We see a brave new world emerging from the combination of modern EM with the rapidly advancing methods for light microscopy. The resulting data should help us to understand the context for functional genomics and elucidate the continuum of structural order that bridges from inanimate molecules to the order of the living cell.

OUTLINE OF ANALYTICAL IMAGING FACILITY CAPABILITIES

The Analytical Imaging Facility provides a comprehensive light and electron microscope imaging facility dedicated to bringing state of the art methods in modern imaging to biomedical scientists with all levels of expertise. The AIF staff has been cross-trained to offer a seamless transition from classical histology, to high resolution light microscope imaging in 3D, to state of the art electron microscopy.

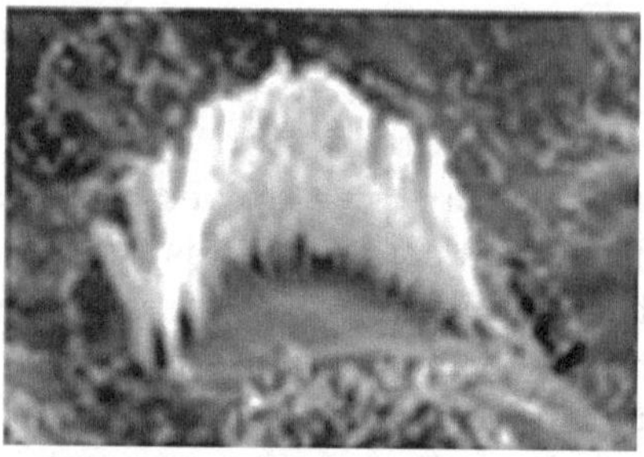

This unified approach facilitates the efficient and appropriate complementary use of these methods in research. For the infrequent user, the AIF provides a completely assisted technical support service. For the trained microscopist, the AIF is an available equipment resource. A significant effort is devoted to training investigators who require microscopy techniques to advance their projects.

The Services

Transmission Electron Microscopy

JEOL 1200EX. This instrument offers the highest basic performance as a 120 kv transmission electron microscope employing a uniquely designed 3-stage 6-lens imaging system. It offers operational ease, excellent image quality and high resolution at low to high magnifications. It is equipped with side entry goniometer stage, minimum dose system, bottom mounted high resolution Gatan video camera and side mounted wide angle Gatan video camera.

JEOL 100CXII. This high performance 100 kv transmission electron microscope features a cool beam electron gun, high image contrast, high-speed cascade differential evacuation, optimum underfocus system using an image wobbler and a side entry goniometer.

Cryo Transmission Electron Microscopy for Single Molecule Imaging

The College, HHMI and NIH (by way of an awarded Shared Instrumentation Grant) have supported establishing a full Cryo EM programme. The technology spans the resolution range from electron microscopy to X-ray crystalography and allows for imaging of single molecules in their hydrated state.

Scanning Electron Microscopy

JEOL 6400. This high performance Scanning Electron Microscope operates with accelerating voltage from 0.2 kv to 35 kv utilizing a high brightness LaB6 filament. It offers full keyboard operation, framestore with digital image processing and digital image capture on a PC running analy SIS software.

Specimen Preparation for Electron Microscopy

The staff of the AIF offer full service sample preparation for many standard and state of the art EM techniques. These include:

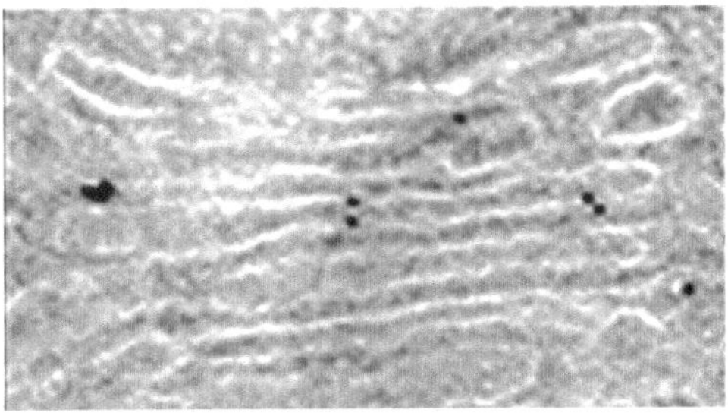

- *Embedding:* utilizing either epoxy or acrylic resins at ambient or low temperatures.
- *Thin Sectioning:* with a Reichert Ultracut E or Leica UCT ultramicrotome.
- Negative Staining.
- *Freeze Fracture* using a Cressington CFE-50 Freeze Etch Unit.
- *Immunogold Labeling* following pre or post embedding protocols.
- *Critical Point Drying* with a Tousimis Samdri 790 Critical Point Dryer and Sputter Coating using a Denton Sputter Coater for preparing cells and tissues for SEM Imaging.
- Cryoultramicrotomy utilizing a Leica UCT cryoultramicrotome for optimizing epitope availability and *morphological preservation for immunogold labeling.*
- Slam Freeze Cryofixation using a Life Cell CF100 Slam Freezer, which can be followed by High Resolution Rotary Shadowing in a Cressington CFE-50 equipped with e-beam guns for platinum or tungsten-tantalum evaporation. Slam Freezing followed by rotary

shadowing is a powerful technique to obtain high resolution images of cells or macromolecules in 3-D, that have been frozen in the hydrated state.

- Slam Freeze Cryofixation using a Life Cell CF100 Slam Freezer followed by *Freeze Substitution* and *Low Temperature Embedding* in a Bal Tec FSU-010 Freeze Substitution Unit. Freeze Substitution is an alternative method for optimizing epitope preservation for immunogold labeling.

Photographic Documentation for Electron Microscopy

The AIF offers a photographic service for producing high quality electron micrographs on conventional photographic paper or high resolution scanning and direct digital printing for poster, lecture, web and journal publication.

Light Microscopy

Routine light microscopy

A Zeiss AxioSkop II with optics for brightfield, darkfield (through the condenser or via true oblique illumination), phase contrast, Nomarski, polarized light and epi-fluorescence with1.25X through 100X objectives serves as the "routine" microscope. Images are recorded with a colour Zeiss AxioCam.

To meet the growing demand for imaging of live material, epecially of eGFP or other fluorescently tagged cells, or of cells in culture dishes, the AIF has Olympus inverted microscopes with a wide array of optics and photography options. On other inverted systems in the AIF researchers continue to use video digitizing technology for imaging motile cells in phase contrast.

Stereo Dissection Microscope

Lower magnification imaging is achieved with a Zeiss SV11 ("STEMI") with a Retiga 1300 digital camera, optical tunable filter for colour imaging and IP Lab for image capture. Reflected light is provided either by a ring illuminator or by two point illumination, transmitted light is provided with a continuously adjustable 100% transmittance to darkfield slider and epi-illumination for fluorescence is provided by a mercury arc lamp with filters for dapi, CFP, GFP, YFP or rhodamine/RFP.

BioRad Radiance 2000 Laser Scanning Confocal Microscope

Thin optical sections of much higher resolution than normal epi-fluorescence can be obtained from live or fixed cultured cells, *vibratome sections,* or intact tissue with colocalization of up to *three different fluorescent probes* and one reflectant probe. Collected images can be *reconstructed in 3D,* enhanced, or analyzed using a variety of techniques. Data can be readily ported to other platforms for analysis or for final presentation.

Leica SP2 AOBS Laser Scanning Confocal Microscope

True spectral imaging with thin optical sections of much higher resolution than normal epi-fluorescence can be obtained from cultured cells, vibratome sections, or intact tissue with simultaneous colocalization of multiple fluorescent probes, reflectance and transmitted light. Collected images can be reconstructed in 3D, enhanced, or analyzed using a variety of techniques.

Data can be readily ported to other platforms for analysis or for final presentation. This instrument also has powerful capabilities for FRAP, *FRET* and time lapse applications

Leica SP5 AOBS Laser Scanning Confocal Microscope

Newer model of SP2. Arriving late December 2007.

Zeiss Live/DUO Confocal Microscope

High speed confocal microscope designed specifically for photoactivation or bleaching via a separate scanner than the imaging path. This confocal with a 100 mW laser at 489 nm and 50 mW lasers at 405 and 561 nm will be used primarily for live cell applications.

PerkinElmer UltraVIEW RS-3 Spinning Disk Laser Confocal Microscope

Preferable for imaging live cell cultures due to reduced phototoxcity, thin optical sections may be imaged as time lapse volumes. The 9 fps full field 12 bits imaging system has laser lines at 488, 568 and 647 nm for exciting three popular ranges of fluorescent probes, a piezo for high speed and reproducable Z axis control, environmental control and Nikon optics. Expected move to new satellite facility on second floor of Michael F. Price Centre for Genetic and Translational Medicine in the Harold and Muriel Block Research Pavilion in January/February 2008.

Multi Photon Confocal Microscopy

Multi-photon microscopy relies on excitation of fluorophores or harmonic generation by femptosecond pulses of highly concentrated long wavelength light. Practically, this allows for imaging multiple fluorescent wavelengths deep in live tissue.

The system reduces bleaching and other problems endemic to epi-fluorescent microscopy, may be more sensitive due to its lack of a confocal pinhole, and solves other problems of light scatter. *The instrument is at the centre of the intravital imaging programme at AECOM.*

FRET

Fluorescence Resonance Energy Transfer, the transfer of energy from a donor fluorophore within 7 nm of an acceptor fluorophore, can be used to measure binding interactions between and within molecules. The AIF provides acceptor bleaching for FRET imaging and measurements on three confocal microscopes and ratio FRET with widefield microscopy. Widefield FLIM in the time domain using a gated cooled CCD camera with LaVision software may be available.

TIRF

Total Internal Reflection Fluorescent microscopy provides excitation of fluorophores only within 100 nm of the subtrate. Therefore, only molecules immediately apposed to the coverslip are excited and imaged. Objective illuminated TIRF is provided on an Olympus IX71 with either a 60X or 100X N.A. 1.45 or a 100X N.A. 1.65 with capability to do TIRF/FRET using probes over the visible spectrum from CFP through red. Image collection and automated shuttering are provided with an Andor EM camera and Uniblitz shutters running under IPLab.

D.I.C. (Nomarski), Darkfield, Phase Contrast, IRM, and Epifluorescence with Digital Imaging

Four inverted microscope stations for high spatial resolution, wide dynamic range (low light to bright light) with *time lapse* and deconvolution capabilities. 12 bit Cooke Sensicam QE cooled CCD cameras mounted on high efficiency throughput Olympus IX70 or IX81 inverted microscopes with state of the art infinity corrected optics.

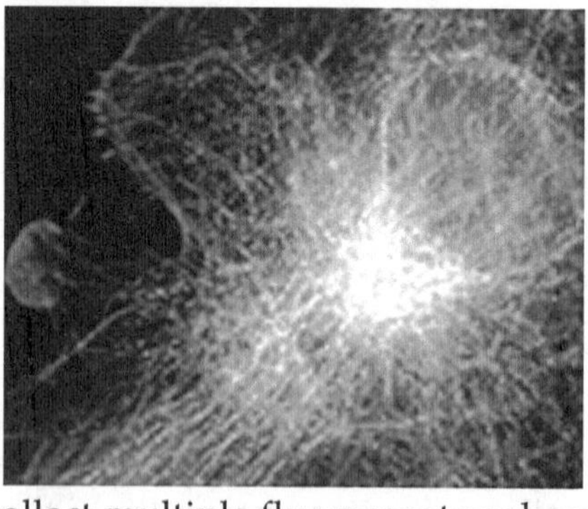

May be used to collect multiple fluorescent probes and transmitted light (brightfield, phase contrast or Nomarski) images with IPLab software running on PCs. Environmental chambers for the Olympus microscopes are available for *in vivo* work. Many applications including spot photometry. Also, focus motors for collection of serial sections for deconvolution. Deconvolution produces images that are confocal-like in their resolution but may have a benefit of imaging weak signal or a wide dynamic range. Standard fluorescent filters include FITC, rhodamine, Cy3, Cy5, Dapi, GFP, CFP, YFP among others more esoteric ones and a 50/50 mirror for *IRM*.

Exhaustive Photon Reassignment (EPR)

EPR deconvolution complements the other deconvolution techniques offered at the AIF by providing preservation of the total energy of the sampled volume for quantitative analysis of very dim specimens. Automated imaging of multiply probed serial optical sections is performed with a piezo controller and a Photometrics 15 bit cooled CCD camera on a high efficiency upright Olympus microscope. This is a specialty technique to image single or few molecules with precise locating within 70 nm.

Microinjection

Microinjection is a method to deliver solutions (proteins, DNA or RNA, other chemicals) directly into individual cells in culture. The AIF has two automated Eppendorf systems for use on any inverted microscopes in the Facility including the confocal, multirphoton, and other digital imaging systems.

Motion Analysis

High speed (200 FPS under bright illumination), real time (30 FPS video), or time lapse (approx. 100 ms to hours) imaging with fluorescence and transmitted light can be performed on inverted microscopes with temperature regulated environmental chambers. Images can be made into movies for video or web presentations. Sophisticated morphometric measurements may be made over time. Quantification of images includes intensity changes of fluorescence, changes in cell or particle velocity, direction, shape and size over time and schematic visualization of such changes. In some cases, volume changes can be measured.

Volume Rendering and 3D Quantitation

For 3D rendering or reconstructions the staff operate and train Imaris Bitplane, Voxx, a number of plugins within ImageJ and Volocity. The staff train investigators in more simple 3D imaging and quantitation via analysis of serial sections with ImageJ and I.P. Lab including the authoring of scripts for automation and result reporting.

Single Photon Uncaging

Uncaging is the activation by removing a photo-labile blocking group from DNA, RNA, protein or small molecules. The uncaging station consists of an Olympus IX70, two Hg arc lamps for UV uncaging and epifluorescence, UV corrected and phase contrast optics for uncaging and viewing cell behaviour, shutters for high speed and timed uncaging and image collection, and a Cooke Sensicam for recording uncaged fluorescence. This system shares a microscope with a *microinjection apparatus* for ease of loading cells for live experiments. A 337 nm laser with has been purchased for the system and is under development on a separate microscope stand in the Biophotonics Innovation Laboratory.

Hard Copy and Presentation

On all imaging platforms, digitized picture files are in standard formats and can be converted easily to other formats; data can be exported to other computer systems or reproduced on a variety of hard copy devices. Adobe Photoshop CS is most widely used for figure preparation and we are happy to assist. A Fujix Pictrography 3000 colour printer makes continuous tone output at 400 PPI indistinguishable from real photographs. Standard laser printing can be used for draft grayscale as well as for crisp graphs and text. Both still images and moving sequences can be prepared for web presentation. The AIF maintains in its inventory tools for video; however, use is by special appointment as video is being phased out.

10

Atomic Force Microscope and Scanning Probe Microscopy

ATOMIC FORCE MICROSCOPY

Atomic Force Microscopy and Scanning Probe Microscopy (AFM/SPM) provide topographic information down to the Angstrom level. Additional properties of the sample, such as thermal and electrical conductivity, magnetic and electric field strength, and sample compliance can simultaneously be obtained using a specialty probe. Many applications require little or no sample preparation.

Principle of Operation

The Atomic Force Microcope uses a physical probe raster scanning across the sample using piezoelectric ceramics. A feedback loop is used to maintain a constant interaction between the probe and the sample. The position of the probe and the feedback signal are electronically recorded to produce a three dimensional map of the surface or other information depending on the specialty probe used.

Data Output: is either a three dimensional image of the surface or a line profile with height measurements. The surface roughness parameters of Ra or RMS are also available with either of the above outputs.

Other types of feature analysis include Partical Grain Size Analysis, Bearing Ratio, Fractal Dimension, Power Spectrum, and Fast Fourier Transform.

Applications Include

- Materials Evaluation
  - Surface roughness on implanted silicon wafers.
  - Thermal properties such as thermal conductivity, glass transition temperature (Tg), and melting temperature of various phases of a blended polymer measured down to the nanometer scale.

- Surface profiles and magnetic field mapping of recording media or reading heads.
- Nanomechanical testing.

- Failure Analysis
  - Rapid hot-spot analysis of powered electronic devices.
  - Defect analysis of compact disk stampers.
- Quality Control
  - Surface profiles of thin film and coatings.
  - Metrology of semiconductor devices and compact disks.
  - Surface finish of substrates for thin film deposition.

SPM Techniques

- *Magnetic Force Microscopy*: Digital Video Disc Surface (10 microns)

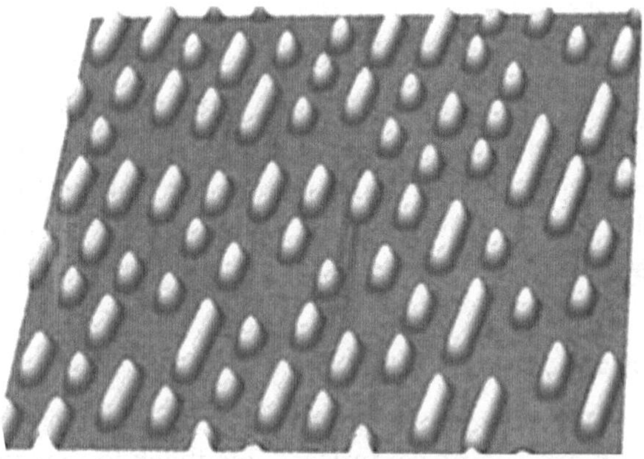

WAVELENGTH DISPERSIVE X-RAY SPECTROMETRY (WDS)

WDS identifies the elemental composition of materials imaged in the *SEM* with an order of magnitude better spectral resolution, sensitivity and ability to determine concentratins of light elements than is achievable with *EDS*. Most elements are detected below 0.1% and some as low as a few ppm.

Applications Include

- Identification of *spectrally overlapped* elements, such as
  - S in the presence of Pb or Mo.
  - W or Ta in Si, or N in Ti.
  - Br in Al, common in semiconductor device failure.
- Detection of *low concentration* species (down to 100 or even 10 ppm)
  - P or S in metals.
  - Contaminants in precious metal catalysts.
  - Trace heavy metal contamination.
  - Performance-degrading impurities in high temperature solder alloys.

- Analysis of low atomic number elements
  - Composition of advanced ceramics and composites.
  - B in BPSG films (sensitivity to 2000 ppm).
  - Oxidation and corrosion of metals.
  - Characterization of biomedical and organically modified materials.

Principle of Operation

The characteristic X-ray photons excited by the electron beam are sorted using a diffracting crystal, whose angular placement relative to the sample and photodetector is a unique measure of their wavelengths. As with EDS, the resulting spectral distribution is automatically compared with those from actual standards or synthetic X-ray fluorescence spectra of material formulations.

WDS vs. EDS

X-ray microanalysis in the scanning electron microscope is accomplished using EDS and/or WDS. EDS is more commonly applied due to its simplicity and speed, while WDS offers an important and often critical refinement of EDS data by providing

- Analysis for light elements with at least an order of magnitude higher sensitivity than available (ultrathin X-ray window) EDS instruments.
- Resolution of severely overlapped spectrum peaks for improved element specificity.
- Lowered detection limits over the entire periodic table.
- More accurate quantitative analyses.

SCANNING AUGER MICROANALYSIS (SAM)

SAM provides elemental and chemical composition for all elements with an atomic number greater than helium. Its sampling depth of 2-3 nm allows films as thin as a few monolayers to be analyzed. Auger also produces images of the distributions of elements along the surface and produces profiles of composition vs. depth from 1 to 2000 nm.

Applications Include

- Materials evaluation and identification
  - Surface contaminants
  - Surface homogeneity
  - Diffusion profiles
  - Particle sizes
  - Catalyst degradation
  - Interfaces

- Failure Analysis
  - Corrosion characterization
  - Stain identification
  - Lifted lead bond evaluation
  - Material delamination analysis
  - Metal embrittlement evaluation
- Quality control screening
  - "Good" to "bad" sample comparison.
  - Material and plating/coating thickness determination.
  - Surface process modification.

Principle of Operation

The sample is scanned with a focused beam of about 5 kV electrons, causing low energy Auger electrons to be ejected from its surface. The kinetic energies of these Auger electrons provide an analysis for the chemical elements present in the top few atomic layers. An auxiliary argon ion beam may be used to remove near-surface layers by "sputtering" to expose a fresh surface for analysis, producing a profile showing the dependence of sample composition on depth.

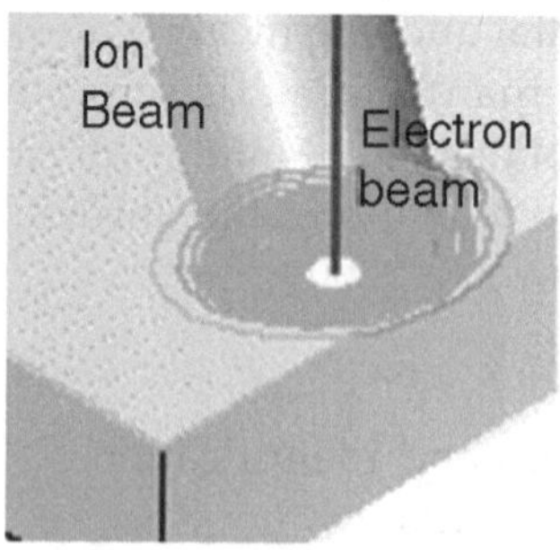

FOURIER TRANSFORM INFRARED SPECTROSCOPY (FTIR)

FTIR spectroscopy is used primarily for qualitative and quantitative analysis of organic compounds, and also for determining the chemical structure of many inorganics.

FTIR analysis applications include:

- Materials Evaluation and Identification
  - Organic compounds
  - Structure of many inorganic compounds
  - Deformulations
  - Forensics
  - Material homogeneity

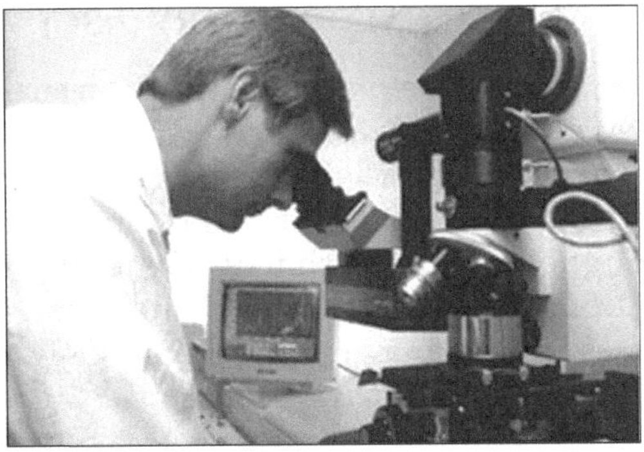

- Failure analysis
  - Micro-contamination identification
  - Adhesive performance
  - Material delamination
  - Corrosion chemistry.
- Quality control screening
  - "Good" to "bad" sample comparison
  - Evaluation of cleaning procedure effectiveness
  - Comparison of materials from different lots or vendors.

Principle of Operation

Because chemical bonds *absorb infrared energy at specific frequencies* (or wavelengths), the basic structure of compounds can be determined by the spectral locations of their IR absorptions. The plot of a compound's IR transmission vs. frequency is its "fingerprint", which when compared to reference spectra identifies the material.

FTIR spectrometers offer speed and sensitivity impossible to achieve with earlier wavelength-dispersive instruments. This capability allows rapid analysis of micro-samples down to the nanogram level in some cases, making the FTIR unmatched as a problem-solving tool in organic analysis.

The FTIR microscope accessory (shown in the photo above) allows spectra from a few nanograms of material to be obtained quickly, with little sample preparation, resulting in more data at lower cost. In some cases, thin films of residue are identified with a sensitivity that rivals or even exceeds electron or ion beam-based surface analysis techniques.

There are few sample constraints; solids, liquids and gases can be accommodated. Many contaminants present on reflective surfaces such as solder pads or printed circuitry are readily analyzed *in situ* using the FTIR microscope in reflectance mode.

More information on infrared spectroscopy:

Basic theory of infrared spectroscopy

Identifying organic structure by FTIR.

DIFFERENTIAL SCANNING CALORIMETRY (DSC)

Differential Scanning Calorimetry, DSC, is a thermo analytical technique in which the difference in the amount of *heat* required to increase the temperature of a sample and reference are measured as a function of temperature.

Both the sample and reference are maintained at nearly the same temperature throughout the experiment. Generally, the temperature programme for a DSC analysis is designed such that the sample holder temperature increases linearly as a function of time. Only a few milligrams of material are required to run the analysis.

Principle of Operation

When a sample undergoes a physical transformation such as a *phase transition,* more or less heat will need to flow to it than to the reference (typically an empty sample pan) to maintain both at the same temperature. Whether more or less heat must flow to the sample depends on whether the process is *exothermic* or *endothermic.*

For example, as a solid sample melts to a liquid it will require more heat flowing to the sample to increase its temperature at the same rate as the reference. This is due to the absorption of heat by the sample as it undergoes the endothermic phase transition from solid to liquid.

Likewise, as the sample undergoes exothermic processes (such as *crystallization*) less heat is required to raise the sample temperature. By observing the difference in heat flow between the sample and reference, differential scanning calorimeters are able to measure the amount of heat absorbed or released during such transitions. DSC may also be used to observe more subtle phase changes, such as *glass transitions.*

Applications

DSC is commonly used to measure a variety of properties in both organic and inorganic materials, from metals and simple compounds to polymers and pharmaceuticals. The properties measured include:

- Glass transitions
- Phase changes
- Melting
- Crystallization

- Product stability
- Cure/cure kinetics
- Oxidative stability
- Heat capacity and heat of fusion measurements

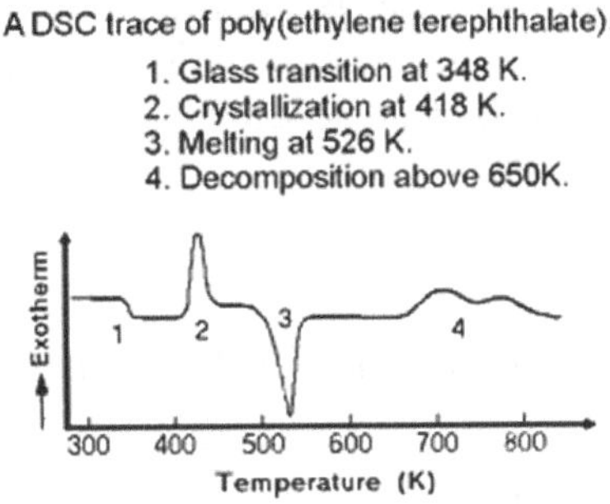

ELECTRON MICROSCOPY IN SITU HYBRIDIZATION

QUESTIONS AND PROBLEMS

For in situ hybridization detection (in the specific example cited, using digoxigenin-labeled RNA probes with mouse anti-digoxigenin antibody, followed by anti-mouse ultrasmall gold-conjugated secondary antibody and enhancement):

- Is a two-step detection method more sensitive than a single-step (gold-labeled anti-digoxigenin primary) method? One review attributes higher sensitivity for two-steps.
- For a two-step method, would gold-conjugated antibodies (against biotin probes) have advantages over streptavidin-gold?
- Will glutaraldehyde levels greater than 0.05% inhibit hybridization? For example, Punnonen et al. labeled rat glioma monolayers using 4% paraformaldehyde and 0.05% glutaraldehyde for 30 minutes, but protocol included 0.1% saponin permeabilization for 30 min.

Success Factors for Effective Labeling

ISH is another form of affinity localization labeling protocol; the probe is a nucleic acid sequence instead of an antibody, and much the same rules apply.

- Two step, sequential detection methods (primary label, followed by secondary with attached visualization probe) are usually more sensitive. This was first observed by Coons (generally considered to be the father of immunoctyochemistry) in the 1950s, and this sequential labeling method can be applied to almost any labeling system.
- If you are using antibodies to digoxigenin, then a secondary antibody-gold probe should be sufficient. If you choose to use Nanoprobes (Fab′ conjugates), it might be difficult to later perform

multiple labeling experiments. If you want to look for biotin labeled nucleic acid probes, it may be worthwhile to try anti-biotin antibodies, and cold water fish skin gelatin as a blocking agent rather than serum.

- Each system has its own unique qualities and sensitivity to fixatives. If it bothers you to include glutaraldehyde, then leave it out. Fix with a strong glutaraldehyde solution after hybridization and silver enhancement.

Ultramicrotomy

The Diatome Handling and Use Manual: Technical Information

Includes information on diamond knives for cryo-ultramicrotomy, low angle diamond knives, as well as knives for light microscopy and histology.

Tips, Tricks, and Data Sheets

Tips and Tricks from the Microscopy Listserver

The site has tips and tricks for biological microscopy. Many of the postings are from the microscopy listserver. Anyone can submit a tip or trick. Please contact Greg Erdos for further information at ICBR EM Core Lab, University of Florida

Electron Microscopy Science's Technical Data Sheets

A collection of almost 100 information sheets for a wide variety of EM plastic resins, fixatives, and other products used in sample preparation.

SCANNING ELECTRON MICROSCOPY (SEM)

SEM Provides Topographical and Elemental Information at Magnifications of 10x to 100,000x with Virtually Unlimited Depth of Field.

Applications Include

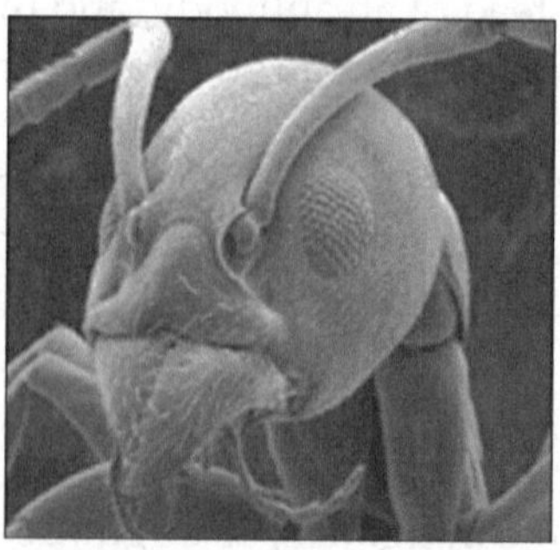

- Materials evaluation
  - Grain size
  - Surface roughness
  - Porosity
  - Particle size distributions
  - Material homogeneity
  - Intermetallic distribution and diffusion
- Failure analysis
  - Contamination location
  - Mechanical damage assessment
  - Electrostatic discharge effects
  - Micro-crack location
- Quality Control screening
  - "Good" to "bad" sample comparison
  - Film and coating thickness determination
  - Dimension verification
  - Gate width measurement
  - Mil Std. screening

Principle of Operation

A finely focused electron beam scanned across the surface of the sample generates secondary electrons, backscattered electrons, and characteristic X-rays.

These signals are collected by detectors to form images of the sample displayed on a cathode ray tube screen. Features seen in the SEM image may then be immediately analyzed for elemental composition using *EDS* or *WDS*.

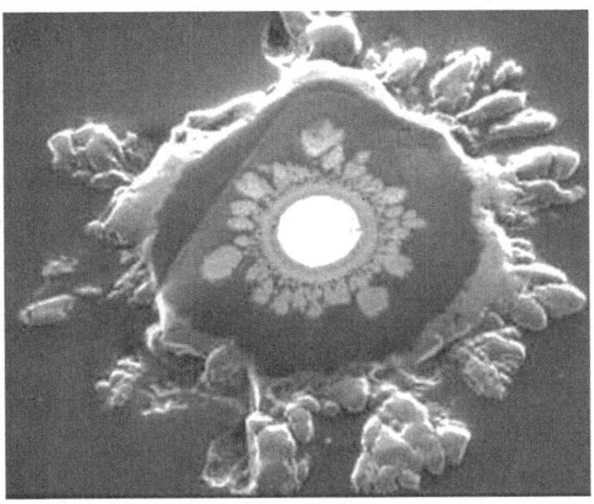

- *Secondary Electron Imaging:* shows the topography of surface features a few nm across. Films and stains as thin as 20 nm produce adequate-contrast images. Materials are viewed at useful magnifications up to 100,000x without the need for extensive sample preparation and without damaging the sample. Even higher magnifications and resolution are routinely obtained by our *Field Emission SEM.*

- *Backscattered Electron Imaging:* shows the spatial distribution of elements or compounds within the top micron of the sample. Features as small as 10 nm are resolved and composition variations of as little as as 0.2% determined.
- *Data Output:* is generated in real time on the CRT monitor. Images and spectra can be printed here, recorded on CD ROM and/or emailed for insertion into your own reports.

Field Emission

Scanning Electron Microscopy (FESEM)

Principle of Operation

A field-emission cathode in the electron gun of a scanning electro-nmicroscope provides narrower probing beams at low as well as high electronenergy, resulting in both improved spatial resolution and minimized samplecharging and damage. For applications which demand the highest magnification possible.

Applications Include

- Semiconductor device cross section analyses for gate widths, gate oxides,film thicknesses, and construction details
- Advanced coating thickness and structure uniformity determination
- Small contamination feature geometry and elemental composition measurement

Why Field Emission SEM?

- FESEM produces clearer, less electrostatically distorted images withspatial resolution down to 1 1/2 nm. That's 3 to 6 times better than conventional SEM.
- Smaller-area contamination spots can be examined at electron acceleratingvoltages compatible with Energy Dispersive X-ray Spectroscopy.
- Reduced penetration of low kinetic energy electrons probes closer tothe immediate material surface.
- High quality, low voltage images are obtained with negligible electrical charging of samples. (Accelerating voltages range from 0.5 to 30 kV.)
- Need for placing conducting coatings on insulating materials is virtually eliminated.
- For ultra-high magnification imaging.

Cross-section of a Laser Window Showing Multiple thin Layers at 50,000x

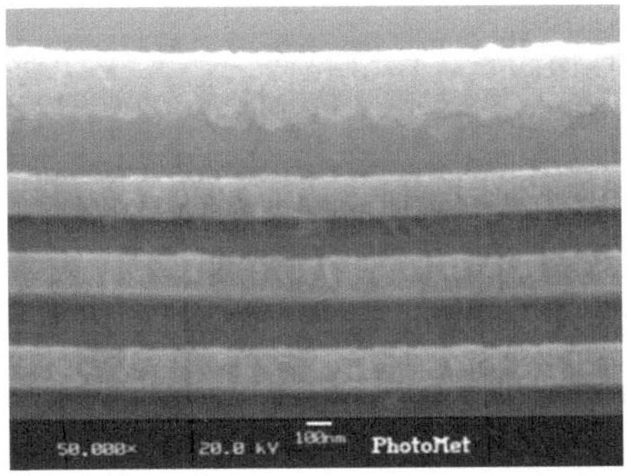

Fig. Cross Section of Contact on Silicon

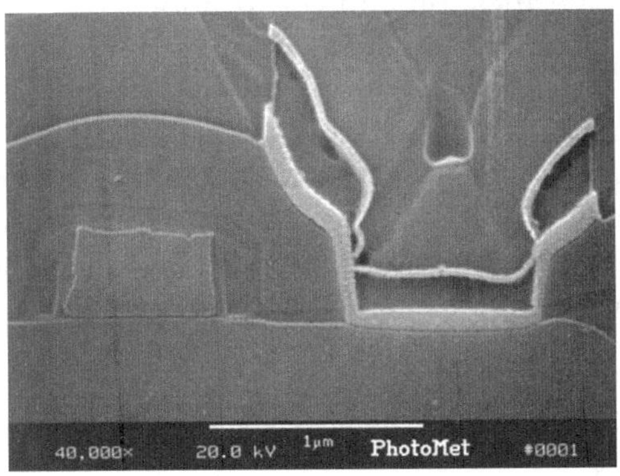

Fig. Cross Section of Via Openings

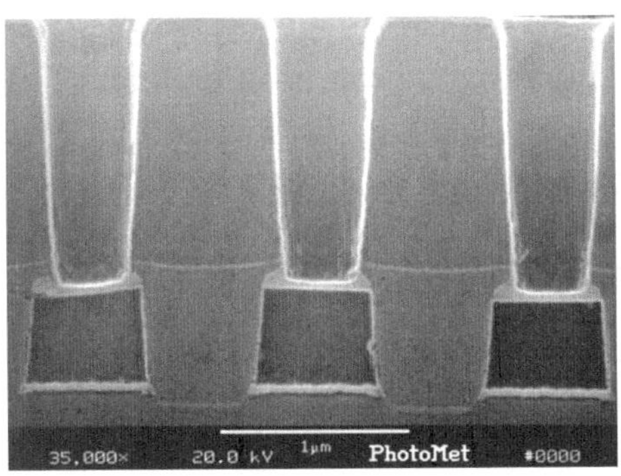

ENERGY DISPERSIVE X-RAY SPECTROSCOPY (EDS)

EDS identifies the elemental composition of materials imaged in a Scanning Electron Microscope *(SEM)* for all elements with an atomic number greater than boron. Most elements are detected at concentrations on the order of 0.1%.

Applications Include

- Materials evaluation and identification
  - Contaminants.
  - Elemental diffusion profiles.
  - Glassivation phosphorus content.
  - Multiple spot analysis of areas from 1 micron to 10 cm in diameter.
- Failure analysis
  - Contamination identification.
  - Unknowns identification.
  - Stringer location and identification.
- Quality control screening
  - Material verification.
  - Plating specification and certification.

Principle of Operation

As the electron beam of the SEM is scanned across the sample surface, it generates X-ray fluorescence from the atoms in its path. The energy of each X-ray photon is characteristic of the element which produced it. The EDS microanalysis system collects the X-rays, sorts and plots them by energy, and automatically identifies and labels the elements responsible for the peaks in this energy distribution.

The EDS data are typically compared with either known or computer-generated standards to produce a full quantitative analysis showing the sample composition.

Data output: Plots the original spectrum showing the number of X-rays collected at each energy, as seen above. Maps of element distributions over areas of interest and quantitative composition tables can also be provided as necessary.

SPECIMEN PREPARATION FOR ELECTRON MICROSCOPY

E.M. Processing Schedule - Epoxy Resin

1. Fix tissue in 2.5% glutaraldehyde in 0.1M sodium cacodylate buffer at 4°C, for a minimum of 4 hours. Tissue should be cut into approx. 1mm cubes for fixing. This may be done in a drop of fix on a sheet of dental wax, using a razor blade. Place in glass processing vials and close with plastic caps. (Tissue may be stored at this stage.)
2. Wash in 0.1M buffer - 1 hour × 2. (Or overnight at 4°C.)
3. Post fix in osmium tetroxide in 0.2M buffer - 1 hour. (Mix equal quantities of 2% aqueous OsO4 and 0.4M buffer and use immediately.)
4. Rinse in 0.2M buffer - 5 mins. × 2.

5. Dehydrate in 70% ethanol - 20 mins. × 2. (May be stored overnight at this stage if absolutely necessary.)
6. Dehydrate in 90% ethanol - 10 mins. × 2.
7. Dehydrate in 100% ethanol - 20 mins. × 2.
8. Propylene oxide (1.2 epoxy propane) - 10 mins. × 2.
9. Propylene oxide/epoxy resin mixture (50/50) - 1 hour.
10. Epoxy resin - overnight - with caps removed from vials. (Allows any remaining propylene oxide to evaporate.)
11. Embed in labelled capsules with freshly prepared resin.
12. Polymerise at 60°C - 48 hours.

NOTES

- All steps must be performed in a fume cupboard and gloves should be worn throughout.
- Osmium tetroxide, propylene oxide and propylene oxide/resin waste should be collected in bottles for safe disposal.
- For steps 2 to 10 the processing vials should be on a rotating mixer.
- It is often beneficial to increase the processing times for specimens with a high connective tissue content such as skin, tendon, cartilage, cornea, etc.. This should be done for all steps and especially for the 50/50 mixture at step 9 (which can even be extended to an overnight treatment). Failure to increase the times may result in a poorly infiltrated block and therefore weak sections which will fail under the electron beam, or even be impossible to cut in the first place.
- For very urgent specimens processing times may be reduced as long as the blocks of tissue are very small and not 'difficult' tissues (see above). Polymerisation can be achieved in 1 hour at 100°C using gelatine capsules (the polythene ones melt at this temperature). Blocks must be cooled in water and are ready to cut in 15 mins..

SUCCESSFUL ELECTRON MICROSCOPY IMMUNOLABELING

Electron Microscopy Immunolabeling

Questions and Issues

Many antibodies don't work for light microscopy, and even more don't work for TEM immunocytochemistry. What experiences and success stories have people had? How does antibody performance compare:

- Between different methods - ELISA, western blot or immunoprecipitation, compared with light microscopy and electron microscopy?
- Between monoclonals and polyclonal antibodies?
- How can we get an antibody which works for light microscopy to work for EM? What proportion of antibodies that work for light microscopy can we expect to work for immunoelectron microscopy?

- How do the requirements for successful immunoelectron microscopy vary between different tissues, antigens, and epitopes?
- What is required to determine a successful EM protocol?
- How is good light microscopy procedure (e.g. acetone whole-mount sections) translated to electron microscopy?

Success Factors for Electron Microscopy Immunolabeling

Introduction and General Approaches

It is sometimes mistakenly assumed that immunoreagents are inert objects, and that their size is the only property to be taken into account. Immunoreagents are biological molecules rather than chemicals; they are charged molecules and have many complex, variable interactive and binding properties.

As a first step in immunoelectron microscopy method development, try the whole preparation and immunolabeling protocol at the LM level first. Use the same reagents and preparation protocols as you plan to use for EM, and expose to silver enhancement for a longer time so that you can see the signal by LM. If you see a signal, then you will be confident that you will see a signal when you go to examine the EM experiment.

Generally, we have found that if there's no signal by LM, the antibodies will never work when used with the same preparation protocols for EM. However, often minor modifications to these protocols can make the system work. General tips for successful EM labeling:

- Get to know the people who make the antibodies and find out as much as possible about the target that the antibodies label, and the biology of the system in which they will be used.
- Screen the antibodies by light microscopy on the system to be used for electron microscopy labeling. Thin sections of Lowicryl-embedded material make good light microscope preparations, and silver enhancement is often particularly good for visualizing protein A-gold.

As a general phenomenon, some antibodies work only for one method. Possible reasons include:

- Antibody binds antigen only in its natural conformation, not in western blots where protein is denatured, or vice versa.
- Strong fixing retains proteins around antigen, making it inaccessible.
- For antibodies to peptides: very few sequences on the target molecule will be recognized by the antibody, and antibody binding will depend upon the position and orientation of these sequences in the protein structure. Interior sequences may be inaccessible, or may be rendered so by fixation; exterior sequences that are part of a soluble component of a complex may be dislodged by washing. Because they recognize only a very small proportion of the target, sensitivity may be a problem as well as accessibility.

- Specificity is often greater for anti-peptide antibodies than for monoclonals, which in turn are more specific than polyclonals.
- Components of blocking agents can bind to the antibody. For example, protein A binds to IgG present in serum, resulting in inhibition of specific signal. Therefore, dilution of protein A in serum is not recommended!

In one very unusual case, autofluorescent organelles with a similar distribution to the target were mistaken for specific labeling in fluorescence labeling experiments; then, when the protocol was repeated for electron microscopy, no labeling was observed. The protein under study was a serum protein, and the blocking agent initially used to dilute the antibody was 10% serum; the antibody reacted with this molecule was removed from labeling solution.

Specimen and Antibody Treatment

- Cultured cells must be grown for long enough for sufficient antigen to be expressed. The time required will vary with the cells under study, so it is important to understand the system thoroughly. For example, some types of cells may need to be grown for 10 or 15 days, while others may express sufficient numbers of specific antigens in only a day or two.
- *Storage issues:* the freezing of antibodies, and especially their treatment after thawing, can affect performance. Freeze-thawing antibodies can cause aggregation, especially for IgM primaries which are large molecules with many branches; if the solution is centrifuged after thawing and there are many aggregates, especially big ones, the antibody can be effectively removed from solution. Therefore, it is worthwhile to check how the antibodies have been stored.

  In LM, even low affinity antibodies can be detected, while at EM they may not. If only about 10% of an IgM primary antibody remains after centrifugation, antibody dilution has to be increased for EM detection; and if there has been any change in the antibody (such as aggregation), although it may still bind to antigens on your section, the secondary antibody may not recognize it. In LM, where there are more antigens in the field of view and amplification methods may be used, the signal may still be detected; however, by EM the signal will be absent.
- Streptavidin-gold is a tricky reagent: when used with silver enhancement, it frequently seems to detect biotin at the LM level, but gives no signal at the EM level. Reasons for this can include:
  - *Blocking agent:* if serum is used, the gold probe will bind to biotin-like molecules in the serum.
  - *The probe is relatively unstable:* colloidal gold separates from the protein after only short storage times. Free streptavidin in the solution may not be sufficient to inhibit labeling for LM, but removes almost all the EM label.

- Protein A-gold can have advantages over secondary antibodies. Since protein A-gold is used by everyone, turnover is faster and quality control and troubleshooting easier (However, it should be noted that protein A does not recognize all classes of IgG, so a bridging antibody may be required).
- *Freund's adjuvant effects:* Freund's adjuvant is a homogenate of ground-up bacterial proteins. Therefore, if the study involves looking at bacteria, some background labeling on bacteria is inevitable. If we know how the antibody was prepared, we can explain this result. If we do not know, we can try many different fixation and blocking solutions without success, only to attibute this to a poor antibody.

Fixation Issues

Different criteria for acceptable morphological preservation for LM and EM can result in differences in labeling. At the LM level, preservation of morphology may be less critical, so less stringent fixation conditions may be used; however, when these protocols are transferred to EM, more stringent fixation may be required for acceptable morphological preservation, which results in lower labeling.

For example, freezing tissue without aldehyde fixation or cryoprotection, then fixing the sections with acetone or methanol and air-drying them before or after this treatment, is used for LM, but it can remove or relocate antigens on the EM scale.

The solution is to begin the development of an EM method by using the same fixation protocol used for LM. This way, labeling will be assured for EM. Then, the protocol may be adjusted until an acceptable compromise is reached between morphological preservation and labeling. The level of acceptance of poor morphology will depend on the result expected. Each system will have its own protocol for getting that particular result.

- If it is possible to label purified (or semi-purified) sub-cellular particles, this may be worthwhile. For example, mitochondria are highly recognizable, so this approach is acceptable for labeling antigens on the outside of mitochondria. If the antigen is inside the mitochondria, then sectioning may be best. If the profile that is labeled is so obviously a mitochondrion, then it doesn't matter if the cristae are not perfectly preserved (unless there is extraction or redistribution of antigen). If specific membrane markers are available, structural morphology is even less important: co-localizing the marker with the test antibody on a membrane or organelle fragment is often acceptable.
- Antigen accessibility can be increased simply by breaking open cells to let out the antigens we are not interested in (unless we are looking for cytoplasmic antigens). For example, one investigator labeled the

lumen of the RER with a very specific antibody. He couldn't do this on well fixed, heathly cells because the antibody could not gain access to the luminal antigens. He noticed that some of the cells in the pellet had obviously died before being fixed, and had poor morphology. However, the RER was still recognizable, and densely labeled.

Antibody and Specimen Characteristics

- Many (monoclonal) antibodies provided today are very well suited for Westerns and sometimes LM because they were developed for the degenerative conditions which are applied in these techniques (up to 90% of the epitopes can be removed by embedding and sectioning).
- In small samples (a few millimeters in square with a section thickness of 50-70 nm) fixation and post-processing can lead to a loss of antigenicity.
- Very few antibodies are suitable for immuno-EM; antibodies specifically generated for immunoelectron microscopy, which recognize their epitope after aldehyde degeneration/fixation and acrylic plastic embedding, are not readily available commercially.

Tips for Developing TEM Immunolabeling Protocols

- Count the signals (specific signals may not be visible at X 20,000 or higher magnification).
- Test different antibodies and manufacturers.

Identifying the Right Fixation Conditions

Immunoelectron microscopy is becoming important for resolving localized GFP transfected proteins which are insufficiently well resolved by confocal microscopy. In pre-embedding experiments with anti-green fluorescent protein (GFP) immunolabeling with Nanogold→ and silver enhancement, it was helpful, before attempting immuno EM, to do a fixation series of formaldehyde and glutaraldehyde dilutions, and see where the fluorescent signal was lost. Fixation vs. penetration is always a compromise.

Other Comments

It was also suggested that pre-embedding labeling with ultrasmall gold conjugates and silver enhancement has made it easier to translate LM success to the EM level. In some cases, the localization is only relevant when the cellular and tissue organization is intact, such as with brain tissue or in the study of cellular RNA/antigen/gene distribution, and the morphology must not be compromised beyond a certain point - which may be beyond the threshold that allows antigen survival or antibody access.

However, ultrasmall gold conjugates and silver enhancement reagents can label antigens in fixed, wet tissue with excellent ultrastructure. It was pointed out that the initial fixation determines the starting point of ultrastructure quality (limited by antigen tolerance level), while the immunolabeling procedure itself is responsible for preventing deterioration of the ultrastructure. Use of Fab' conjugates allows less harsh permeablization of the tissue, and a near neutral pH silver enhancement reagent reduces the harm of pre-osmium enhancement on morphology.

Mixed results were described; one reply gave a success rate of 5 to 10 % for immunoelectron microscopy labeling (TEM thin sections) with antibodies which worked for fluorescence confocal microscopy. It was also suggested that investigators using fee-for service facilities may be unwilling to take the time necessary to develop an effective method for their project.

FRICTION FORCE MICROSCOPY

Soon after the introduction of the AFM, it was realized that the same instrument could be used to also measure forces in the direction(s) parallel to the surface, i.e. the *friction* forces.

For this application, the AFM usually detects not only the deflection of the cantilever perpendicular to the surface, but also the torsion of the cantilever, resulting from one component of the lateral force. Again, the sensitivity can go down to the atomic level, so that this instrument allows us to investigate the atomic-scale origin of the friction force.

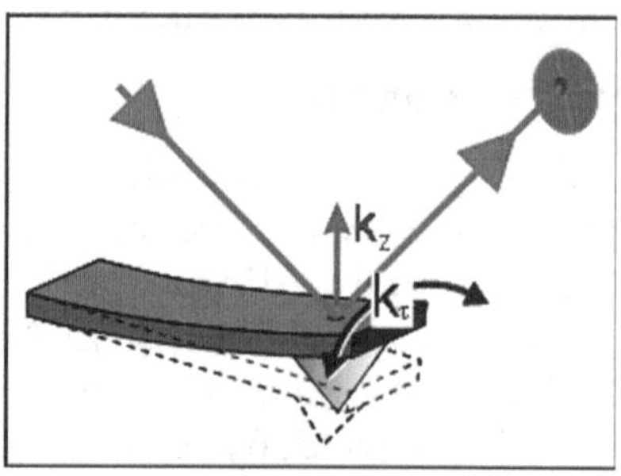

This field of research is known as *nanoTribology*. Unfortunately, the geometry of a traditional AFM cantilever is not very sensitive to friction, unless the cantilever is made very thin, in which case it becomes 'over-sensitive' to the forces normal to the surface.

In order to overcome this problem, the *Interface Physics Group* (The Interface Physics group uses various types of *Scanning Probe Microscopy* to investigate the structure and dynamic behaviour of surfaces and interfaces.

The microscopes used in this research are all home-built, special-purpose instruments. Topics that we are working on at present include *surface diffusion, surface phase transitions, gas-surface interactions, nanotribology,* and *nanobiology.*

Click on any of the items below to find out more about our most recent results.) has designed and constructed a special Friction Force Microscope (FFM), which is highly sensitive to both components of the force parallel to the surface, while it has the regular sensitivity, typical for a traditional AFM, to the perpendicular force. More about atomic-scale friction and our research on friction can be found on the research pages of our group about *nanoTribology* (The Nano-Triboscope is an instrument designed to measure friction and wear down to nanometer and nanonewton scales. The term tribology derives from the Greek "tribos" meaning "to rub." Peter Jost formally named this field in 1966 in a report that he presented to the H.M.

Stationary Office in London. Naturally, friction and wear—and the study of friction and wear—have been around a lot longer than that. Still, people refer to tribology as a "young field" for it has only been in recent years that scientists have truly started to quantify the study of friction and wear.*)*.

OVERVIEW OF SCANNING PROBE MICROSCOPY TECHNIQUES

Scanning Probe Microscopy has enabled researchers to image surfaces at the nanometer scale. Rather than using a beam of light or electrons, SPM uses a fine probe that is scanned over a surface (or the surface is scanned under the probe). By using such a probe, researchers are no longer restrained by the wavelength of light or electrons.

The resolution obtainable with this technique can resolve atoms, and true 3-D maps of surfaces are possible. Scanning Probe Microscopy is a general term, used to describe a growing number of techniques that use a sharp probe to scan over a surface and measure some property of that surface. Some examples are STM (scanning tunneling microscopy), AFM (atomic force microscopy), and NSOM (Near-Field Scanning Optical Microscopy).

Basic Idea of Scanned Probe Techniques:

The number of techniques is constantly growing, as the tip can be modified in many ways to investigate surface properties. In addition, the use of a tunneling current, simple force feedback or light through an aperture in the probe enable different ways to interact with the surface.

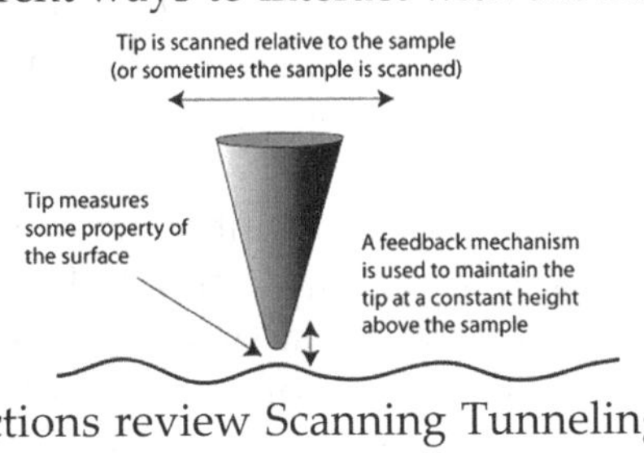

The following sections review Scanning Tunneling Microscopy, Atomic Force Microscopy & its multiple variations, and NSOM (or SNOM).

SPM on the Market

Although several academic research groups, such as the Interface Physics Group in Leiden, develop new types of SPMs, or improve existing types of SPM, most SPM investigations in academic and industrial research groups are carried out with commercial SPMs nowadays. There is a wide collection of general-purpose or more specialized scanning probe microscopes, which can be obtained from a growing number of companies. Go to our *links* page, in order to find links to several SPM-suppliers.

IMAGING, MEASURING AND MANIPULATING NATIVE BIOMOLECULAR SYSTEMS WITH THE ATOMIC FORCE MICROSCOPE

Introduction

Biological systems can only be fully understood if their structure is known. This underscores the importance of structural biology, the science investigating the structure and function of the components of living systems.

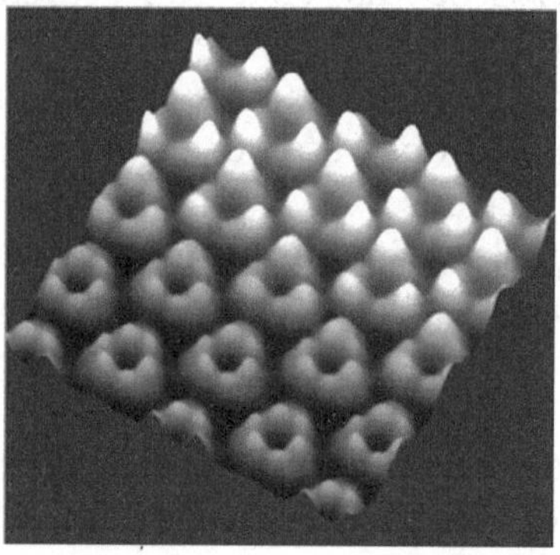

In contrast to other fields of molecular biology, it is an area of research where progress depends critically on sophisticated instruments. The atomic force microscope (AFM) is one of the most powerful tools for determining the surface topography of native biomolecules at subnanometer resolution.

Unlike X-ray crystallography and electron microscopy (EM), the AFM allows biomolecules to be imaged not only under physiological conditions, but also while biological processes are at work. Because of the high signal-to-noise (S/N) ratio, the detailed topological information is not restricted to crystalline specimens.

Hence single biomolecules without inherent symmetry can be directly monitored in their native environment. The AFM can also provide insight into the binding properties of biological systems. In order to determine the specific interaction between two kinds of molecules (e.g., avidin and biotin) in an AFM, one kind (i.e., avidin) is bound to the tip of a cantilever and the other kind (i.e., biotin) covers the surface of the sample support.

The adhesion force upon separation is then a measure of the binding strength. This method allowed the intermolecular forces between individual ligand-receptor pairs, complementary DNA strands, cell adhesion proteoglycans, and the specific antigen-antibody interaction to be determined.

PRINCIPLE

The AFM works in the same way as our fingers which touch and probe the environment when we cannot see it. By using a finger to "visualize" an object, our brain is able to deduce its topography while touching it. The resolution we can get by this method is determined by the radius of the fingertip. To achieve atomic scale resolution, a sharp stylus (radius ~1-2 nm) attached to a cantilever is used in the AFM to scan an object point by point and contouring it while a constant small force is applied to the stylus.

With the AFM the role of the brain is taken over by a computer, while scanning the stylus is accomplished by a piezoelectric tube. This simple technique provides a peep into the microscopic world, and it enables us to understand how the "smallest bricks" (i.e., the biomolecules) of biological systems might function.

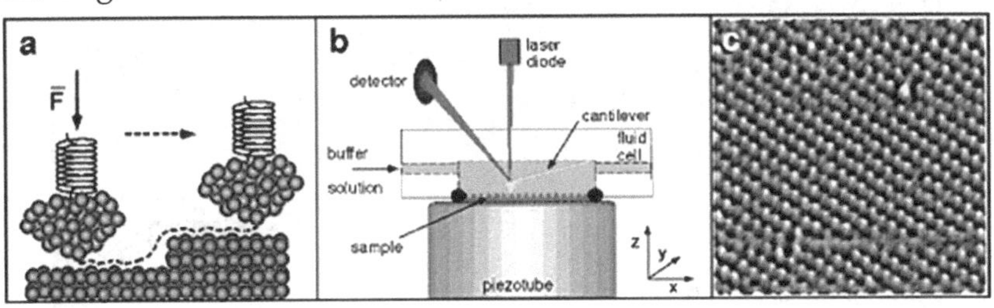

Principle of the AFM.

(a) A fine stylus is mounted on a cantilever spring and scanned over the surface. At sufficiently small forces the corrugations of the scanning lines represent the surface topography of the sample.

(b) The vertical deflection of the cantilever is detected by reflecting a laser beam onto a 2-segment photodiode.

The photodiode signal is used to drive a servo system which controls the movement of the piezo xyz-translator. In this manner the applied force between the stylus and the sample can be kept constant within some tens of a piconewton. The imaging process can be performed in a liquid cell filled with buffer solution. This ensures that the biomolecules remain hydrated. (c) Atomic resolution of a mica surface recorded in aqueous solution. The distance between adjacent protrusions is 5.4 E.

Applications: From Cartilage to Collagen, and to the Structure and Function of Individual Proteins

With the AFM, native tissue can be directly observed without prior dehydration. This can be important as demonstrated with articular cartilage which has to be kept in physiological buffer to preserve its ultrastructure. Surface irregularities observed by scanning electron microscopy (SEM) are absent on AFM inspection. Occasionally, the cartilage surface exhibits local discontinuities where an underlying fibrous network is distinguishable.

Digestion of the cartilage surface with chondroitinase AC exposes this fibrous network more systematically so that the individual fibers are visualized with great clarity by AFM.

When imaged at higher magnification, these distinct fibers exhibit a 60 nm repeat, indicating that they are assembled from collagen fibrils. In addition, the micromechanical properties along the articular surface can be measured with the AFM.

Hence this instrument provides new possibilities for the characterization of both surface structure and mechanical properties of freshly excised articular cartilage under physiological conditions. Preliminary data confirm the existence of an amorphous, non-fibrous articular surface which is expected to be vital for lubrication and wear of articular cartilage surfaces *in vivo.*

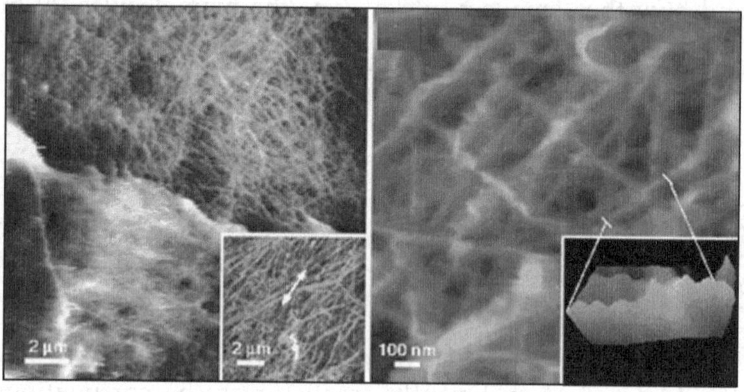

Surface and sub-surface of fresh articular cartilage from a bovine humeral head.

(a) The most superficial layer, typically 200-500 nm thick, consists of acellular and non-fibrous tissue. Occasionally, this exhibits local discontinuities where the underlying network of collagen fibrils oriented parallel to the surface becomes visible. Treatment with chondroitinase AC removes the majority of the articular surface and exposes the superficial collagen fibrils.

(b) Higher magnification view of the collagen fibrils which exhibit a mean diameter of ~35nm. Tangential fibrils reveal a characteristic periodic banding pattern of 60±5 nm.

AFM topographs of purple membrane from Halobacterium salinarium.

Purple membrane consists of 25% lipids and 75% bacteriorhodopsin. This light driven proton pump comprises 7 transmembrane a-helices which surround the photoactive retinal.

(a) Imaged at forces of about 3 × 10 –10 N two of the three loops connecting the a-helices are visible on the cytoplasmic surface.
(b) When the applied force is reduced during imaging (from 3 × 10 –10 N at the beginning (bottom) to 1 × 10 – 10 N at the end of the scan), the proteins undergo a conformational change.
(c) The most prominent loop connecting the a-helices E and F is imaged at 1 × 10 –10 N, but is bent toward the membrane surface at higher forces.

As a result of their high S/N ratio unprocessed AFM images provide atomic detail of solids. In contrast, biomolecules such as proteins which are designed to undergo conformational changes and form flexible supramolecular assemblies, are mechanically "soft", i.e., they are best compared with a "sponge". Hence their surface topography cannot be probed at atomic detail by AFM.

Nevertheless, state-of-the-art specimen preparation and instrumentation now allow the surface topography of native proteins to be imaged at subnanometer resolution.

As documented in figure, by scanning the cytoplasmic surface of the purple membrane of the *Halobacterium salinarium* with a very low cantilever force (i.e., ~10 – 10 Newton), the most prominent loop connecting the transmembrane a-helices E and F of the bacteriorhodopsin molecule can be imaged in its extended conformation.

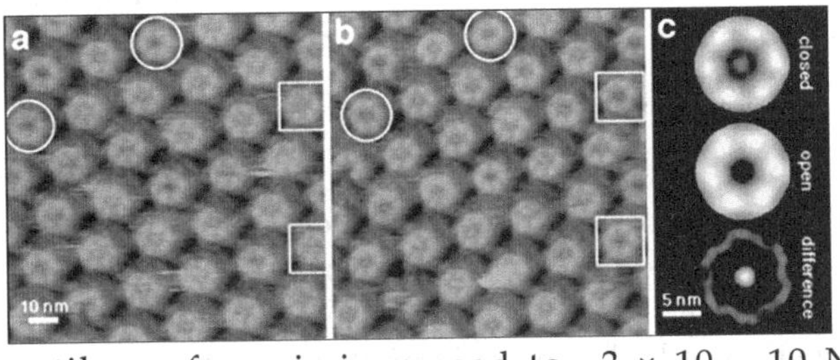

As the cantilever force is increased to ~3 × 10 – 10 Newton, this prominent loop is mechanically bent toward the membrane surface. The energy required for this reversible structural change is approximately 6kJ/mol, much too small to disrupt single chemical or electrostatic bonds. Conformational change of the hexagonally packed intermediate (HPI) layer of Deinoccocus radiodurans.

The HPI layer consists of units which display a large central pore and are assembled from six protomers. The function of the surface layer is not understood, but it is thought to protect the cell from hostile factors of the environment, and might also be responsible for the uptake and release of nutrients and cellular signals.]

(a) Single pores of the inner surface occur in "open" (circled) and "closed" (boxed) conformations.
(b) Imaging the same surface area after 5 min, some of the pores have changed their conformation. This conformational change is fully reversible and can be observed over hours.
(c) Averages of open and closed pores together with a calculated difference image.

As illustrated in figure, at nanometer scale resolution AFM images of the cytoplasmic surface of the hexagonally packed intermediate (HPI) layer of the bacterium Deinoccocus radiodurans reveal protruding protein cores. Being formed from six protomers these are conected by emanating arms and contain a central pore. The pores exhibit two conformations: a "closed" with, and an "open" without a central plug. Most exciting, over time a stochastic opening and closing of individual pores is monitored with the AFM.

The extracellular and periplasmic surface topographies of 2-D crystalline lipid-protein arrays of the E. coli outer membrane protein OmpF have been determined with the AFM, and the accuracy of the corresponding surface contours has been assessed by comparison with the atomic structure of the OmpF trimer. The atomic model of the 2-D lipid-OmpF crystals which has been built so as to optimally fit the surface contours measured with the AFM, illustrates how membrane proteins interact with the lipid bilayer.

Identification by AFM of individual amino acid residues at protein surfaces appears now to be feasible. However, this requires a better understanding of the interactions between stylus and biomolecule, and further developments of AFM techniques. For high resolution imaging both the radius and the geometry of the stylus, and the sensitivity of the deflection detector are critical and require improvement. In addition, biological specimens must be immobilized onto ultraflat supports to decrease their lateral flexibility.This can be achieved by chemisorption of the sample to activated substrates, or by its physisorption to atomically flat layered crystals.

Our results emphasize that AFM is an useful technique in structural biology. Today, AFM images provide information on the surface structure of biomolecular systems which is complementary to other established techniques such as light and electron microscopy, nuclear magnetic resonance and X-ray crystallography.

The advantage of directly observing biomolecular systems in their native environment opens the exciting possibility to analyse their structural and functional properties at the submolecular level.

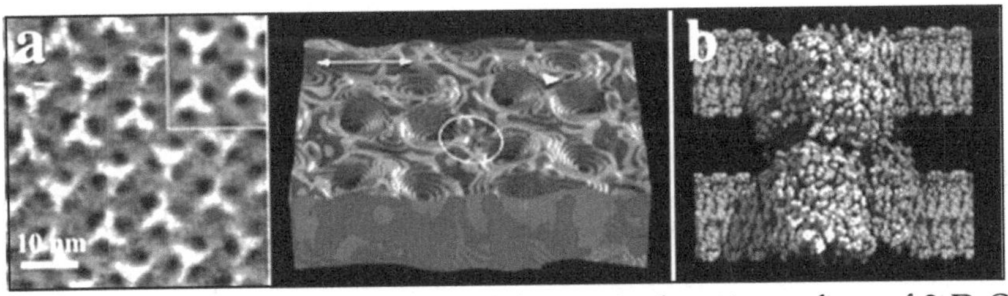

(a) High resolution topograph of the perisplasmic surface of 2-D OmpF porin crystals. The average shown was calculated from 25 translationally aligned subframes.

To allow comparison, the electron density based on the X-ray structure of the OmpF trimer was rendered at a lateral resolution of 15 E. Zebra-like contours (dark and light blue) on the overlaid transparent AFM topograph mark zones of identical altitude, with a height difference between contours of 1 E. (b) Atomic model of protein-protein and lipid-protein interactions based on the X-ray structure of the porin trimer.

USE OF SEMI-THIN CRYOSECTIONS FOR LIGHT MICROSCOPY

Semi-thin sections can be obtained from frozen blocks of cryoprotected biological material by sectioning at -90°C. The advantages for using these sections for light microscopy are that the subcellullar morphology is preserved and the optical resolution is improved.

Chemically fixed biological material is cryoprotected by immersion in 2.3 M sucrose and frozen onto specimen pins by immersion in liquid nitrogen. The sectioning is performed in a cryo-ultramicrotome with the thickness control of the ultramicrotome adjusted to cut thicker sections (0.3 to 1μm). Once the sections have been obtained, they are removed from the knife using a sucrose droplet but instead of being placed on specimen grids they are placed on glass slides.

The glass slides should be marked with a diamond pencil on the side where the sections will be placed (mark the glass, wash and then coat the glass).

After cleaning with detergent and water they may be coated with either gelatin, poly-L-lysine or alcian blue. For the first of these, a drop of 1% gelatin is smeared onto the glass slide using a second slide and air dried. For the other two methods, see below. Coating the slides may ensure that the sections stick to the slides during the labeling procedure.

Immunolabeling of Cryosections on Glass Slides

1. Immerse the slide, with sections on, into a Coplin jar containing PBS. This will wash away the sucrose.
2. Immerse the slide into PBS containing 50mM ammonium chloride in order to quench free aldehyde groups.

3. Immerse the slide into PBS containing a blocking agent (eg 1% gelatin). Remove the slide and dry the surface of the slide such that the sections and the area around them remain wet. It is important that the sections do not dry out.
4. Put 10μl -20μl of diluted, centrifuged antibody onto the sections and leave in a moist atmosphere for 15 min to 24 hr.
5. Wash the antibody away with fresh PBS (in a Coplin jar).
6. Repeat the antibody incubation and washing steps (steps 4 and 5) with the secondary fluoresceinated antibody.
7. Mount the slides in 90% glycerol in PBS or wash with water and mount in Moviol or Permount containing an anti-fade compound. This will produce a permanent mount where the fluorescent signal is not easily bleached when exposed to UV light.

Problems with Autofluorescence

If the material under study was fixed with glutaraldehyde then autofluorescence will be present. This can be removed by treating the sections with 0.1% sodium borohydride in PBS (pH 8). Borohydride stock solutions can be stored indefinitely at pH 12. At pH 7 the molecules have a half life of 10 seconds. At pH 8 the half life is 100 seconds.

If the autofluorescence is from the tissue, then gold- or enzyme-conjugated antibodies can be substituted for the secondary fluorescent antibodies. In this way the antibody binding can be visualized using silver enhancement to reveal the gold, or enzyme cytochemistry to produce a colored reaction product in the tissue slice.

Poly-L-lysine Coating of Glass Slides

1. Dissolve 10 mg of poly-L-lysine in 100 ml of distilled water.
2. Leave the clean, marked slides in this solution for 30 min.
3. Either air dry without washing or rinse briefly in distilled water.

Alcian Blue Coating of Glass Slides

1. Make 0.5% alcian blue stock solution
2. Dilute this solution 1:100 with distilled water
3. Dip the clean, marked slides in the dilute solution for 10 min.
4. Briefly rinse the slides in distilled water. They will remain slightly blue after this wash.
5. Dry before use and only use freshly prepared slides.

Sample and Preparation

Goals

- Obtain a highly purified, homogeneous sample at a concentration and thickness appropriate for TEM or cryoEM.

- Preserve the structure of the specimen for examination in a TEM at the desired resolution. For cryoEM this involves rapid freezing or vitrification, which preserves the specimen in a native, hydrated state.
- Mount the specimen on grids with a thin carbon film or, ideally for cryoEM, suspended in the holes of a holey support film to reduce background noise.

Preliminary Considerations/Questions Regarding the Nature of the Specimen

In order to successfully study any specimen one must first gather all existing information regarding the sample. This is not only to aid in understanding the specimen but to also protect all involved from potential hazards. Some of this information (i.e. biohazard level, vaccines required etc.) should have already been discussed at the onset of the collaborative process.

How Labile is the Specimen?

- For how long and under what conditions can it be stored?
- What temperature, solutes, ionic strength, pH, metal ions, etc. are tolerated or beneficial?
- Does it withstand freezing or storage at 4°C or room temperature?
- Does it need to be prepared for microscopy immediately?
- If yes and it is prepared outside the Purdue campus, what is the appropriate way to ship the sample to assure its preservation in transit?

How Concentrated is the Specimen?

- Virus samples of ~0.5-2 mg/ml generally give good results for cryoEM.
- Virus samples of <1.0 mg/ml generally give good results for negative staining.
- If the sample is too dilute, how can it be concentrated?
- If the sample is too concentrated, how can it be diluted?

How is the Specimen Currently Stored?

- What temperature, buffer, solutes, ionic strength, pH, metal ions, etc. are used?

How Precious is the Specimen?

- Should unused sample be returned, saved for additional experiments, or destroyed?
- How should it be returned or destroyed?

Is the Virus Complexed with other Macromolecules/Ligands/Labels?

- What ratio of virus and ligand are required to form the complex?
- How long should they be incubated to assure saturation?
- Does the complex need to be further purified to remove excess ligand?
- How should this be accomplished? (Column gradient?)
- Will gold labeling be used? How will it be performed?
- Is the ligand large enough to see with negative stain?

Is the Specimen Prepared off Site and Being Shipped?

- What are the guide lines for sending samples?
- Questionnaire for collaborators to give specimen specifics (*e.g.* B1b.-g.).

EM Grid Preparation

The EM grid is to the TEM what the glass slide is to the light microscope. It is actually a 3 mm diameter, thin-foil mesh, and it can be made from a number of different metals (copper, gold, nickel etc.) and have various spacing patterns.

Often an electron lucent film is deposited upon the grid to aid in specimen support. The TEM techniques that are most often used to investigate macromolecular structures require films made of a thin layer of carbon which may or may not have numerous small perforations. The macromolecules are then examined while adhered to the carbon or while suspended in the holes.

Preliminary Considerations

- The type of grids and films used are usually determined by the desired specimen preparation technique and sample characteristics.
- Copper grids are usually used because they are conductive, stable in the beam and inexpensive. But when treating the sample on the grid with some substances such as acids, bases, some salts, etc.; more expensive non-reactive grids such as gold or platinum may be necessary.
- Both holey films and regular carbon films are mounted on grids with a small mesh such as 400 bars per inch for stability in the beam at higher magnifications.
- Vitrified samples are best examined when suspended in the holes of a holey carbon film, while negative stained samples are examined on regular carbon films.
- Purchased films or in house prepared films

Manufactured holey (e.g. Quantifoil, Sources: *Electron Microscopy Sciences, Quantifoil*) and regular films

- Advantages:
  - Big time saver
  - Specified hole sizes/spacings

- Disadvantages:
  - Expensive
  - Unreliable quality
  - "Freshness" issue (hydrophobicity increases with age)

In house preparation

- Advantages:
  - Age of grids is known
  - Price
  - Vary film thickness and hole size to suit specimen
  - Availability
- Disadvantages:
  - Difficult technique
  - Variability
  - Time

Preparation of Regular Carbon Films

Goals

- Provide an electron translucent substrate for the collection and examination of macromolecules and other particulate samples.
- Produce a substrate that is durable, simple to produce and easy to handle.
- Provide a substrate minimizes background noise.

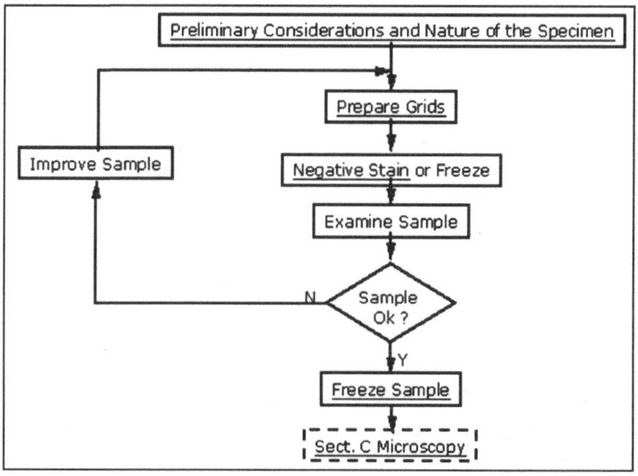

Preparation of Holey Carbon Films

Goals

- Provide a secondary mesh over the copper grid for the suspension of a thin film of vitrified ice forming an unsupported layer within the holes of that mesh.

- Produce holes within that mesh that have appropriate diameters and depth (depth = film thickness) to contain the correct amount specimen solution for freezing to the appropriate ice thickness.
- Produce a support that has the necessary surface characteristics to allow for the quick freezing of thin films of hydrated specimens.

B2d. Carbon Vacuum Evaporation (using the Emitech K950)

Goals

- Evaporate a uniform carbon layer over plastic films or freshly cleaved mica to be used as a stand alone support film.
- Produce very thin support films (~250 E) for use in negative staining.
- Evaporate a slightly thicker carbon layer over holey films for use in CryoEM.

Negative Staining

Scientists have for years used the technique of negative staining to help them examine the morphology of macromolecules. It is a fast and relatively easy procedure whereby a small amount of a sample (i.e. virus suspension) is adhered to the surface of a carbon film on an EM grid and a solution (stain) containing an electron dense material is applied, essentially embedding the sample in a thin film. The electron scattering of the sample is much lower than that of the stain thus resulting in high contrast in the images.

Goals

- To add contrast to low contrast biological samples.
- To add definition to poorly defined details.
- To quickly characterize and assess new sample quality prior to CryoEM.

Preliminary Considerations

- Most negative staining in this lab involves using 2% Uranyl Acetate in water. It is undesirable for the sample to react directly with the stain. This can cause positive staining or precipitation of the stain obscuring rather than revealing details of the specimen. When this happens it may be necessary to try other stains such as Aurothioglucose or Phosphotungstic Acid.
- Negative staining is usually done on thin (~100E) carbon films to reduce background noise, which are mounted on 400 mesh copper grids. This combination is stable at high magnifications while providing adequate viewing area for sample surveys.
- Carbon films are hydrophobic and become more so with age. They tend to repel water and biological samples unless glow discharged just before use. Grids several months old may be used if glow discharged.

Glow Discharging

Carbon coated grids are exposed to ions to change the surface charge of the film. Many macromolecules have positive charges exposed at their surface. Exposing the carbon film to a reduced atomosphere under a high voltage renders the films hydrophilic and negatively charged. This allows the sample to spread more evenly on the surface of the grid.

Negative Staining Procedure

Goals

- To add contrast to low contrast biological samples.
- To enhance details at 15-20E resolution.
- To quickly characterize and assess new sample quality prior to CryoEM.

Index